THE GENE ILLUSION

THE GENE ILLUSION

Genetic Research
in Psychiatry and Psychology
Under the Microscope

Jay Joseph, Psy.D.

Algora Publishing
New York

Library of Congress Cataloging-in-Publication Data —

Joseph, Jay.

The gene illusion : genetic research in psychiatry and psychology under
the microscope / by Jay Joseph.

   p. cm.

Includes bibliographical references and index.

ISBN 0-87586-344-2 (alk. paper) — ISBN 0-87586-343-4 (pbk. : alk.
paper) — ISBN 0-87586-345-0 (ebook)

1. Mental illness—Genetic aspects 2. Genetic psychology. 3. Twins--
Psychology—Research—Methodology. I. Title.

RC455.4.G4J67 2004

616.89'042—dc22

2004017619

Front Cover: Twin Sisters Touching Noses and Smiling

© Anna Palma/CORBIS

ACKNOWLEDGMENTS

Professor Steve Baldwin of the University of Teesside, UK, my friend and colleague, was killed in a March, 2001 train crash in Selby, England. Steve was always available to provide helpful advice and support. I miss him dearly, and I dedicate this book to his memory.

I would like to thank the people who helped make this book possible. In doing so, I want to be very clear that the views expressed in this book are entirely my own, and do not necessarily reflect the opinions of those who helped me along the way. In addition, I take full responsibility for any errors found in the text. The origins of this book lie in a 1998 work analyzing schizophrenia twin and adoption research. I have published several papers based on this research over the past few years. I thank Raymond C. Russ, Editor of the *Journal of Mind and Behavior*, and David Cohen, Associate Editor of JMB. They believed in my work early on and were instrumental in helping me get started in the publishing arena. I would like to thank those who provided helpful feedback at various stages in the preparation of this book. These include Jonathan Beckwith, Ty Colbert, Galina Gerasimova, Leon J. Kamin, Jonathan Leo, Loren Mosher (deceased), Sarah Norgate, Alvin Pam, Ken Richardson, and Al Siebert. I also thank my family for the support they provided. Daniel Ramseyer, Annette Schoch, Ingrid Stapf, and especially Esther Rosen helped greatly with German translations. I am also grateful to the University of California library system and the people who work hard to keep its treasures easily accessible for research. I want to acknowledge Maggie Taylor-Sanders and Pete Sanders of PCCS Books for publishing the UK edition of this book in 2003, as well as the Editors of the PCCS Critical Psychology division, Craig Newnes and Guy Holmes. Finally, I want to thank James H. Joseph and Algora Publishing for helping make the present edition possible.

PREFACE TO THE NORTH AMERICAN EDITION

This book was first published in the United Kingdom in March, 2003. I have taken the opportunity provided by the publication of this book in North America to update the text where necessary. In addition, I have revised every chapter in an effort to correct errors, make the reading more accessible, and to sharpen formulations. Some chapters, particularly Chapters 6 through 8, were greatly revised. The basic issues, however, remain the same, and genetic theories remain entrenched in the mainstream of psychiatry and psychology.

Jay Joseph
Berkeley, California

ADHD	Attention-Deficit Hyperactivity Disorder
APA	American Psychiatric Association
B1	Chronic Schizophrenia
B2	Acute Schizophrenia
B3	Borderline Schizophrenia
C	Schizoid or Inadequate Personality
CBCL	Child Behavior Checklist
D1	Uncertain Chronic Schizophrenia
D2	Uncertain Acute Schizophrenia
D3	Uncertain Borderline Schizophrenia
DSM	Diagnostic and Statistical Manual
DZ	Dizygotic Twins
DZA	Dizygotic (Fraternal) Twins Reared Apart
EEA	Equal Environment Assumption
IQ	Intelligence Quotient
MMPI	Minnesota Multiphasic Personality Inventory
MZ	Monozygotic Twins
MZA	Monozygotic (Identical) Twins Reared Apart
NAS-NRC	National Academy of Sciences-National Research Council
N	Number of Subjects
PKU	Phenylketonuria
SES	Socioeconomic Status
SPD	Schizotypal Personality Disorder
SSD	Schizophrenia Spectrum Disorder
TDI	Thought Disorder Index

TABLE OF CONTENTS

Chapter 1. Introduction

In 1925, when the eugenics movement's influence was at its height and a belief in the overriding importance of genes was widespread, psychiatrist Abraham Myerson observed, "We often hear of hereditary talents, hereditary vices, and hereditary virtues, but whoever will critically examine the evidence will find that we have no proof of their existence."[1] The evidence of Myerson's era consisted of family pedigrees, preconceived notions, and prejudice. Today it consists mainly of family studies, adoption studies, studies of twins reared-together, studies of twins reared-apart, and molecular genetic research. Although it is widely believed that the results of these investigations converge on the importance of genetics, I will argue in this book that conclusions in favor of genetics based on family, twin, and adoption studies are faulty. And as we will see in Chapter 10, molecular geneticists have failed to find postulated genes for the major psychiatric disorders. Thus, Myerson's 1925 observation is more relevant to today's evidence than is commonly believed.

In 1996, twin researcher Irving Gottesman wrote, "no educated person . . . can be oblivious to the fact we are in the midst of a genetic revolution." He added that the younger generation of genetic researchers could hardly imagine "the uphill battle that had been fought for the past 45 years."[2] With what weapons was this "battle" fought? Gottesman identified the "old-fashioned strategies" of family, twin, and adoption studies. And he is correct that these methods helped pave the way for the ascendancy of the genetic position as articulated by the fields of behavior genetics and psychiatric genetics.

---

1. Myerson, 1925, p. 23.
2. Gottesman, 1996, p. 52.

1

Stories in the media often report claims that a specific gene has been linked to a psychiatric disorder, trait, or behavior, although most of these claims are subsequently retracted or fail replication attempts. Several books have been published popularizing behavior genetic research, which include William Wright's *Born That Way*, Lawrence Wright's *Twins*, Hamer and Copeland's *Living With Our Genes*, and Judith Harris's *The Nurture Assumption*. Most have focused on the results of twin and adoption studies which, it is claimed, call forth a radical reevaluation of the "nature-nurture" debate on the side of nature.

In this book I offer an alternative view of this body of literature. Far from establishing the importance of genes, we will see that family, twin, and adoption studies are plagued by researcher bias, unsound methodology, and a reliance on unsupported theoretical assumptions. I examine research methods, theories, and specific studies in the only way they should be examined — at their roots. I devote special attention to twin research because of its central position in the case for genetics.

OUTLINE OF THE CHAPTERS

Chapter 2 looks into the history of twin research, for a simple reason: The full story of this history has never been told. Unfortunately, twin researchers cannot be relied upon to provide details on the more unsavory aspects of this history. In fact, the origin of twin research as a tool of eugenics, "racial hygiene," and Nazism is rarely mentioned by twin researchers. Beginning with Galton, I discuss the various ways that twins have been used for research purposes, as well as some of the methodological problems discussed by critics. We will see that interest in twin research waned in the late 1940s and 1950s, but began a revival in the late 1960s that has continued to the present day.

Chapter 3 is devoted to an analysis of the theoretical underpinnings of the most commonly used method in twin research, the "classical twin method." The twin method compares reared-together identical and reared-together same-sex fraternal twins, and its proponents claim that the greater correlation or concordance of identical twins points to the operation of genetic factors on the trait in question. However, this finding depends on the validity of a critical theoretical assumption which holds that identical and fraternal twins experience equal environments. Interestingly, most people — including most leading twin researchers — recognize that identical twins experience physical and social environments that are much more similar than those of fraternals. Nonetheless, twin researchers continue to uphold the validity of the twin method on the basis of a new definition of the equal environment assumption. I shall, however, demonstrate that this new definition is as untenable as its predecessor. Thus, contrary to the views expressed in countless textbooks and popular reviews, identical-fraternal comparisons do not constitute scientifically acceptable evidence in support of genetic influences on psychiatric disorders and psychological trait differences.

Chapter 4 turns to studies of twins whom the investigators have called "reared-apart." Although these investigations might appear to eliminate the environmental factors confounding the twin method, we will see that they suffer from important methodological problems, not the least of which is that few pairs were actually reared apart without knowledge of and contact with each other. I focus on the well-known Minnesota studies because they have been instrumental in strengthening the genetic position. I also look at cases of individual pairs who have been reported by journalists. Although their stories are interesting, there are many reasons why they don't tell us anything important about genetics.

In Chapter 5, I explain the reasons why heritability figures are discussed only briefly in this book. The heritability concept was designed for use in agriculture to predict the results of a program of controlled breeding. Unfortunately, it has been falsely promoted as a "nature-nurture ratio" of the relative influences of genes and environment on particular traits.[3] As I argue in this chapter, there is little reason to believe that heritability estimates serve this function.

Chapter 6 begins a two-part examination of the evidence supporting the widely held view that schizophrenia is a genetic disorder. The genetic basis of schizophrenia is nowadays so widely accepted that the question is almost never debated. Amazingly, however, the evidence supporting this position is weak. The main evidence, we will discover, is derived from biased, methodologically unsound, and environmentally confounded research. In this chapter, I point to trends and data from schizophrenia twin research to further illustrate problems with the twin method.

The schizophrenia adoption studies of the 1960s and 1970s are largely responsible for closing the "genetics of schizophrenia" debate. In Chapter 7, I perform an in-depth critical analysis of these studies, which were carried out in the United States, Denmark, and Finland. We will discover that schizophrenia adoption research was plagued by methodological problems and bias. Most

---

3. Hirsch, 1997.

importantly, all schizophrenia adoption studies were performed in regions which had eugenic sterilization laws in the era when adoptees were placed. This makes it unlikely that genetically stigmatized children were placed into the same types of homes as other adoptees. Therefore, the basic assumption of adoption studies — that index and control adoptees had the same chance of being placed into available adoptive homes — was violated. On the basis of the potentially invalidating problems with twin and adoption studies, I call for reopening the debate on the role of genetics in schizophrenia and other psychiatric disorders.

Chapter 8 examines twin and adoption studies of criminal and antisocial behavior as an example of how genetic research has been carried out in the area of human behavioral differences. Genetic theories of criminality have been regaining the foothold they had before they were discredited by their association with eugenics, Nazism, and German "criminal biology." In keeping with one of the main themes of this book, I argue that the reported higher identical twin concordance for criminality found in some of the studies is plausibly explained on environmental grounds. Like schizophrenia, adoption studies have been put forward as important evidence in support of the role of genetics. Also like schizophrenia, there are several invalidating features of these studies, which I document.

In Chapter 9, I examine the argument that intelligence (as allegedly measured by standardized IQ tests) has an important genetic component. I look at studies cited in support of this position only briefly, and place more emphasis on the intelligence tests themselves. Specifically, we will see that assumptions about the genetic inferiority of lower classes and subordinated races are *built into* IQ tests. Thus, it is astonishing that anyone who knows how these tests are constructed could argue that the lower IQ scores of blacks versus whites, or working class versus upper class, are the result of genetic differences.

Chapter 10 examines the failure of molecular geneticists to find genes for psychiatric disorders and psychological trait variation. The belief that these genes exist is based on the results of family, twin, and adoption studies. As I argue throughout this book, these studies do not show what they are believed to show, and this may explain the failure to find genes, which may not even exist. Unfortunately, the general public is sometimes led to believe that genes have been found. Sensational accounts of the original "findings" are followed by brief reports of replication failures. Behavior genetics and psychiatric genetics may have arrived at a blind alley, as they attempt to show us what they believed all along actually exist. Unlike family, twin, and adoption studies, in this case unsound methodology and glossed-over implausible assumptions cannot produce the desired results of finding actual genes.

In Chapter 11, I summarize and integrate the main arguments from the preceding chapters, and propose that the entire body of evidence produced by behavior genetics and psychiatric geneticists be reevaluated. I note further that biased and unsound research is not limited to these fields, but is a widespread problem. Finally, I provide an optimistic view of human progress; a view which stands in contrast to the pessimism and limits of genetic theories in psychiatry and psychology.

FAMILY STUDIES

Although twin and adoption studies are a major focus of this book, researchers sometimes cite family studies in support of the genetic position. The family (or consanguinity) method of study constitutes a systematic attempt to determine whether a condition clusters in families, thereby laying the basis for the possibility of finding a genetic component with other methods. Family studies locate persons manifesting a particular trait or condition and attempt to determine whether their biological relatives are similarly affected more often than members of the general population or a control group. If a condition is found to cluster or "run" in families, it is said to be familial. Note that "familial" is not the same as "genetic." Unfortunately, many people view these terms as being synonymous, when in fact they are not. As most genetic researchers now recognize, the aggregation of a trait in families is consistent with genetic or environmental causation. Psychiatric geneticists Steven Faraone and Ming Tsuang, for example, noted that a family study can provide only "the initial hint that a disorder might have a genetic component," because "disorders can 'run in families' for nongenetic reasons such as shared environmental adversity, viral transmission, and social learning." They concluded that "Although family studies are indispensable for establishing the familiality of disorders, they cannot, by themselves, establish what type of transmission."[4] However, this has not always been the prevailing view.

The first schizophrenia family study was published in Germany during World War I by Ernst Rüdin, and Franz J. Kallmann published the most influential in 1938. (Kallmann, as we will see, also studied twins.) Most of the early

---

4. Faraone & Tsuang, 1995, pp. 88-89.

family study authors were strong proponents of the genetic position, and most did not perform blind diagnoses. Kallmann believed that the familiality of schizophrenia *proved* that the condition was genetic in origin: "The principal aim of our investigations was to offer conclusive proof of the inheritance of schizophrenia and to help, in this way, to establish a dependable basis for the clinical and eugenic activities of psychiatry."[5] Psychologist Alvin Pam noted Kallmann's faulty logic and commented further on the family study method,

> The most serious breach in inductive logic committed by Kallmann was his use of kinship concordance rates to determine genetic transmission of psychopathology. We have already noted that no family inheritance study can control for environment in human research; such data, therefore, are nowhere near "suggestive" — they are at best inconclusive and at worst misleading. . . . This inferential limitation holds with respect to any consanguinity finding, even if the design and technique employed in the investigation were scientifically impeccable.[6]

Today, most behavior genetic and psychiatric genetic researchers recognize that family studies cannot establish the existence of genetic factors, and have cited twin and adoption studies as the main evidence in favor of the genetic basis of schizophrenia and other conditions.

In spite of their formal pronouncement that family studies cannot be used as evidence of genetic transmission, some researchers continue to cite these studies in support of genetic conclusions. For example, Russell Barkley, in his authoritative 1998 handbook for the diagnosis and treatment of attention-deficit hyperactivity disorder (ADHD), wrote, "Family aggregation studies find that ADHD clusters among biological relatives of children or adults with the disorder, strongly implying a hereditary basis to this condition."[7] And schizophrenia researchers Frangos and colleagues concluded their 1985 family study by writing, "We consider the results to provide evidence that even narrowly defined schizophrenia . . . has a genetic component . . ."[8] Whereas Faraone and Tsuang correctly viewed the results from family studies as providing only an "initial hint" that genetic factors might be operating, investigators such as Barkley and Frangos et al. believed that these results provide evidence in favor of genetics.

Even before the first family studies were performed, researchers saw patterns in the family pedigrees of affected individuals as definitive proof of genetic influences. The case of pellagra provides an example. Pellagra is a disease that

---

5. Kallmann, 1938a, p. xiv.
6. Pam, 1995, p. 19.
7. Barkley, 1998, p. 36.
8. Frangos et al., 1985, p. 385.

ravaged poor people in the southern part of the United States during the first half of the 20th century. Before then, it had been known in southern Europe for almost 200 years. The often fatal disease, still found among the world's poor, is characterized by severe skin rash, gastrointestinal problems, and mental disturbance. Between 1730 and 1930, pellagra claimed over half a million lives, including tens of thousands of poor blacks and whites in the southern United States. The pioneering work of Joseph Goldberger and others in the early part of the 20th century firmly established that pellagra is caused by a dietary deficiency linked to malnutrition. In other words, pellagra was (and is) a disease of hunger and poverty. But it has not always been viewed this way.

The pioneers of the American eugenics movement believed that pellagra had an important hereditary component. For example, in 1916 Charles Davenport wrote, "Pellagra is not an inheritable disease in the sense in which brown eye color is inheritable. The course of the disease does depend, however, on certain constitutional, inheritable traits of the affected individual."[9] Davenport based his conclusions on the results of his family pedigrees, then known as "eugenical family studies."[10] Yet even before the advent of eugenical family studies, Davenport believed that "many physical, mental and moral traits have been proved to have an hereditary basis, and it seems probable that in practically all there is an hereditary factor of more or less importance." He saw eugenical family studies as "afford[ing] the means of studying this [proven] hereditary factor."[11]

In support of his position, Davenport produced 15 pages of family pedigrees for people diagnosed with pellagra. These diagrams showed that pellagra clustered in families at a rate far greater than expected in the general population. Although Davenport mistakenly believed that the condition was communicable, he believed that "the constitution of the organism must be held to be the principal cause of the diversity which persons show in their reaction to the same disease-inciting factors. This constitution of the organism is a racial, that is, hereditary factor."[12] In Davenport's paper one searches in vain for any mention of the fact that most pellagrins lived in dire poverty. Conspicuously absent from his study are words such as "poor," "poverty," "hunger," or "malnutrition," in spite of the fact that several researchers had pointed to nutritional factors in pellagra and

---

9. Davenport, 1916, p. 13.
10. Davenport & Laughlin, 1915.
11. *Ibid.*, p. 3
12. Davenport, 1916, p. 4. Davenport mistakenly believed that blacks were constitutionally less susceptible to pellagra than whites.

that there were no reported cases of health care providers, working in close contact with pellagrins, who had developed the disease. Above all, pellagra was known to be correlated with poverty and the consumption of corn.

That Davenport and others were blinded by hereditarian views and failed to pay serious attention to possible environmental confounds in pellagra family pedigrees is hardly an original observation; indeed, it is the classic example of the potential fallacy of reaching conclusions about genetic factors when studying people who share a common environment as well as common genes.[13] This is true for virtually all psychiatric conditions, psychological traits, behaviors, and medical conditions. Thus, it is now generally agreed that family study results can be explained on either genetic or environmental grounds.

TERMS USED IN THE BOOK

I would like to define some terms used in the book. I use **human genetics** as a shorthand for "behavior genetics and psychiatric genetics," although several aspects of human genetic research do not fit into these categories. I frequently discuss the underlying **assumptions** of genetic research. In this sense, an assumption is something that is taken to be true. Human genetic researchers base many of their conclusions on questionable assumptions, some of which the researchers identify and discuss, and many others which they do not identify. **Concordance rates** are used in psychiatric twin studies. Twins are said to be concordant for a condition when both members of a pair are affected, and *dis-cordant* when only one is affected. Unless otherwise noted, all discussions of twins in this book refer to pairs reared together in the same home. **Correlation coefficients** quantify the degree of relationship or association between two variables. A positive correlation is expressed as a coefficient ranging from 0.0 to 1.0. A "relationship" means that two occurrences vary together, not necessarily that they are similar. In addition, correlation says nothing about the *cause* of the association. A classic example is that ice cream sales increases are often correlated with an increase in the crime rate. However, this does not mean that eating ice cream causes people to commit criminal acts. This correlation is spurious because a third factor, warm weather, contributes to the causes of both higher ice cream sales a higher crime rate. Although the position that correlation does not imply cause is a basic statistical principle, it is frequently misunderstood or

---

13. See Chase, 1975, 1980.

ignored by people attempting to use correlational data to imply cause and effect. When I say that a study is **methodologically** unsound, I mean that it violates a set or system of methods and principles used in psychological or psychiatric research. The word **environmental** refers to all non-genetic factors influencing a trait or contributing to the causes of a condition. Examples of environmental factors include family or social milieu, viruses, prenatal complications, air and water pollution, and socioeconomic status. When discussing environmental and genetic influences, I should emphasize that they frequently interact. For example, human height is influenced by heredity, but the social environments of adults in Western societies who are three feet tall, and those who are six feet tall, will be vastly different.

Another word I employ frequently is **confound**. A basic study tests the effect of a factor or factors (the independent variable) on another factor (the dependent variable.) The investigators attempt to eliminate (control for) factors other than the independent variable which could influence (confound) the results. For example, a serology laboratory takes precautions to keep foreign substances out of its samples, since contamination could lead to false results. In human genetic research, conclusions favoring genetic factors are dependent upon being able to control for environmental confounds. Human genetic investigators usually state that the results of family studies are confounded by environmental factors, but that twin and adoption studies are not.

In discussions of psychiatric adoption studies, the terms **index (experimental) group** and **control group** are used frequently. An index group consists of subjects identified because they are, or because they are related to, people diagnosed with the trait under study. Control groups consist of people matched with members of the index (experimental) group, but who are not diagnosed with the condition under study. The investigators draw their conclusions from comparisons between the index group and the control group. The **genetic theory of schizophrenia** refers to the view that, although environmental factors might be important, genetic factors are equally if not more important.

It is also important to distinguish between genetic influences on a **trait** versus genetic influences on **trait differences**. A trait is a distinguishing characteristic or quality of a person. Clearly, there are many inherited physical and mental traits that distinguish humans from other species. Behavior genetic research takes this basic fact as a starting point and seeks to establish that genetic factors contribute to differences among people for human traits such as personality and intelligence. As behavior geneticist Robert Plomin has written, "It is critically important to recognize that behavioral genetic research is limited to the investigation of the genetic and environmental origins of individual differences within a species, not species-typical development."[14] Still, the critical differentiation between genetic influences on behavior versus genetic influences on behavioral *differences* is often conflated. In a 1994 article in *Science*, Plomin added to this problem by writing, "Just 15 years ago, the idea of genetic influence on complex human behavior was anathema to many behavior scientists."[15] While hardly anyone in 1979 doubted that genes influence human behavior, many disputed the contention that genes were involved in human behavioral *differences*.

I also discuss various aspects of the way that the **eugenics** movement impacted genetic research. This movement was founded by Francis Galton in the 19th century. Eugenics refers to the belief that human beings can be improved by means of selective breeding. "Positive eugenics" refers to measures used to increase the reproduction of those seen as genetically "fit," whereas "negative eugenics" refers to attempts to curb or eliminate the reproduction of the "unfit." Examples of negative eugenic measures include marriage restrictions, laws against miscegenation (racially mixed marriages), sterilization, and in the case of the Nazi Germany, genocide. The eugenics movement was very active and influential in the United States in the first four decades of the 20th century, until the crimes of the Nazis in the name of eugenics and "racial hygiene" demonstrated what the logic of eugenics could ultimately lead to. The mainstays of eugenic research were family and twin studies. Although we have seen that family studies are no longer seen as conclusive evidence for genetic factors, twin studies remain "the workhorse of behavioral genetics."[16] The use of twins in an attempt to prove that heredity is important has a long and yet largely unknown history. In the next chapter, I attempt to give readers a better sense of this history.

---

14. Plomin, 1996, p. 30.
15. Plomin, 1994, p. 187.
16. Plomin & DeFries, 1998, p. 64.

CHAPTER 2. TWIN RESEARCH: MISUNDERSTANDING TWINS, FROM GALTON TO THE 21ST CENTURY

> Why shall we only deprive these persons, of no use to society or even for them-
> selves, the ability of reproduction? Is it not even kinder to take their lives? This
> kind of dubious reasoning will be the outcome of the methods proposed today.
>
> — Swedish member of Parliament Carl Lindhagen, speaking in opposition
> to a proposed eugenic sterilization law in 1922.[17]

> That man should be one of the few species that produce an abundant supply of
> identical twins is so fortunate that it might be regarded as a direct intervention
> of Providence!
>
> — Eugenicist J. A. Fraser-Roberts in 1934.[18]

The history of twin research is usually told by twin researchers. These typically brief accounts are written from the standpoint that twin studies are a
valuable tool for assessing the influence of genetic factors on human trait variation. They tend to portray the history as a linear path from necessarily crude
methods and uncertain biology to the scientifically precise twin studies of the
current period. Because one of the central themes of this book is the problem of
drawing genetic inferences from the results of twin research, I will focus more
closely on the theoretical and methodological shortcomings of this research. This
chapter is a relatively brief overview of a large body of work, and I will cover only
what I view as the most important points.

The story cannot be told without an integration of the social and political
views and motivations of twin researchers, as well as the social and political
environments in which they carried out their research. As we will see, most pioneers of twin research believed strongly in the importance of heredity, which
often led them to advocate the use of selective breeding programs for humans. To
leave this out is to fail to understand the history of twin research. It would be
like writing a history of the US Civil War without mentioning slavery.

---

17. Quoted in Broberg & Tydén, 1996, p. 104.
18. Fraser-Roberts, 1934, p. 31.

Galton

The history of twin research usually starts with Galton, which is where I also shall begin. Francis Galton (1822-1911), born to a well-off Birmingham, England family, was a pioneer of the fields of statistics and psychology, and is widely recognized as the father of twin research. Galton was also the founder of the eugenics movement and coined the word "eugenics," meaning "well-born." In his 1869 book *Hereditary Genius*, he wrote that "it would be quite practical to produce a highly-gifted race of men by judicious marriages during several consecutive generations."[19] As for the others, Galton wrote in 1883 that "There exists a sentiment, for the most part quite unreasonable, against the gradual extinction of an inferior race."[20]

In *Hereditary Genius*, Galton demonstrated that British judges, commanders, scientists, and the members of other highly regarded professions tended to produce high-achieving offspring. Although Galton attributed the success of fathers and sons to heredity, the resemblance also can be explained by the vastly superior socioeconomic status of these fathers. Galton's awareness of this objection led him to seek other methods to demonstrate the importance of heredity. In an article published in 1875 (and reproduced in a revised form in his 1883 book, *Inquiries into Human Faculty*), Galton discussed the need for

> some new method by which it would be possible to weigh in just scales the respective effects of nature and nurture, and to ascertain their several shares in framing the disposition and intellectual ability of men. The life history of twins supplies what I wanted.[21]

Galton sent 600 inquiries to persons he knew to be twins, or the relatives of twins, but received only 159 responses.[22] From these responses he obtained 35 cases of twins "of close similarity" and 20 cases where twins displayed "sharply contrasted characteristics, both of body and of mind." Contrary to popular belief, Galton did not compare identical and fraternal twins for the purpose of assessing hereditary influences, and thus he cannot be credited with inventing

---

19. Galton, 1881, p. 1.
20. Galton, 1883, p. 308. For a discussion of Galton's racist perspective, see Fancher, 2004.
21. Galton, 1875, p. 391.
22. Burbridge, 2001.

what we now know as the twin method.[23] Galton did, however, believe that there were two types of twins. In his view, one type "is derived from a separate ovum, while the other is due to the development of two germinal spots in the same ovum."[24]

Galton viewed his 35 closely similar pairs as being alike because of their shared heredity, and noted that he had not heard of a case where "growing dissimilarity" was the result of "firm freewill of one or both of the twins." He argued that physical illness or accidents were the only circumstances, apart from heredity, that would have a "marked effect on the character of adults."[25] Of his 20 dissimilar pairs, none was reported to have become more similar due to having received identical nurture. He concluded, "There is no escape from the conclusion that nature prevails enormously over nurture when the differences of nurture do not exceed what is commonly to be found among persons of the same rank of society and in the same country."[26]

Clearly, Galton came to important conclusions on the basis of what today we would consider sketchy, responder-biased, and second-hand reports, and like many subsequent twin researchers he studied twins for the purpose of proving what he already believed to be true. Galton saw the histories of his similar twins as evidence of the importance of their common heredity, or as he put it, "the clocks of their two lives move regularly at the same rate, governed by their internal mechanism."[27] Galton based this conclusion on reports in which twins were said to have become sick and recovered at the same time. He viewed cases describing twins who simultaneously became ill from contagious diseases as evidence of an "intimate . . . constitutional resemblance,"[28] when today we could say that close physical proximity was the likely cause. In another case, both members of a pair of female twins developed "the defect of not being able to come downstairs quickly" at the age of 20, which Galton also attributed to heredity.[29]

Turning to behavioral and psychological traits, Galton's explanations become even more implausible. He described a twin pair who "make the same

---

23. See Bulmer, 1999; Burbridge, 2001; Rende et al., 1990.
24. Galton, 1875, p. 392.
25. *Ibid.*, p. 402.
26. *Ibid.*, p. 404.
27. Galton, 1883, p. 237.
28. Galton, 1875, p. 397.
29. *Ibid.*, p. 396.

remarks on the same occasion" and "begin singing the same song at the same moment," as if they were genetically programmed to do so independently of one another.[30] Galton's belief in genetic predestination is illustrated by his interpretation of the history of a pair of French twins.[31] François and Martin were 50-year-old twin railroad contractors living about six miles apart from each other in small French towns. On January 15th, someone robbed the box where the brothers had deposited their savings. On the night of January 23rd, both twins, though living separately, were said to have been awakened at 3 a.m. by the same dream, at which point they shouted, "I have caught the thief! I have caught the thief! They are doing mischief to my brother!" Martin attacked his grandson and would have strangled him had he not been stopped. He went outside and attempted to drown himself in the River Steir, but was prevented from doing this by his son. Martin was then picked up by gendarmes and was taken to an asylum, where he died three hours later. According to the account, François had calmed down by the morning of January 24th and attempted to solve the robbery. Then,

> By a strange chance, he crossed his brother's path at the moment when the latter was struggling with the gendarmes; then he himself became maddened, giving way to extravagant gestures and using incoherent language (similar to that of his brother). He then asked to be bled, which was done, and afterwards, declaring himself to be better, went out on the pretext of executing some commission, but really to drown himself in the River Steir, which he actually did, at the very same spot where Martin had attempted to do the same thing a few hours earlier.[32]

Galton apparently viewed these twins as being constitutionally programmed at birth to attempt to drown themselves in the same river on the same day. According to the story, Martin had regained his composure until he came across his twin brother who was struggling with the gendarmes. It was only then that he began to make "extravagant gestures" and use "incoherent language ... similar to that of his brother." Clearly, there was a strong bond between these twins which likely manifested itself in similar behavior, and explains François's reaction to what was happening to his twin brother. This story describes the mutual interaction and association of twins, which Galton completely misun-

---

30. *Ibid.*, p. 400.
31. *Ibid.*, p. 399.
32. *Ibid.*, p. 399.

derstood as evidence that the genetically set "clocks of their lives" moved at the same rate.

In studying twins, Galton saw the opportunity to strengthen his case for a hereditary behavioral blueprint stamped on people from conception. He interpreted his case histories as showing that twins who closely resembled each other in childhood were "keeping time like two watches, hardly to be thrown out of accord except by some physical jar."[33] Although Galton understood that common upbringing could be a factor, he was blind to the influences of identity confusion and emotional closeness. Galton saw the twins' histories as evidence that heredity has mapped out a behavioral life plan for each person in the same way that it might influence, for example, baldness, menstruation, or menopause. He was arguing that psychological traits are under as much hereditary control as physical traits, failing to recognize that twins' behavioral similarity could be the result of their similar environment and the emotional bond of the pairs. In another example, Galton wrote that "no less than nine anecdotes have reached me of a twin seeing his or her reflection in a looking-glass, and addressing it in the belief it was the other twin in person."[34] Galton did not understand that these stories could describe the identity confusion of twins who — because of similar appearance, mutual interaction, and societal expectations that they see themselves as two halves of one whole — found it difficult to become individuals separate from each other. "Identity formation," wrote Ricardo Ainslie in his 1985 book on twinship, "is often considered the cornerstone of any discussion of the psychology of twinship. The idea that twins encounter difficulties in the process of identity formation is as pervasive in scientific writings on twinship as it is in popular culture."[35]

From a historical perspective, the most important aspect of Galton's twin study was his failure to understand the causes of the observed behavioral similarities of twins. This error, as we will see, was repeated by subsequent twin researchers similarly blinded by a belief in the importance of eugenics and genetics.

After 1875, Galton did not perform twin research and he turned his attention to other questions. It would be another thirty years before the publication of the next twin study, this time by an American investigator. Never-

---

33. *Ibid.*, p. 402.
34. Galton, 1883, p. 221.
35. Ainslie, 1985, p. 50.

theless, Galton is rightly considered the father of eugenics and twin research — and these "fatherhoods" are by no means independent of each other. Interestingly, Galton was one of the first to mention reports of adoptees in the context of genetic research. Using stories supplied to him, in 1865 Galton attempted to show that "a wild, untamable restlessness is innate with savages."[36] His evidence?

> I have collected numerous instances where children of a low race have been separated at an early age from their parents, and reared as part of a settler's family, quite apart from their own people. Yet, after years of civilized ways, in some fit of passion, or under some craving, like that of a bird about to emigrate, they have abandoned their home, flung away their dress, and sought their countrymen in the bush, among whom they have subsequently been found living in contented barbarism, without a vestige of their gentle nature.[37]

Thorndike

The first twin study to use psychological tests was published in 1905 by the noted American psychologist E. L. Thorndike, who sought to assess the relative influences of nature and nurture on mental abilities. Thorndike selected 50 pairs of same-sex twins from the New York City school population. He was interested in using twins to test whether training and common environment had an important role in mental development. If this were the case, he reasoned, then (1) twins should grow more alike as they get older, and 13- to 14-year-old twins should be much more alike than 9- to 10-year-old twins; (2) the similarity of ordinary siblings not more than 4–5 years apart should resemble the similarity of twin pairs; and (3) if training influences mental traits, twins should be more similar for abilities that require training as opposed to those less subject to training. Thorndike believed that his data did not support the effects of training, and concluded that he had found "conclusive evidence" that "mental likenesses . . . are due, to at least nine tenths of their amount, to original nurture."[38]

Thorndike did not compare correlations between identical and fraternal twins. He believed that "The evidence . . . gives no reason for acceptance of the hypothesis of two distinct types of twins."[39] He viewed the evidence as sug-

---

36. Galton, 1995, p. 406.
37. *Ibid.*, p. 406.
38. Thorndike, 1905, p. vii.
39. *Ibid.*, p. 44.

gesting that all twins were the result of two separate ova. Thorndike's data revealed a unimodal distribution of the twins' test scores, but he expected that if there were two types of twins, there would be a bimodal distribution. Although by this time many scientists had come to accept the existence of two types of twins, the authority of Thorndike (and later Fisher[40]) delayed the widespread acceptance of this idea until the mid-1920s.

In Thorndike's monograph one is struck by the certainty of his belief in the importance of genetics on the basis of what today we would view as weak evidence. Like Galton, Thorndike used twins as a backdrop for expressing his pre-existing belief in the importance of heredity:

> The facts are easily, simply and completely explained by one simple hypothesis: namely, that the natures of the germ cells — the conditions of conception — cause whatever similarities and differences exist in the original natures of men, that these conditions influence body and mind equally, and that in life the differences in modification of body and mind produced by such differences as obtain between the environments of present-day New York City public school children are slight.[41]

Also like Galton, who acknowledged that some would doubt his conclusion "that nurture should go for so little,"[42] Thorndike discussed his and others' "repugnance to assigning so little efficacy to environmental forces as the facts of this study seem to demand."[43]

Following Thorndike's study, almost two decades would pass before twins would again be used in psychological research. In the 1920s, a convergence of events led to a rapid growth in twin research.

Origins of the Twin Method

It became firmly established in the mid-1920s that there are two types of twins: identical (also known as monozygotic, MZ, one-egg, duplicate, or uniovular) and fraternal (also known as dizygotic, DZ, two-egg, or binovular) twins, although several investigators had been certain of this years earlier.[44] In 1924,

---

40. Fisher, 1919.
41. Thorndike, 1905, p. 9.
42. Galton, 1875, p. 404.
43. Thorndike, 1905, p. 10.
44. For example, Newman, 1917.

Hermann W. Siemens of Germany became the first to propose that the similarities of reared-together identical and reared-together fraternal twins be compared as a way of determining the role of genetic factors for diseases or traits.

In his book *Die Zwillingspathologie* (Twin Pathology), Siemens presented the results of a study using this new method, which came to be known as the "classical twin method" or "twin method."[45] He invented the twin method in the context of the study of racial biology [rassenbiologie], family biology, and twin biology. According to Siemens, the main problem with previous racial pathology research [rassenpathologische Forschung] was that its findings were not widely accepted due to the existence of plausible alternative explanations. The new method of comparing identical and fraternal twins offered, according to Siemens, a more definitive research method in support of racial hygienic ideas.[46] Siemens, a member of the famous German industrialist family, was a leading figure in the German racial hygiene movement in the early to middle parts of the 20th century, and was later a supporter of Nazi racial policies (see below).

Three contemporary behavior geneticists have challenged the claim that Siemens deserves full credit as the discoverer the twin method. In 1990, Rende, Plomin, and Vandenberg[47] argued that American psychologist Curtis Merriman should share this distinction on the basis of his 1924 twin study published in *Psychological Monographs*.[48] However, the authors acknowledged that, unlike Siemens, Merriman "did not indicate that the comparison between identical and fraternal twin resemblance could be used to assess hereditary influences on a trait."[49]

Merriman believed that identical twins should show a greater resemblance than fraternals, but he made this observation in the context of demonstrating the existence of two types of twins. As Rende et al. acknowledged, Siemens is usually recognized in Europe as the discoverer of the twin method. But this view was also held by early American twin researchers. For example, in 1934 twin researcher Paul Wilson named Siemens, Dahlberg, and Holzinger as the developers of the twin method,[50] and in 1937 Newman, Freeman, and Holzinger wrote, "Siemens in 1924 published a book entitled *Zwillingspathologie*, in which he

---

45. Siemens, 1924b, pp. 1-3.
46. These views are discussed in Siemens, 1924b, pp. 1-2.
47. Rende et al., 1990.
48. Merriman, 1924.
49. Rende et al., 1990, p. 283.
50. Wilson, 1934.

proposed a new method of studying pathology and human heredity in general."[51] I should add that Italian twin historian Luigi Gedda named Siemens as deserving "credit for opening a new approach to twin studies."[52] I am not aware of any twin researcher naming Merriman as a discoverer of the twin method prior to Rende and colleagues' 1990 article. Having raised Merriman to the status of a co-founder of the twin method 65 years after the fact and contrary to the position of all other twin researchers, in 1997 behavior geneticists Plomin, DeFries, McClearn, and Rutter named Merriman as the *sole* discoverer of the method: "The first real twin study in which identical and fraternal twins were compared in order to estimate genetic influence was conducted in 1924 (Merriman, 1924)."[53] Interestingly Plomin, as a co-author of the 1990 Rende and colleagues article, had written that Merriman *did not* study twins for this purpose (see above).

Siemens: The unknown pioneer. Hermann Werner Siemens (1891-1969) holds the distinction of being one of the most unknown and unheralded inventors of a widely used research technique. Contemporary twin researchers rarely mention him, and his 103-page founding document of the twin method, *Die Zwillingspathologie*, has not been translated into English. One explanation for twin researchers' reluctance to talk about Siemens is the nature of his views on racial hygiene and German National Socialism.

The German Society for Racial Hygiene (Gesellschaft für Rassenhygiene), the world's first eugenics organization,[54] was founded in 1905 by Alfred Ploetz, Anastasius Nordenholz, and Ernst Rüdin. Its goals were to promote the repro-duction of the genetically "fit," and to prevent the reproduction of those viewed as unfit. The racial hygienic and eugenics movements elsewhere shared similar goals and ideas, but many in the German movement preferred the name "racial hygiene" because they wanted to stress the racial nature of their movement.[55] The German racial hygiene movement grew slowly through the years of World War I and emerged in the mid-to-late 1920s as an influential force in German intellectual circles. In 1916, Siemens published the first edition of his book *Grundzüge der Vererbungslehre, Rassenhygiene und Bevölkerungspolitik* (Foundations of

---

51. Newman et al., 1937, p. 19.
52. Gedda, 1961, p. 24.
53. Plomin, DeFries, et al., 1997, p. 81.
54. Weindling, 1999.
55. Proctor, 1988.

Genetics, Racial Hygiene, and Population Policy). This was followed by at least twelve more editions spanning more than 35 years.

An English edition of Siemens's book was published in 1924 under the title *Race Hygiene and Heredity.*[56] In this book, Siemens argued strongly for the importance of genetic influences and called for the implementation of racial hygienic measures. He warned that European civilization was in danger because "fertility among the inferior is greater than that among those of more average capacity,"[57] claiming that Rome and Greece had fallen because "care was not taken to maintain a sufficient fertility among the 'capable,' who were the real girders of civilization."[58] Siemens believed that the state must take action in order to stop the reproduction of the "unfit":

> It cannot be denied that the slightly feeble-minded, those who lack self-control and those with feeble will, which make up the largest contingent of confirmed criminals, tramps and prostitutes, represent a certain danger from the standpoint of race hygiene, for through their lack of inhibition they tend to multiply rapidly. The sterilization of such pathological persons at their own request should, therefore, as soon as possible, become a matter of legal regulation. . . . To render pathological natures permanently harmless and to prevent the reproduction of other miserable creatures should consciously become the goal of our courts.[59]

It is unclear whether Siemens's desire for the courts to act was a call for the compulsory sterilization of "pathological persons." This was published in the atmosphere of Weimar Germany but, as we will see, the atmosphere changed radically after the National Socialist seizure of power in 1933. Siemens invented the twin method in order to create more credible evidence in favor of the importance of heredity and to help eugenicists and German racial hygienists identify the carriers of genetic defects, so that these "miserable creatures" would be convinced not to reproduce, or would be prevented from reproducing.[60]

After the Nazis took power in 1933, most German racial hygienists called for compulsory eugenic sterilization. For Siemens, Hitler's regime transformed

---

56. Siemens, 1924a.
57. *Ibid.*, p. 110.
58. *Ibid.*, p. 98.
59. *Ibid.*, pp. 125-127.
60. Galton also believed in sterilization. In a 1907 letter to his collaborator Karl Pearson, Galton wrote, "except by sterilization I cannot yet see any way of checking the produce [sic] of the unfit who are allowed their liberty and are below the reach of moral control" (quoted in Kevles, 1985, p. 94).

his racial hygienic ideas from a "utopian dream" into state policy. Siemens's forward to the 1937 (8th) edition of his book on racial hygiene contains this passage:

> Since the National Socialist seizure of power the political goals that we, the racial-hygienists, are in favor of, have now become a part — and not the least important one — of the German government program. "Racial hygiene as a utopian dream" became "Racial hygiene as political program". . . . Our future will be governed by racial hygiene — or it will not exist at all.[61]

Later in the 8th edition, Siemens discussed how Galton's ideas were being put into practice in Germany:

> Galton already saw the possibility of integrating racial-hygienic ideals — just like a new religion — into the national conscious. The national [völkische] state, however, is now called on to be really serious about it. According to its *Führer*, it is the obligation of the national state "to declare children as a people's most precious commodity" so that "it will one day be considered reprehensible to withhold healthy children from the nation." [62]

Here, Siemens praised the "Führer" while dropping the mask of the apolitical scientist. He spoke of his field by its proper designation: "Racial hygiene as political program."

Siemens lived in the Netherlands during the Nazi period, having been named the chair of dermatology at the University of Leiden in 1929,[63] so the views expressed in his writings cannot be attributed to the pressures of living under the regime. Although postwar editions of his book on racial hygiene and population policy removed statements in support of Hitler's policies, the 1952 edition of Siemens's book "continued to advocate the forcible sterilization of inferior stocks."[64] In the 1930s and early 1940s, Siemens was a co-editor of Rüdin's *Archiv für Rassen- und Gesellschaftsbiologie* (Archive for Racial and Social Biology). The *Archiv* began publication in 1904 as the journal of the Society for Racial Hygiene. By the mid-1930s, it had become an official organ of the Nazis' Reich Committee for Public Health.[65]

---

61. Foreword to Siemens, 1937 (my translation).
62. *Ibid.*, p. 180 (my translation).
63. Wiener, 1958.
64. Proctor, 1988, p. 306.
65. Weindling, 1989.

DEVELOPMENTS IN TWIN RESEARCH, 1924-1945

Early Criticism of the Twin Method

The twin method (Figure 2.1) rests on five basic theoretical assumptions, which have changed little since the mid-1920s. The most important is that the environments of identical and fraternal twins are roughly equal.[66] Today, this is known as the as the "equal environment assumption." The questionable validity of this assumption has been the most common criticism of the twin method, and it is therefore useful to look at the writings of some early critics as a prelude to a more in-depth discussion. (I examine the equal environment assumption in detail in Chapter 3.)

Figure 2.1

The Classical Twin Method and its Assumptions

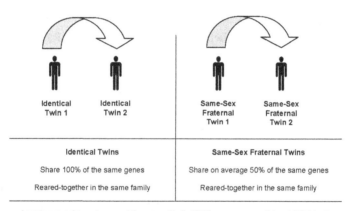

Identical Twins	Same-Sex Fraternal Twins
Share 100% of the same genes	Share on average 50% of the same genes
Reared-together in the same family	Reared-together in the same family

A greater concordance rate or correlation among identical (MZ) versus same-sex fraternal (DZ) twins is attributed to genetic factors and is generalizable to the non-twin population, assuming **ALL** of the following are true:

(1) There are only two types of twins, identical and fraternal.
(2) Investigators are able to reliably distinguish between these two types of twins.
(3) The risk of receiving the diagnosis is the same among twins and non-twins (generalizabilty).
(4) The risk of receiving the diagnosis is the same among identical and fraternal twins.
(5) Identical and fraternal twin pairs experience equal environments (the "equal environment assumption" or "EEA").

Early critics of the twin method pointed out that identical twins experience a more similar environment than fraternals, which might cause identicals to resemble each other more than fraternals. Because this objection often was

---

66. Plomin, DeFries, et al., 1990.

raised by people who believed strongly in the importance of heredity, it was sometimes argued that identical twins created more similar environments for themselves on the basis of their more similar inherited personalities and temperament. For example, in 1930 American eugenicist S. J. Holmes wrote that "the environmentalist may contend that ordinary, or fraternal, twins by virtue of their differences soon get into different relations with their environment and come to experience very unlike reactions from their associates."[67] In the same year, British medical statistician Percy Stocks argued that the twin method "may involve a fallacy" because

> many dizygotic twins are very different in general body build, healthiness, tastes and temperament so that they naturally tend to subject themselves, or be subjected, to differences in nurture to a greater degree than monozygotic twins who have usually the same needs, tastes and inclinations and are rarely seen apart during childhood.[68]

In 1934, Wilson conducted his own twin study and concluded that "The greater dissimilarities in environment of the fraternal pairs may be attributed in part to the influence of their different hereditary makeup."[69] Wilson therefore believed that the assumption of equal environments was "unjustified."[70] In the same year, Helmut von Bracken published the results of a study which found that identical twins experience a higher degree of attachment than fraternals. He attributed this to identical twins' greater genetic similarity, which reduces "slight misunderstandings which are an important cause of conflicts between brothers and sisters . . ."[71] Von Bracken also noted, "where identicals live together, there is a characteristic tendency towards uniformity which must inevitably exert a powerful causal influence upon the properties of such twins."[72] Although von Bracken was the first person to study differing levels of attachment and association among identical and fraternal twins, in his 1934 article he did not discuss the implications his findings might have for the twin method. American psychologist R. S. Woodworth performed a critical review of the literature in 1941 and concluded that twins reared in the same home "may have very different effective environments, for the reason, at bottom, that they

---

67. Holmes, 1930, p. 247.
68. Stocks, 1930, p. 104.
69. Wilson, 1934, p. 352.
70. *Ibid.*, p. 353.
71. von Bracken, 1934, p. 305.
72. *Ibid.*, p. 306.

differ genetically."[73] According to Woodworth, "we cannot derive much information from a comparison of the results from the two classes of twins."[74]

In a 1940 summary of his ten years of twin research, University of California researcher Harold Carter wrote that "the assumption that the nurture influences are approximately equal for fraternal and identical twins. . . . seems untenable to anyone who has had much contact with twins in their own social environment."[75] He further observed that

> Identical twins obviously like each other better; they obviously have the
> same friends more often; they obviously spend more time together; and they
> are obviously treated by their friends, parents, teachers, and acquaintances
> as if they were more alike than fraternal twins are.[76]

We have seen that most of the early critics believed that the twin method was flawed because the greater genetic dissimilarity of fraternal twins led them to create more dissimilar environments than identical twins. Interestingly, in Chapter 3 we will see that contemporary twin researchers *uphold* the validity of the twin method on the basis of the very same argument.

Developments in North America

Apart from Germany, the United States was the leader in twin research in the 1920s and 1930s.[77] I have already mentioned Merriman's 1924 study. C. E. Lauterbach published his twin study of intelligence in 1925.[78] Like Merriman, he did not make direct identical-fraternal correlational comparisons, but instead gathered information in order to answer other questions about the features and abilities of twins. One of his findings was that, in accord with Merriman, twins were not intellectually handicapped when compared with non-twins. This was seen as an important question, as it related to whether IQ twin studies are generalizable to the single-born population.

Two additional early IQ twin studies were published by the Canadian Wingfield in 1928 and the American Hirsch in 1930.[79] Both attempted to make

73. Woodworth, 1941, p. 12.
74. *Ibid.*, p. 21.
75. Carter, 1940, p. 247.
76. *Ibid.*, p. 247.
77. A description of the methods, results, and flaws of the studies I discuss will not be
    made here, other than to point out that most studies found significantly higher identical vs. fraternal similarity, which was attributed to genetic influences.
78. Lauterbach, 1925.

identical-fraternal comparisons, although critics have noted that these studies suffered from inadequate methods of separating identical and fraternal twins.[80] Wingfield believed, as he wrote in *Eugenics Review*, that "intelligence is an inherited trait."[81] Hirsch claimed that his data showed that "heredity was about five times as significant as environment in determining I.Q. differences between twins . . . ,"[82] and concluded, "We could not better end our essay than by quoting from Professor Conklin," who wrote that civilization was threatened by "hereditarily primitive men," and that "either the responsibilities of life must be reduced and the march of civilization stayed, or a better race of men, with greater hereditary abilities, must be bred."[83] Four years earlier, Hirsch had written a lengthy monograph in which he contrasted the abilities of the "Nordic," "Alpine," and "Mediterranean" "races" of Europe, while advocating eugenic measures. According to Hirsch, "It would really be worse than miscegenation for millions of our citizens of many of our Natio-Racial groups to breed freely among themselves or to blend with others of our Natio-Racial groups. Both of these are contingencies that will *naturally* occur unless intelligence and foresight intervene."[84]

In the 1930s, Carter published twin studies of personality, vocational interests, and emotional traits,[85] and Rosanoff and colleagues studied schizophrenia, criminality, and delinquency.[86] Aaron Rosanoff was an early American psychiatric twin researcher and eugenicist who served on the committee on insanity of the American Breeders Association.[87] In the 1938 edition of his *Manual of Psychiatry and Mental Hygiene*, Rosanoff, although recognizing that social and economic factors influence criminal behavior, wrote,

> In all probability, even under ideal social and economic conditions, there will still be many cases of antisocial personalities in which rehabilitation or a tolerable social adjustment will prove unattainable. For such cases, *permanent segregation* and, possibly, *sterilization* will have to be arranged, for the double purpose of protection of society and eugenic effect.[88]

---

79. Wingfield, 1928; Hirsch, 1930.
80. Newman et al., 1937.
81. Wingfield, 1930, p. 186.
82. Hirsch, 1930, p. 148.
83. Conklin, quoted in Hirsch, 1930, p. 150.
84. Hirsch, 1926, pp. 399-400.
85. Reviewed in Carter, 1940.
86. For schizophrenia, see Rosanoff, Handy, Plesset, & Brush, 1934; for criminality and delinquency, see Rosanoff, Handy, & Rosanoff, 1934.
87. Haller, 1963.

Rosanoff wrote that "for mental disorders which are determined solely by hereditary factors the only prevention is through eugenics,"[88] although "the problem is more complicated in connection with mental disorders which are of complex and variable etiology."[90] Given that Rosanoff was a leading voice in psychiatry for eugenics and sterilization in the 1920s and 1930s (particularly in California), it is intriguing that Edward Shorter, in his 1997 *A History of Psychiatry*, cited Rosanoff as an example of a benevolent early psychiatric geneticist. According to Shorter, "Rosanoff was not a marginal right-wing figure, as some historians of psychiatry have tried to characterize these early psychiatric geneticists."[91] My point is not that Rosanoff was "right-wing," but rather that he, like most other twin researchers of his era, was motivated by eugenic beliefs.

For most eugenicists, a demonstration that conditions or traits run in families was sufficient evidence to implicate heredity.[92] As twin researcher and eugenicist Horatio H. Newman wrote in his 1932 book, *Evolution, Genetics, and Eugenics*, the three main "methods of research in eugenics" were "the pedigree method," "the statistical method," and "the twin method."[93] Today, twin studies are the most frequently cited evidence in support of genetic influences because the results of family studies are *now* recognized as explainable on environmental as well as on genetic grounds.

Twins Reared Apart

Although there is controversy about who invented the twin method, according to contemporary twin researchers Bouchard and Pedersen there is no controversy about who invented the twins reared apart method, because "apparently, no one knows."[94] Galton had received a report on a reared-apart identical pair, but did not mention this in his published writings.[95] The mid-1920s witnessed the first published reports of identical twins reared apart. The first reported pair, "B" and "J," were separated at two weeks of age and were brought

---

88. Rosanoff, 1938, p. 642. Emphasis in original.
89. *Ibid.*, pp. 749-750.
90. *Ibid.*, p. 750.
91. Shorter, 1997, p. 243.
92. For example, see Davenport, 1911.
93. Newman, 1932, pp. 444-445. Newman was a member of the Advisory Council of the journal *Eugenics* (for example, see *Eugenics, III* [1930]).
94. Bouchard & Pedersen, 1999, p. 71.
95. Burbridge, 2001.

up in different US states. These twins were first reported in 1922 by Paul Popenoe, who was also a prominent champion of compulsory eugenic sterilization in the United States and was a co-author of *Applied Eugenics.* This was followed by an investigation by the subsequent Nobel laureate geneticist and prominent eugenicist, H. J. Muller.[96] B and J have thus gone into history as the "Popenoe-Muller pair." Muller found that the twins scored similarly on IQ tests, but on other psychological tests their scores were no more similar than randomly selected individuals. Other pairs were subsequently identified, and in 1937 the first account of a study of several reared-apart twins was published by Newman, Freeman, and Holzinger (see Chapter 4). Various single-case studies of reared-apart pairs concordant or discordant for psychiatric disorders have been published over the years. Reports of schizophrenia diagnoses among these pairs are discussed in Chapter 6.

The Co-twin Control Method

Another method of studying twins is known as the "co-twin control method," which was developed in the late 1920s by Gesell and Thompson. This method performs an environmental manipulation on one member of an identical twin pair in order to test the effect of the manipulation. For example, it might be possible to test the efficacy of a particular teaching method by applying it to one twin and not the other. Because genetic factors are controlled for on the basis of the genes shared by identical twins, it is reasoned that test scores would reveal whether the method improved learning. The co-twin control method differs from other types of twin research because it assesses environmental influences, whereas most other types of twin studies are used to detect genetic influences. Still, a finding that environmental influences do not lead to large behavioral differences can be used to support the genetic position, and Gesell, like other pioneer twin researchers, was influenced by eugenic ideas.[97] The co-twin control method is not used widely today for several reasons, which include ethical problems and its history of abuse (see below).

---

96. Muller, 1925.
97. Gesell, 1942; Gesell & Thompson, 1929. For a discussion of Gesell's view on eugenics, see Marchese, 1995.

Twin Research in Germany

Germany was the world center of twin research in the late 1920s and 1930s. It is commonly known that biology and genetics played a major role in the "Third Reich" and helped provide a "scientific" justification for the murder of mental patients and, ultimately, the Holocaust. What is not well known is that the intellectual seeds of the misuse of genetics were sown long before Hitler was named Chancellor in January, 1933.

An early investigator interested in twin research was Wilhelm Weinberg (1862-1937), who also served as chairman of the Society for Racial Hygiene's Stuttgart branch.[98] Weinberg was a popularizer of the "differential method," which uses a simple calculation to estimate the number of expected identical pairs in a given population of twins. He also created the first "morbidity tables" used in medical statistics, as well as an age-correction method subsequently used by several twin researchers. In a 1938 obituary, F. J. Kallmann wrote that Weinberg's most well-known publications "deal with the frequency of taint in the families of tuberculosis and schizophrenia patients, the relationship between tuberculosis infection and heredity, the racial importance of accurate fertility and mortality statistics, and the consideration of selective factors and age distribution in biological studies."[99] According to geneticist Curt Stern, Weinberg worked alone without colleagues or students, although other investigators (such as Rüdin) sought his advice from time to time. "Only when he was old," wrote Stern, "did a new generation again begin to explore the areas in which he had achieved much, a nearly lone worker."[100]

In 1928, Hans Luxenburger published the first schizophrenia twin study. Another study performed in Germany was published by Johannes Lange in 1929, which appeared in English the following year under the title *Crime and Destiny* (the German title of Lange's book, *Verbrechen als Schicksal*, translates as "Crime *as* Destiny"). Lange studied 13 identical and 17 fraternal pairs, finding a much higher identical twin concordance rate for criminality. He concluded that genetic factors "play a predominant part"[101] in criminality, and that "We must try to make it impossible for human beings with positive criminal tendencies to be born."[102]

---

98. Stern, 1962.
99. Kallmann, 1938c, p. 263.
100. Stern, 1962, p. 5.
101. Lange, 1930, p. 46.

1933 was a pivotal year in world history and also began the golden age of twin research in Germany. Hitler's regime gave a prominent place to biology and genetics. Indeed, Rudolf Hess called the National Socialist movement "applied racial science."[103] It was therefore natural that Germany would take the lead in racial hygienic measures, and in July, 1933, the regime passed the "Law for the Prevention of Genetically Diseased Offspring." This law (which was not officially abolished until 1988[104]) established genetic courts having the power to involuntarily sterilize people labeled feebleminded, schizophrenic, manic-depressive, epileptic, genetically blind of deaf, or alcoholic. The law was justified on the basis of genetic studies, which included twin research. In an official commentary on the new sterilization law, Gütt, Rüdin, and Ruttke wrote that natural laws applicable to plants and animals also applied to humans, and that this view was "fully and completely . . . confirmed during the last three decades both through family research and through the study of bastards and twins."[105] The commentary included family pedigree charts and illustrations showing how the sterilization procedure is surgically performed.[106]

The focus of Nazi-era genetic studies included criminality, epilepsy, manic depression, tuberculosis, mental retardation, and abnormal personality. These studies were performed in order to help justify the regime's racial hygienic measures. The authors of two criminal twin studies published in 1936, Heinrich Kranz and Friedrich Stumpfl, both called for the compulsory sterilization of criminals. According to Kranz,

> One could ascertain so far on the basis of twin concordance rates that have been found that the imbecile criminals are undesirable in terms of racial hygiene [rassenhygienisch unerwünscht]; furthermore, some types of criminal psychopaths are borderline psychotics and severe alcoholics. All of these are already being recorded to a large extent through the sterilization law.[107]

Kranz wrote that the genetic impairment of sex offenders "can hardly be questioned," while adding that "the castration law is simultaneously fulfilling the racial hygienic task [rassenhygienische Aufgabe]"[108] Stumpfl went on to

---

102. *Ibid.*, p. 242.
103. Quoted in Proctor, 1988, p. 64.
104. Weindling, 1999.
105. Quoted in Weinreich, 1946, p. 33.
106. Gütt et al., 1934.
107. Kranz, 1936, pp. 250-251 (my translation).
108. *Ibid.*, p. 251 (my translation).

work in the Innsbruck *Intitut für Erb- und Rassenbiologie* (Institute for Genetics and Racial Biology) in Nazi occupied Austria from 1939-1945.[109]

Twins were regarded as a precious resource in Hitler's Germany. As the prominent racial hygienist Eugen Fischer observed, twin research was the single most important tool in the racial hygiene field.[110] Summer camps for German "Aryan" twins were established in 1936 for the purpose of scientific study. As described in 1939 by Germany's leading twin researcher, Otmar Freiherr von Verschuer, "the latest development of the methodology of twin research" are "the so-called 'twins camps'":

> Twin children of Berlin and other towns who were in need of a holiday were brought together in two camps, for a 6 weeks' summer holiday. Several doctors, psychologists, and trained nurses lived with the children during this time, kept them under constant observation, and took notes. Every experience of the twins from early morning until bedtime, the daily rhythm of life, the succession of moods, the reaction to difficulties, to successful and unsuccessful experiences and to conflicts were exactly noted. This kind of observation was supplemented by experimental psychological tests introduced unobtrusively into the daily life. These investigations have led to essentially new conclusions about the hereditary nature and development of normal psychological tendencies, about which we cannot give a detailed account here.[111]

According to the Reich Research Council records, twin camp research was funded until November, 1944.[112]

A visitor to Germany in the winter of 1939-40 wrote of twin studies which "oblige some ten thousand pairs of twins, as well as triplets and even quadruplets, to report to a scientific institute at regular intervals for all kinds of recordings and tests."[113] He described the setting where twins were tested,

---

109. Weindling, 1989, p. 516.
110. Proctor, 1988, p. 43.
111. von Verschuer, 1939, p. 79.
112. Weindling, 1989, p. 562. Further information on German twin camps can be found in Ash, 1998.
113. Ellinger, 1942, p. 141.

> I was particularly interested in their laboratory for the study of the inherit-
> ance of behavior and mental capacities. For this purpose the twins were
> placed in two identical rooms, separated by a narrow corridor for the
> observer, who had a free view over both rooms through big windows in the
> walls. . . . If the observer wants a permanent record of the behavior of the
> twins, he can operate two concealed movie cameras with the lenses nicely
> camouflaged in the wall paper.[114]

Rüdin. The psychiatric genetics field was born in Germany in the early
20th century. The founder of the movement was the Swiss-German psychiatrist
Ernst Rüdin (1874-1952), who (as we have seen) was also one of the founding
members of the German Society for Racial Hygiene. The primary goal of psychiatric genetics was to establish the genetic basis of psychiatric conditions for the
purpose of promoting racial hygienic measures for those diagnosed with these
conditions. German psychiatric genetics played a central role in training people
to conduct twin studies, and researchers from other European countries came to
Munich to study under Rüdin and his associates at the Genealogical-Demo-
graphic Department of the Kaiser-Wilhelm Institute of Psychiatry. Among these
students were several people who went on to have long and influential careers in
psychiatry, such as Franz Kallmann (Germany and the United States), Eliot
Slater (UK), Eric Strömgren (Denmark), and Erik Essen-Möller (Sweden).

Today, one can still find positive references to the "pioneering work" of
Rüdin's "Munich School."[115] Some contemporary psychiatric geneticists claim
that the findings of Rüdin and his associates were merely *used* by the Nazis to
justify murder,[116] or that psychiatric genetics suffered from a "guilt by association" with eugenics.[117] These positions overlook the role of German scientists as
collaborators with the Nazi regime and as people who provided the intellectual
justification for many of its practices.[118] To understand this more clearly, we
must look at the views and deeds of Rüdin himself.

Rüdin was the founder of psychiatric genetics and has been referred to as
one of its "great pioneers."[119] We have seen that Rüdin was a popularizer of the
Nazi eugenic compulsory sterilization law, under which 300,000-400,000

---

114. *Ibid.*, p. 141.
115. For example, see Rosenthal, 1971a; Zerbin-Rüdin & Kendler, 1996; Faraone & Bied-
    erman, 2000.
116. For example, see Faraone, Tsuang, & Tsuang, 1999; Faraone & Biederman, 2000.
117. Farmer, 2003, p. 428.
118. For documentation, see Burleigh, 1994; Lifton, 1986; Müller-Hill, 1998; Proctor, 1988.
119. Faraone & Tsuang, 1995, p. 124.

people were forcibly sterilized.[120] In 1934, Rüdin (like Siemens) recognized that "only through the political work of Adolf Hitler, and only through him, has our more than thirty-year-old dream become reality: to be able to put race hygiene into action."[121] In 1935, Rüdin proposed that the sterilization law be extended to include "valueless individuals . . . all who were socially inferior psychopaths on account of moral confusion or severe ethical defects."[122] Rüdin thereby remained true to the views he had put forward at the beginning of his career in 1903, when he wrote that in order to obtain "biologically fit members of the race," it would be necessary to promote "maximum propagation of those who are healthy, robust and . . . ethically superior," while preventing "the weak, ill, unfit, and morally reprehensible from reproduction by artificial selection. . . . by instruction and by private and government force." [123] Thus, before the passage of compulsory eugenic sterilization laws anywhere in the world, the 29-year-old Rüdin was calling for the use of "private and government force" to prevent the reproduction of the "unfit."

According to his biographer, Rüdin "played a major role in the propagation of racial hygiene doctrines in the 'Third Reich.'"[124] Hitler awarded Rüdin the Goethe medal for art and science in 1939, which was accompanied by a telegram from Nazi Interior Minister Wilhelm Frick, which read, "To the indefatigable champion of racial hygiene and meritorious pioneer of the racial-hygienic measures of the Third Reich I send . . . my heartiest congratulations."[125] In the same year Rüdin, who by now had joined the Nazi party,[126] gave a talk which ended with a long quotation from the "Führer, spoken at the 1937 Party Day in Nuremberg." Rüdin ended his speech by calling on psychiatrists and neurologists to "not only hear [Hitler's] serious words, but to take them to heart and act accordingly!"[127] Rüdin's university lecture notes help illuminate his racial ideas. He instructed his students that "Nordics [are] tall, slim, narrow, and long-headed. Blue-eyed, creative, brave, heroic spirit, self-restrained, cautious, self-respecting, [and] honest." Rüdin's notes for "Jews" read, "racial mixture, pre-

---

120. Proctor, 1988, p. 108.
121. Quoted in Weingart, 1989, p. 270.
122. Quoted in Müller-Hill, 1998, p. 33.
123. Quoted in Weber, 1996, p. 325.
124. Weber, 1996, p. 328.
125. Quoted in Weinreich, 1946, pp. 32-33.
126. Proctor, 1988, p. 292.
127. Rüdin, 1939, pp. 16-17.

dominantly Near Eastern, oriental and negroid. Characterized by persuasiveness and importunacy [*sic*]; early maturity, quickness of understanding but not creative; *thiebhaft* [*sic*] (tendency to follow baser instincts with emphasis on sexual crimes); dishonest; tendency to agitate (Karl Marx, Kurt Eisner)."[128]

In the late 1930s, the regime took "racial hygiene" and eugenic sterilization to its logical conclusion and instituted an extermination campaign against mental patients and "defectives." This Nazi extermination campaign was euphemistically called "euthanasia" (the program was code named "T4"), and resulted in the murder of approximately 70,000 people.[129] It is known that Rüdin discussed "euthanasia" as a type of "therapeutic reform."[130] A memorandum on euthanasia co-authored by Rüdin and others suggests that Rüdin supported the murder of mental patients:

> Even the euthanasia measures will meet with general understanding and approval, as it becomes established and more generally known that, in each and every case of mental disease, all possible measures were taken either to cure the patients or to improve their state sufficiently to enable them to return to work which is economically worthwhile, either in their original professions or in some other occupation.[131]

In 1942, Rüdin stressed "the value of eliminating young children of clearly inferior quality,"[132] and in the same year he published an article celebrating the 10th anniversary of National Socialist rule, while the Holocaust was already underway:

> The results of our science had earlier attracted much attention (both support and opposition) in national and international circles. Nevertheless, it will always remain the undying, historic achievement of Adolf Hitler and his followers that they dared to take the first trail-blazing and decisive steps toward such brilliant race-hygienic achievement in and for the German people. In so doing, they went beyond the boundaries of purely scien-

---

128. See Bernstein, 1945, p. 5. The author interviewed Rüdin after the war, writing, "I am sure that Prof. Rudin never so much as killed a fly in his 74 years. I am also sure he is one of the most evil men in Germany" (p. 5).
129. Proctor, 1988, p. 177. More recent estimates have put the number of murdered mental patients at 300,000 (Peters, 2001, p. 297).
130. Quoted in Weber, 1996, p. 329.
131. Quoted in Müller-Hill, 1998, p. 46.
132. Quoted in Weber, 1996, p. 329. The word "stressed" is Weber's.

tific knowledge. He and his followers were concerned with putting into practice the theories and advances of Nordic race-conceptions . . . the fight against parasitic alien races such as the Jews and Gypsies . . . and preventing the breeding of those with hereditary diseases and those of inferior stock.[133]

Rüdin wrote favorably about Nazi racial laws which led to a "progressive reduction of Jewish influence and especially the hindrance of further intrusions of Jewish blood into the German gene pool."[134] Rüdin saw World War II as being the result of "the Jewish-plutocratic and Bolshevik directed forces," and regretted that the racial hygienic agenda had been slowed by the war.[135] This setback, he wrote, "will only spur us on to multiply our racial hygienic efforts after the victory."[136] Readers may wonder why someone of Rüdin's stature would put these views into publication, but the answer is clear: at the time these lines were written, Rüdin, like most Germans, expected Germany to win the war.

Nonetheless, Rüdin's contemporary supporters continue to claim that he opposed the Nazi T4 euthanasia program and that he helped Jews and was an opponent of anti-Semitism. Gottesman and Bertelsen claimed that Rüdin "was not directly involved" in the euthanasia program, and that when he found out about it "in 1940, he declared it to be murder."[137] Gottesman and Bertelsen also claimed that Rüdin, "rationalizing after the war," stated that his reason for joining the Nazi party in 1937 was the necessity "to preserve his life's work, his research institute, and its funding, and to protect his staff, including Jews, for as long as possible."[138] Many Nazi criminals rationalized their actions after the war. In common language, it is called trying to save your skin. Former Munich student Erik Strömgren wrote about a 1936 funeral for a Jewish researcher which "everybody" at the institute, braving bad weather, attended because they "wanted to pay him the last tribute." The implication was that Rüdin and his colleagues were not anti-Semitic.[139]

Luxenburger. Rüdin's history is discussed here as an important example; there were many other figures in German psychiatry who held similar views and performed similar functions. Some were prominent twin researchers. Rüdin's associate Hans Luxenburger (1894-1976) published the first schizophrenia twin study in 1928 and wrote extensively on twin research and psychiatric genetics

133. Rüdin, 1942, p. 321. Quoted in Müller-Hill, 1998, p. 67.
134. Rüdin, 1942, p. 321 (my translation).
135. *Ibid.*, p. 322 (my translation).
136. *Ibid.*, p. 322 (my translation).

and racial hygiene throughout the 1930s, yet the unsavory aspects of his back-ground are left unmentioned by contemporary twin researchers.

For Luxenburger, the "core significance of human twin and multiple research" was to provide a scientific basis for the "unity of racial hygienic and medical thinking."[140] Luxenburger supported and helped implement the 1933 eugenic sterilization law, although he viewed the sterilization of identified "unfit" people as an inadequate measure to achieve racial hygienic goals, *before* Hitler's rise to power. At the 25th annual conference of Bavarian psychiatrists, held in 1931, Luxenburger discussed calls for sterilization,

> Firstly, occasional mistakes are unavoidable, but they can be tolerated in view of the grave crisis in which our race currently finds itself. It is clear that we cannot fully prevent the transmission of recessive hereditary prop-erties through sterilization; but they will certainly be considerably con-tained, and it is impossible to see why one should sit back and do nothing only because a radical eradication of degenerate hereditary properties is still impossible today.[141]

---

137. Gottesman & Bertelsen, 1996, p. 320. Another example of historical distortion is found in Gottesman & Bertelsen's description of Rüdin's role in the forcible steriliza-tion of children who were the offspring of black African French occupation troops and native Germans (known pejoratively in Germany as the "Rhineland bastards" ["Rheinlandbastarde"]). According to Gottesman & Bertelsen,

> A committee on which Rüdin served considered the ["Rheinlandbastarde"] ques-tion again in 1935 and recommended either sterilization after parental "persuasion" and permission or, preferably, forced emigration.... The only detailed report of the subsequent Gestapo action in 1937 is provided by Pommerin [1979], not translated into English. (Gottesman & Bertelsen, 1996, p. 320)

The sources of this information were listed as Proctor (1988), and Weindling (1989). Gottesman & Bertelsen's description is indeed taken from Weindling and from page 113 of Proctor's book. Interestingly, they chose to ignore Proctor's description of Rüdin's activities in 1937 (in English, I might add), which is found on page 112:

> The 1933 Sterilization Law made no provision for sterilization on racial grounds. (Jews, for example, were never specifically targeted by the law.) In 1937, however, on secret orders from Hitler, the 500 or so *Rheinlandbastarde* ... were sterilized in a joint Gestapo/genetic health court operation organized by Ernst Rüdin, Fritz Lenz, and Walter Gross, among others. (Proctor, 1988, p. 112)

138. Gottesman & Bertelsen, 1996, p. 320.
139. Strömgren, 1994.
140. Luxenburger, 1934b, p. 315.
141. Quoted in Burleigh, 1994, p. 41.

Apparently, even in 1931, compulsory sterilization was for Luxenburger only a temporary method to contain "degenerate hereditary properties" until a future time when they could be "radically eradicated."

During this period Luxenburger published articles expressing his outrage over the lack of legal support to eugenic practices. In 1931 Luxenburger wrote about the "misdeed" of allowing "the most dangerous people . . . the improved or cured mentally ill" from helping to create the next generation. "We are forced," he lamented, "to watch without the ability to act."[142] In another article he expressed misgivings about therapeutic interventions with psychiatrically diagnosed people because "if the germ cells are left untouched," the "hereditarily ill" recovered patients "become a serious danger for the race" because they will pass along their predispositions to their offspring.[143] Luxenburger warned that, "If psychiatric practice succeeds in widely preventing the manifestation of a mental illness or in the reduction of the time spent in an asylum by the mentally ill, this would undoubtedly lead to a disaster for the race."[144] He concluded by stating that eugenics will support therapeutic interventions only on the assurance that, in addition to the health of the individual, the "health of the race" is also considered.[145]

Luxenburger's hard-line racial hygienic views are seen clearly in a 1931 article in *Eugenik*. There, he demanded the sterilization of "every securely diagnosed schizophrenic" before discharge from an institution, if the person was of an age where he or she could father a child or become pregnant. Additionally, he demanded that non-institutionalized people diagnosed with schizophrenia be institutionalized if they did not submit to sterilization.[146] Turning his attention to other diagnoses, Luxenburger wrote that every person who (after being tested) was judged to be hereditarily feebleminded "must be sterilized."[147] Not only the mentally retarded but also their entire families were candidates for racial hygienic measures. According to Luxenburger, "The biological, social, and cultural standard among the families of the hereditarily feebleminded [erblich Schwachsinniger] is in most cases so low that the complete elimination [völlige Ausschaltung] of these families from the population would be a clear gain."[148]

---

142. Luxenburger, 1931a, p. 755 (my translation).
143. Luxenburger, 1931c, p. 53 (my translation).
144. *Ibid.*, p. 55 (my translation).
145. *Ibid.*, p. 55 (my translation).
146. Luxenburger, 1931b, p. 119 (my translation).
147. *Ibid.*, p. 122 (my translation).

He wrote that every criminal "whose criminality is mostly hereditary" has to be "stopped from procreation," eventually through force, even though the "social causes of criminality have to be studied and completely eliminated."[149] Regarding "sexual perversion," Luxenburger warned that "attempts to heal the sexually perverse, especially exhibitionists or homosexuals, through marriage or normal sexual activity without prior sterilization are not only ineffective, they are a virtual crime against the race."[150]

Luxenburger worried that the procreation of the "unfit" would lead to a situation where "the fit will be overtaken and annihilated" by a "huge army of dull inferiors." It is "the responsibility of eugenics," he wrote, "to make sure that this horrible vision will never become a reality in our culture."[151] The situation was so grave, Luxenburger warned, that urgent measures were required:

> The preservation of the fit by means of elimination of the unfit [Ausmerze des Untüchtigen] — the highest goal of eugenics — will never be fully achieved. The eugenicist is in the position of a surgeon who, with the intention of getting rid of a cancerous growth, will prefer to cut off some of the healthy substance in order to be sure to get rid of the disease.[152]

In a chilling forecast of events to come, in the early 1930s Luxenburger wrote that doctors would soon become "the executors of the eugenic will of the nation."[153] Although Luxenburger saw the "elimination of the unfit" as the "highest goal of eugenics," he stopped short of calling for the "destruction of lives unworthy of living," proposed by Hoche and Binding in 1920. In 1931, Luxenburger wrote that the discussion of "social euthanasia" (killing the "unfit") "is unworthy of a civilized people."[154]

In a discussion of how the 1933 sterilization law should be carried out, Luxenburger argued that medical and racial hygienic interventions should be

---

148. *Ibid.*, p. 121 (my translation).
149. *Ibid.*, p. 122 (my translation).
150. *Ibid.*, p. 122 (my translation).
151. *Ibid.*, p. 124 (my translation). Similar ideas were expressed in American medical circles. In a 1933 *New England Journal of Medicine* editorial, physicians were warned that the differential reproduction rate "threatens to replace our population with a race of feeble-minded; we must study its causes and the sources from which it springs. If we wait too long, this viper that we have nourished may prove our undoing" (quoted in Sofair & Kaldjian, 2000, p. 316).
152. *Ibid.*, p. 123 (my translation).
153. Quoted in Burleigh, 1994, p. 41.
154. Luxenburger, 1931a, p. 755 (my translation).

thought of as operating on a continuum.[155] For Luxenburger, "the healthy identical co-twin of a schizophrenic is genetically a schizophrenic," and "We should demand an extension of the definition of sickness according to the Law for the Prevention of Genetically Diseased Offspring [the 1933 sterilization law] to apply to both identical twins, including the twin who is not ill."[156] Thus, from Luxenburger's racial hygienic perspective, there was no difference between a person diagnosed with schizophrenia and his or her "well" identical co-twin. Luxenburger listed four indications that an individual should be the subject of racial hygienic measures: (1) The person manifests a condition deemed hereditary; (2) the person is the identical twin sibling of an affected person; (3) for recessive conditions, the person has two affected parents; or (4) the person manifests latent symptoms of the (supposedly) hereditary condition. Luxenburger was particularly concerned that these "latent" individuals would escape sterilization because they were not clinically identified.[157]

Luxenburger's position that the identical twin of a person diagnosed with schizophrenia is also "schizophrenic" demonstrated his view of individuals more as genotypes than as people. That Luxenburger did not similarly target other biological relatives (as we will see Kallmann attempted to do) was due only to his inability to definitively identify them as carriers. In 1940, Luxenburger wrote that a close relationship between race and hereditary illness was "likely,"[158] and that "hebephrenia," depression, and hysteria "are rather widespread among the Jewish people living in Europe."[159] Thus, after Kristallnacht and the passage of the Nuremberg Laws, and on the eve of the Holocaust, Luxenburger found it necessary to remind his readers that Europe's Jews were the disproportionate carriers of "hereditary illness."

Luxenburger published several articles in the mid-1930s in von Verschuer's influential journal *Der Erbarzt* [The Genetic Doctor]. This journal, which first appeared in 1934, was dedicated to the promotion of racial hygiene, the sterilization law, and the maintenance of the German "Volk's" racial purity.[160] In a 1939 edition of the Nazi-controlled *Archiv für Rassen- und Gesellschaftsbiologie*, Lux-

---

155. Luxenburger, 1934b.
156. *Ibid.*, p. 306.
157. See Luxenburger, 1934b.
158. Luxenburger, 1940a, p. 256 (my translation).
159. *Ibid.*, p. 257.
160. Proctor, 1988. Luxenburger published several articles in *Der Erbarzt* (for example, see Luxenburger 1934a, 1936a, 1936b).

enburger reviewed a book on racial hygiene by the Romanian eugenicist G. Banu. There, he took Banu to task for not going far enough in pursuit of the racial hygienic agenda.[161] In the late 1930s Luxenburger ran afoul of the regime because he publicly criticized Julius Streicher's theory that the children of an "Aryan" woman would not be "Aryan" if she had previously had intercourse with a "non-Aryan." Still, Luxenburger was one of the Nazi era's leading psychiatric genetic and racial hygienic ideologues, and played a role in creating the climate that culminated in mass murder.

Schulz. Bruno Schulz (not to be confused with Bruno K. Schultz) was another important associate of Rüdin. Of the leading Munich twin researchers, Schulz is most often portrayed as an opponent of the National Socialist regime. It also is said that he, and not Rüdin, was the person most responsible for training students in psychiatric genetic methods. In 1991, Gottesman wrote that Schulz was a "star member of Rüdin's Munich school."[162] According to British twin researcher Eliot Slater, a student of Schulz's in 1934, "Schulz's political integrity was the equal of his scientific honesty. He could not stand the Nazi ideology and never compromised with it."[163] (Slater did admit that Klaus Conrad, another Munich twin researcher, was "an enthusiastic Nazi."[164]) However, it was not possible to be a public figure (and publish) under the Nazi totalitarian regime without "compromising" with it. Of course, it would have taken tremendous courage to speak out under the conditions of Hitler's Germany, and one cannot be condemned for not doing so. Still, Slater's account describes the impossible scenario of Schulz living under a fascist regime, as an opponent, and not compromising with it. In 1997, American psychiatric geneticist Kenneth Kendler portrayed Schulz in a positive light, writing that "Schulz never joined the Nazi party to the detriment of his academic career."[165] According to Strömgren, Schulz "was deeply disgusted with the Nazi regime."[166]

In fact, Schulz aided the regime and supported its early racial hygienic measures. In 1934, Schulz published an article entitled "Racial Hygienic Marriage Counseling" in *Volk und Rasse* ("People and Race"; the edition was a

---

161. Luxenburger, 1939. In 1937-38, Banu held a position in the far-right anti-Semitic Goga-Cuza Romanian regime (Bucur, 2002, p. 43).
162. Gottesman, 1991, p. 94.
163. Slater, 1971, p. 18.
164. *Ibid.*, p. 19.
165. Kendler, 1997, p. 461.
166. Strömgren, 1994, p. 406.

Festschrift to Ernst Rüdin on the occasion of his 60th birthday) in which he supported racial hygienic measures and the sterilization law.[167] *Volk und Rasse* had
been taken over by Hitler's SS (Schutzstaffel) in 1933 and was used to promote
racial theories and Nordic superiority.[168] The editorial board of the 1934 edition
of *Volk und Rasse* containing Schulz's article (Volume 9, Number 5) featured Nazi
"Reichsminister" Walter Darré, "Reichsführer SS" Heinrich Himmler, and other
members of the SS. According to a historian, *Volk und Rasse* "became more and
more a mouthpiece for Himmler's Nordic programme."[169] Although according to
Slater's account, Schulz was an uncompromising opponent of the political
regime he "detested,"[170] Schulz "compromised" enough to write an article on
racial hygiene and sterilization which earned the warm approval of "detested"
Nazis Heinrich Himmler, Walter Darré, and associates.

Schulz wrote in this 1934 article that the sterilization law was insufficient
to eliminate all "hereditary illness," writing, "It is without question that the
reproduction of another large number of persons who do not fall under the law
are undesirable in terms of racial hygiene." In this group he included "diabetics . .
. hereditary neurotics, certain types of psychopaths, etc." For those whom Schulz
was willing to grant a reprieve from the sterilizer's knife, he recommended
"racial hygienic marriage counseling."[171] According to Schulz, the "Children of
someone who later becomes schizophrenic have to be considered qualitatively
equal to children of obvious schizophrenics," because the "illness" serves merely
to identify the nature of the genes.[172] In 1939, Schulz published an article in von
Verschuer's *Der Erbarzt* in which he wrote of empirical genetic prognostic
research in the context of support for the 1933 sterilization law and other
German racial hygienic statutes.[173]

Rüdin's biographer, Matthias Weber, claimed that in 1936 Schulz "questioned Nazi sterilization laws under the protection of his mathematical appa-
ratus, which only specialists could understand."[174] Although Weber performed a
public service by exposing some of Rüdin's criminal activities, in this case he

---

167. Schulz, 1934.
168. Weindling, 1989, p. 500.
169. *Ibid.*, p. 500.
170. Slater, 1971, p. 18.
171. Schulz, 1934, pp. 138-139 (my translation).
172. *Ibid.*, p. 142 (my translation).
173. Schulz, 1939.
174. Weber, 1996, p. 328.

failed to mention Schulz's public support of the sterilization law in his 1934 *Volk und Rasse* article. After the war, Schulz was a founder of the Max Planck Institute in Munich, which was the successor to Rüdin's institute.

* * *

Psychiatric geneticists Rüdin, Luxenburger, Schulz, Conrad and others studied twins and popularized twin research in order to provide what they viewed as additional evidence that people with "undesirable" traits were "hereditarily tainted," and deserved to fall under the sterilizer's knife or worse. This was the main reason that they studied twins, and this is how Rüdin's Munich school should be remembered.

But the full history of the Munich school continues to be denied. According to Kendler, Luxenburger was "an active Catholic," who "deeply disliked the Nazis."[175] It is true that Luxenburger was an active Catholic, although this did not prevent him from also being an active eugenicist and sterilization proponent. The claim that Luxenburger "deeply disliked" the Nazis, while containing an element of truth, is highly misleading. From the standpoint of educated elitists such as Luxenburger, the Nazi leaders and their plebeian followers were "uneducated rabble." As a historian correctly noted, "Many geneticists initially had a disdainful attitude toward nazism, as being a vulgar rabble-rousing movement."[176] Strömgren wrote that Luxenburger "did not hesitate to make jokes about the government and the Nazis."[177] In the same spirit, an aristocratic Wehrmacht Field Marshal may have looked down upon (and made jokes about) "Corporal Hitler" and his modest origins and educational background. To cite this as evidence that the same general was an opponent of Nazism, however, would be a gross distortion. Undoubtedly, many racial hygienists who recognized Hitler as enabling their "utopian dream" to become a reality were at the same time repulsed by the plebeian origins and methods of German National Socialism. But to paint them as anti-Nazi is very far from the truth. In *The History of Psychiatry*, Shorter acknowledged that Luxenburger "served as an apologist for

---

175. Kendler, 1997, p. 461. Kendler also wrote that "In 1935," Luxenburger "was banned from future public lectures during the Nazi regime" for his remarks against Streicher (Kendler, 1997, p. 461). According to the historian Paul Weindling, however, "Rüdin, Luxenburger and Lenz continued to hold large numbers of lectures until 1937" (Weindling, 1989, p. 502). In addition, we have already seen that Luxenburger was published in Hitler's Germany as late as 1940 (as another example, see Luxenburger, 1940b).
176. Weindling, 1993, p. 643.
177. Strömgren, 1994, p. 406.

racist genetics for a few years under the Third Reich," but then mistakenly claimed that Luxenburger became a critic of the regime in subsequent years.[178] Getting into trouble for publicly criticizing the ravings of one particularly outrageous Nazi figure is not the same thing as being a critic of the regime in general.

In a 1971 account by American psychologist David Rosenthal, who apart from Kallmann was the most important popularizer of psychiatric genetics in the United States, we read,

> Psychiatrists founded a research institute in Munich, Germany, in the first decade of the century, and the field of psychiatric genetics was born there, fathered by Ernst Rüdin, a man little known in our own country. From this institute emerged all of the pioneering psychiatric geneticists, some of whom are still alive and active. They concerned themselves with many psychiatric disorders, especially the new diagnostic entities, dementia praecox (schizophrenia) and manic depressive psychosis. They devised systematic research strategies based on sound principles of human genetics, and generated a large body of data, never doubting that they were dealing with genetic diseases. Unfortunately, concomitant with the development of psychiatric and other aspects of human genetics, there grew in Germany an ugly, racist view of man, in which the Aryans were deemed to be the master race. Many German scientists voiced their approval of these ideas, offering pseudoscientific statistics and generalizations in support of them, while the Nazi *Herrnvolk* [master race] went about the inhuman business of committing genocide.[179]

Thus, according to Rosenthal, there was a group of dedicated psychiatric researchers, headed by Rüdin in Munich, who performed pioneering genetic studies "based on sound principles." At the same time presumably *other* German scientists supported the master race idea and helped set the stage for genocide. Rosenthal's account is even more inexcusable when we realize that he was very well acquainted with the Nazi-era German psychiatric genetic literature.[180]

Auschwitz. The "twin research" at Auschwitz conducted by Josef Mengele in the midst of the Holocaust has been ably described in Robert J. Lifton's 1986 book, *The Nazi Doctors*. Mengele was a committed member of the SS and had studied under von Verschuer in Frankfurt. Mengele's teacher was perhaps the most famous twin researcher in the world during the 1930s.

---

178. Shorter, 1997, p. 242.
179. Rosenthal, 1971a, p. 7.
180. For example, see the bibliography in Rosenthal, 1970b.

According to a 1937 account by American twin researchers Newman, Freeman and Holzinger,

> The most active and influential of modern students of human genetics using this twin method is O. von Verschuer, who heads a large group of students of twinning at the Kaiser-Wilhelm Institut für Anthropologie . . . This group of investigators has used the twin method for the study of anthropological, clinical, serological, and physiological problems and has contributed extensively to our knowledge of these subjects and, incidentally, to our general knowledge of twins.[181]

Mengele arrived at Auschwitz in the spring of 1943 for the purpose of conducting studies on twins. His task was to identify twin inmates from those at the camp who had not yet been put to death, as well as to select twins from among the new arrivals. Mengele frequently performed this task personally by wading into the stream of new arrivals and shouting "Zwillinge heraus!" (twins, get out!).

Mengele performed gruesome and murderous "experiments" on twins, at times killing them personally and performing post-mortem examinations. His technique of choice was the co-twin control method. Sometimes he would send the body parts of his victims to von Verschuer in Berlin, with whom he was in close contact.[182] Thus, the world's "most active and influential" twin researcher was an accomplice to the murder of twins in the name of science. In 1942, von Verschuer wrote that "The policies of National Socialism have finally put an end to the last threat to the race — that which comes from the Jews,"[183] and in 1944 he proclaimed that the alleged dangers posed by Jews and Gypsies had been "eliminated through the radical-political measures of recent years."[184]

For Mengele, von Verschuer and others, Auschwitz presented an unprecedented opportunity to conduct twin research. "It would be a sin, a crime," Mengele is reported to have said, "not to realize the possibilities that Auschwitz had for twin research. There would never be another chance like it."[185] It is estimated that 3,000 twins were "examined" by Mengele, but only about 200 were left alive by the end of the war.

---

181. Newman et al., 1937, p. 19.
182. See Lifton, 1986, for an account of von Verschuer's role as accomplice to Mengele's crimes at Auschwitz.
183. Quoted in Müller-Hill, 1998, p. 52.
184. Quoted in Proctor, 1988, p. 211.
185. Quoted in Lifton, 1986, p. 358.

* * *

Twin research and human genetic investigation were dealt a severe blow by the crimes committed by the Nazis and their willing scientific accomplices. Still, some contemporary psychiatric geneticists persist in denying these crimes. In response to my modest statement in 2000 that "history has shown that the results of genetic studies have often been used to stigmatize individuals and groups,"[186] American psychiatric geneticists Stephen Faraone and Joseph Biederman responded by criticizing me for presenting "no data" supporting this claim, while admitting only that "psychiatric genetic data had been used by the Nazis to justify eugenic sterilization and murder."[187] In a similar fashion, psychiatric geneticists Zerbin-Rüdin and Kendler wrote in 1996 about the Nazis'' "possible [sic] misuse and abuse of findings from the field of psychiatric genetics."[188] Statements such as these ignore the *collusion* of German psychiatric genetics and Hitler's regime. In their 1999 psychiatric genetics textbook, Faraone, Tsuang, and Tsuang wrote that the "findings from psychiatric genetics were abused by scientists and politicians to kill rather than to cure."[189] But these scientists had names, and among them was the "great pioneer" of psychiatric genetics, Dr. Ernst Rüdin.

---

186. Joseph, 2000c, p. 562.
187. Faraone & Biederman, 2000, p. 575.
188. Zerbin-Rüdin & Kendler, 1996, p. 336.
189. Faraone, Tsuang, & Tsuang, 1999, p. 222. The authors also wrote (p. 223) that "Adolph Hitler and his Nazi regime began a systematic program first to sterilize and then to kill 'genetically defective' people." The role of Rüdin and other psychiatric geneticists in popularizing the sterilization law is not mentioned, in spite of Faraone & colleagues' citation of two articles (Gottesman & Bertelsen, 1996; Weber, 1996) which documented Rüdin's role. Also on page 223 they wrote, "Contemporary researchers in psychiatric genetics are especially disturbed to learn that the Nazis' used this research to justify their eugenics policies regarding the mentally ill . . ." Thus, Faraone et al. portrayed eugenic sterilization as something advocated by the Nazis, but made no mention of the role of psychiatric geneticists. The false implication is that eugenic sterilization was "their" (the Nazis') program, but not "ours" (the psychiatric geneticists). Also overlooked is the fact that several psychiatric geneticists were themselves members of the Nazi party, such as Rüdin himself who joined in 1937.

TWIN RESEARCH FROM 1945-1960

In the years after World War II, twin research was performed by a small group of people, many of whom were motivated by eugenics, who believed that their research continued to have an important role to play is assessing the relative weight of nature and nurture. This was a difficult time for twin research, as Germany was seen as an example of what eugenics and racial hygiene could ultimately lead to. In addition, behaviorist and Freudian explanations for psychiatric disorders were widely accepted, although few psychiatrists believed that conditions such as schizophrenia had no genetic component at all. For those who believed that psychiatric disorders were strongly influenced by heredity, the flame was kept burning by researchers who had studied at Rüdin's institute in Munich.

One of these researchers was Munich-trained Eliot Slater, who published the results of his British schizophrenia twin study in 1953. Along with his colleague James Shields, Slater played an important role in promoting psychiatric genetic ideas in the English-speaking world. Another Munich graduate, Erik Essen-Möller, performed twin research in Sweden and published his own schizophrenia twin study in 1941.

Kallmann

In the United States, the message of psychiatric genetics was brought from Rüdin and Munich in the person of Franz J. Kallmann (1897-1965). Kallmann published his schizophrenia twin study in 1946, in which he claimed to have investigated and diagnosed 691 pairs of American twins. He reported age-corrected pairwise concordance rates of 85.8% identical, and 14.7% fraternal. Kallmann concluded that schizophrenia required a genetic predisposition "which is probably recessive and autosomal"[190] (see Chapter 6 for a critique of schizophrenia twin research). He had published his famous schizophrenia family study eight years earlier. This investigation was based on the families of German patients diagnosed with schizophrenia.

Kallmann was well known in the United States because of his large reported twin samples and the high concordance rates he reported. He was also controversial, in part because of the impact that his strong eugenic views may have had on the way he diagnosed twins. In most cases, Kallmann was aware of

---

190. Kallmann, 1946, p. 321.

the co-twin's diagnosis and whether a twin was identical or fraternal. Critics Lewontin, Rose, and Kamin noted the irony of the fact that, while Kallmann's data is less accepted as time goes on, a belief in the genetic basis of schizophrenia, which his data helped establish, becomes more accepted.[191]

Like his former Munich colleagues, Kallmann's primary interest was to produce evidence in support of policies that would eliminate the genes he believed cause psychiatric disorders. Kallmann's background and activities in Germany and the US in the 1930s are seldom discussed by his contemporary supporters. When they are, he is typically portrayed as a "Jewish refugee" who "fled Hitler's Germany."[192] These portrayals grossly misrepresent Kallmann's history.

In 1935, while still active in Germany, Kallmann called for the compulsory sterilization of people diagnosed with schizophrenia and their "heterozygous" relatives as well. According to Kallmann,

> It is desirable to extend prevention of reproduction to relatives of schizophrenics who stand out because of minor anomalies, and, above all, to define each of them as being undesirable from the eugenic point of view at the beginning of their reproductive years.[193]

Kallmann was forced to leave Germany in 1936 because he had a Jewish father. It is ironic that Kallmann, ever in search of "hereditary tainting" in others, would be forced to flee his homeland with little more than the clothes on his back and his family study manuscript, on the basis of his own "taint."

So perhaps, upon arriving in the United States, Kallmann's personal experiences would compel him towards a greater sensitivity to those with "bad blood"? He hardly missed a beat. In a 1938 article in *Eugenical News*, Kallmann called for the use of "negative eugenic measures" against the "carriers of any mental disease" meeting certain criteria.[194] He continued,

> From a eugenic point of view, it is particularly disastrous that these patients not only continue to crowd mental hospitals all over the world, but also afford, to society as a whole, an unceasing source of maladjusted cranks, asocial eccentrics and the lowest types of criminal offenders. Even

191. Lewontin et al., 1984.
192. Gottesman & Shields, 1982, p. 42. This theme is consistent in Gottesman's writings on Kallmann's background.
193. Quoted in Müller-Hill, 1998, p. 11.
194. Kallmann, 1938b, p. 105.

the faithful believer in the predominance of individual liberty will admit that mankind would be much happier without those numerous adventurers, fanatics, and pseudo-saviors of the world who are found again and again to come from the schizophrenic genotype.[195]

Kallmann presented data from his family study which, he claimed, showed that "schizophrenia is a single-recessive trait and a specific genetic entity."[196] In order to protect society, Kallmann advocated "eugenic prophylactic measures" to deal with schizophrenia "taint-carriers."[197]

Another misleading account of Kallmann's history was published shortly after his death by his associate John Rainier. According to Rainer, Kallmann was a "medical man who takes a humanitarian interest in the psychological and emotional problems of his patients."[198] Regarding Kallmann's activities in the 1930s, Rainer wrote,

> By 1934, [Kallmann] was becoming increasingly anxious to leave Germany, where he found the political abuse of genetics for racial purposes, including sterilization laws for the mentally ill, not at all consistent either with the Hippocratic responsibility of the physician or with the genetic data on schizophrenia which he was beginning to collect. In 1936 Kallmann came to the United States . . .[199]

And Kallmann himself was not above describing his motivations in ways that suited the times. Let's compare two accounts. The first is from his 1938 schizophrenia family study. In the foreword to the book describing this study, Kallmann wrote,

> The present study of the genetic and reproductive conditions in schizophrenia families was undertaken in 1929, in collaboration with the German Research Institute for Psychiatry in Munich (Prof. Dr. Rüdin). . . . *The principal aim of our investigation was to offer conclusive proof of the inheritance of schizophrenia* and to help, in this way, to establish a dependable basis for the clinical and eugenic activities of psychiatry [emphasis added].[200]

However, in 1953 Kallmann provided this account:

---

195. *Ibid.*, p. 105.
196. *Ibid.*, p. 112. For those interested in Kallmann's family study, see Kallmann, 1938a.
197. *Ibid.*, p. 109.
198. Rainer, 1966, p. 414.
199. *Ibid.*, p. 415.
200. Kallmann, 1938a, p. xiv.

> As a young psychiatrist, I divided my time between clinical service and work in histopathology. On visiting days, the head nurse . . . often announced a group of waiting visitors with the statement: "Doctor, the familial taint is here." It was just because that nurse was so domineering that *I undertook my first family study to "disprove" the inheritance of schizophrenia* [emphasis added].[201]

Kallmann's account therefore shifted 180 degrees, from someone who studied families in order to provide "conclusive proof" of the genetic basis of schizophrenia in order to "aid the eugenic activities of psychiatry," to someone seeking to "disprove" the idea that schizophrenia was the result of a "familial taint." A parallel process occurred in Germany, where post-World War II editions of the works of leading German scientists were edited to remove positive references to Hitler and National Socialism.[202] In the 1950s and 1960s, Kallmann wrote frequently about eugenics and was active in the American Eugenics Society until his death.[203]

In spite of his history of collaboration with Rüdin and other German racial hygienics, Kallmann has been called upon by twin researchers Gottesman and Bertelsen as a posthumous character witness for Rüdin.[204] After mentioning a few of Rüdin's relatively minor crimes while ignoring the most heinous, Gottesman and Bertelsen concluded, "perhaps 50 years is not yet enough time to render a final evaluation" of Rüdin.[205] They ended with a lengthy quote from a 1947 letter written by Kallmann in support of Rüdin's successful attempt to escape justice in the latter's post-war "denazification" trial (Rüdin received a 500-mark fine). In his letter, Kallmann claimed that Rüdin "is no criminal, of course." Amazingly, Gottesman and Bertelsen were "content to let Kallmann have the last word for now"![206] (In 2003, British psychiatric geneticist Anne Farmer wrote that "it remains to be seen" whether the denazification court's decision that Rüdin was not a "major contributor" to Nazi atrocities is true).[207]

---

201. Kallmann, 1953, p. 18. At the end of this quotation, Kallmann cited his "first family study" as "Kallmann, 1938."
202. Proctor, 1988.
203. Ludmerer, 1972.
204. Gottesman & Bertelsen, 1996.
205. *Ibid.*, p. 320.
206. *Ibid.*, p. 321. Psychiatric geneticist Erik Strömgren (1994, p. 406) denied Rüdin's criminal role, writing that Rüdin was released after the war "following very conscientious investigations performed by the Americans."
207. Farmer, 2003, p. 427.

Israeli genetic investigators Lerer and Segman, objecting to Gottesman and Bertelsen's "whitewash," wrote bluntly that Rüdin "was not only a willing accomplice to the most abhorrent crimes against humanity but an enthusiastic theorist who provided the intellectual basis for many of these crimes."[208] For American geneticist Elliot Gershon, Rüdin was a "Nazi psychiatrist and geneticist" who subscribed to "a racist/eugenist conception of human genetics which is now universally abhorrent."[209]

In a rare discussion of the embarrassing aspects of their field's history and origins, twin researchers Torrey, Bowler, Taylor, and Gottesman wrote in 1994 about the popularity of twin research in Nazi Germany, and about Mengele's crimes at Auschwitz in collaboration with von Verschuer.[210] At the same time, they said nothing about Rüdin's criminal role, and they discussed Luxenburger in a favorable light. Finally, Kallmann was portrayed in his usual role as a German researcher forced to flee his homeland because of his Jewish ancestry, and his racial hygienic views are left unmentioned.

Selected Topics in Twin Research

Zygosity determination. Zygosity determination refers to researchers' necessity of accurately assessing whether a given pair is identical or fraternal. The ability to make this distinction is critical for twin research and is more difficult to accomplish than many people realize. Zygosity determination by physical appearance alone can lead to mistakes. It was once thought that zygosity could be determined at birth on the basis of whether twins arrived in one or two chorion sacks, but it has been known for many years that about a one-third of identical twins also have two chorions. Additionally, there is the problem of determining the number of chorions many years after the births took place.

One of the first ways that zygosity was determined was the "similarity method" discussed by Siemens in 1927, which compared traits such as eye color, hair color, skin color, palm prints, ear form and body build.[211] Siemens believed that by comparing a sufficient number of traits, he would be able to determine zygosity with a very high level of accuracy. In 1937, Newman and colleagues

---

208. Lerer & Segman, 1997.
209. Gershon, 1997, p. 457.
210. Torrey et al., 1994.
211. Siemens, 1927.

agreed with Siemens that the chances of making a mistake with this method were negligible. At the same time, several twin researchers were seeking better methods.

Essen-Möller introduced the method of blood typing in 1941, where samples of twins' blood were compared and an accurate determination of zygosity could be made. Blood typing is more expensive than most other methods, and the ability to obtain samples is also required. Subsequent methods have combined blood typing with physical characteristics, and errors are said to be made less than one time in a thousand.[212] Even more recently, researchers have determined zygosity very accurately by DNA analysis.[213]

Another technique is the "questionnaire method." Several researchers have developed questionnaires asking twins or their parents about the similarity of the twins' physical characteristics, whether they were mistaken for each other as children, and whether they were considered as "two peas in a pod." In general, the questionnaire method has the advantage of being relatively inexpensive and fairly accurate (about 95%). It is also used in cases where the investigators are unable to examine twins personally, such as studies where twin psychological information is obtained by mailed questionnaires and tests.

Interestingly, the differentiating criteria found in these questionnaires could be interpreted as a demonstration of the dramatically different childhood experiences of the two types of twins. For example, Cohen and colleagues' zygosity determination questionnaire asked mothers of twins (identical, N = 94; fraternal, N = 61) the following questions (which are accompanied by the results): "Does either mother or father or ever confuse them?" (identical = 78%, fraternal = 10%); "Are they sometimes confused by other people in the family?" (identical = 94%, fraternal = 15%); and "Is it hard for strangers to tell them apart?" (identical = 99%, fraternal = 16%).[214] The irony of using dissimilar identical and fraternal childhood environments as part of the twin method's zygosity determination process was not lost on Cohen and colleagues, who wrote, "the impact of such repeated confusion on individual twinships, or the effect of these differences between MZ and DZ twins is not known with certainty. However, such information must cast doubt upon the assumption of environmental equivalence."[215] It is a telling indictment of the twin method's logic that a procedure

212. Lykken, 1978.
213. See, for example, Thompson et al., 2001.
214. Cohen et al., 1973, p. 467.
215. Dibble et al., 1978, p. 247-248.

twin researchers use to distinguish identical and fraternal twins is based on a demonstration of the dissimilarity of their childhood environments, which, paradoxically, they have assumed (at least until recently) does not exist.

There are several other problems associated with zygosity determination in twin studies. Some researchers have given vague descriptions of how they determined zygosity. For example Franz Kallmann, in his large 1946 schizophrenia twin study, stated that he determined zygosity "on the basis of personal investigation and extended observation" using the similarity method.[216] This means that Kallmann, a strong believer in the genetic basis of schizophrenia, made personal non-blinded diagnoses. It is reasonable to suspect bias because, when there was doubt, Kallmann might have seen concordant pairs as identical, and discordant pairs as fraternal. Because Kallmann also made the schizophrenia diagnoses, bias becomes a bigger problem.

Rosenthal discussed some of the problems with zygosity determination in schizophrenia twin research in 1962. He noted that in some cases the investigator had to make a determination for a twin who was not available or had died. If the investigator decided to eliminate questionable cases, he or she may have introduced a bias in the sampling procedure. Due to the problem of small sample sizes, Rosenthal described the tendency "to retain as many pairs as possible by getting as much information as one can about all pairs of twins and making one's best judgment about zygosity and concordance for each pair."[217] Because of the large size of Kallmann's sample, Rosenthal believed that it was difficult for Kallmann to personally investigate all of the pairs and that he sometimes had to rely on "lay testimony," which had been shown to be inaccurate much of the time. Rosenthal suggested that future investigations assign one or two people the task of determining zygosity, and two (presumably other) people the task of making diagnoses. Although zygosity determination has been a problem in past twin studies, it is not a major problem today as long as investigators use modern techniques such as blood typing or DNA analysis.

Sampling issues. In psychiatric twin studies there are three main ways in which twin subjects have been obtained: the resident hospital method, the consecutive admissions method, and the register method. In the resident hospital method, the researcher examines the records of psychiatric hospital patients who have been diagnosed with the disorder in question. The researcher then

---

216. Kallmann, 1946, p. 312.
217. Rosenthal, 1962b, p. 121.

attempts to discover if any of these patients are members of a twin pair. After a subject is found in this way, the researcher attempts to locate the co-twin of the affected person. The zygosity of the pair is determined, after which a diagnosis of the co-twin is made. Typically, the same researcher makes the diagnosis for both members of the pair. Sometimes a twin pair will be located because the researcher has been told of the existence of a diagnosed twin (or pair of twins) by a person connected with a particular hospital. The bias this creates is clear, as identicals' more similar physical appearance would make it more likely that hospital workers would know about a pair of concordant identical twins than fraternal twins. In addition, the co-twin of a hospitalized identical twin would be easier to locate.

The resident hospital method was employed in several early schizophrenia twin studies (Kallmann's 1946 study being the most prominent). Since the 1960s, there has been a consensus among twin researchers that this method creates a sample biased in favor of concordance. Rosenthal discussed some of the problems in obtaining samples for schizophrenia twin studies from resident hospital populations.[218] He observed that many people diagnosed with relatively mild cases of schizophrenia are never hospitalized. This skews the sample towards the most severely affected, who are known to be more concordant than those with milder conditions. In addition, the studied hospitalized twins were already diagnosed with the condition when they were seen by the investigator (although this problem exists in all studies sampling from hospital populations).

Another problem is that in countries and eras where a disorder is believed to have an important genetic component, it is more likely that an identical twin will be diagnosed with a psychiatric disorder if his or her co-twin has already been diagnosed with the disorder. Studies making blind diagnoses independent of hospital diagnoses can reduce this bias. Luxenburger argued in his 1928 study that samples should be obtained from twins identified from lists of people consecutively admitted to a hospital for a specific period of time. This consecutive admissions method avoids the problems associated with selecting twins who are already known to be affected with the condition in question, although the problems with sampling from hospital populations remain.

The third method identifies subjects from twin registers. Beginning in the 1940s, national registers were created in Denmark, Norway, and Sweden for the purpose of scientific research. Although twin registers existed before this time,

218. *Ibid.*

it is not clear whether they were created with scientific research in mind. The first attempt to create a national twin register for this purpose may have been 1939, when Hitler's Interior Minister Wilhelm Frick "ordered the registration of all twins, triplets, or quadruplets born in the Reich, for the express purpose of research to isolate the effects of nature and nurture in the formation of the human racial constitution."[219] A register allows researchers to sample from an entire population of twins and makes it easier to locate both members of a pair. If the investigator is able to locate and evaluate both twins independently of any other source, bias is reduced. However, most psychiatric twin studies check the names of those identified by the twin register with people whose name appeared on psychiatric records or registers. Thus, the old problems of relying on hospital records remains.

Criticism of the twin method by geneticists. In 1950, geneticist Bronson Price published a paper entitled "Primary Biases in Twin Studies." Price began with a provocative statement: "In all probability the net effect of most twin studies has been [the] underestimation of the significance of heredity in the medical and behavior sciences."[220] Contrary to most previous critics, Price argued that the twin method *under*estimated genetic contributions. The reason, according to Price, was the unique prenatal experience of twins. As examples, he listed the effects of mutual circulation and lateral inversions. Price concluded that the authors of individual case reports of identical twins tended to exaggerate the effects of the postnatal environment, and that interpretations of larger twin studies "have been about as mistaken."[221] He advised future twin researchers to determine, two or more months before birth, whether identical twins are monochorial or dichorial and to factor this information into their findings. Even so, Price believed that "such a study might show that the twin method, as ordinarily applied, is too crude for the purposes of modern nature-nurture studies."[222]

Geneticists James Neel and William Schull criticized the twin method in their 1954 textbook *Human Heredity*. They noted that concordant identical twins are more often reported and are therefore more likely to come to the investigator's attention than discordant pairs. They also cited Price's criticism of the

---

219. Proctor, 1988, p. 43. I am not aware of any twin researcher mentioning the Nazis' creation of a national twin register.
220. Price, 1950, p. 293.
221. *Ibid.*, p. 343.
222. *Ibid.*, p. 343.

twin method's "primary biases." Regarding postnatal influences, Neel and Schull discussed the assumption that identical and fraternal twins share similar environments, stating, "It has been clearly shown that rarely is this the case," because "Identical twins are treated more similarly than are fraternal twins by both their families and their acquaintances."[223] Neel and Schull concluded that "twins have contributed little which may be extrapolated to other genetic situations," and that "In its present context, the twin method has not vindicated the time spent on the collection of such data."[224]

TWIN RESEARCH FROM 1960 TO 1980

The end of the 1950s was not a good time for twin research for several reasons. The memories of the Holocaust and collaboration of genetic researchers with the Nazi regime were still fresh. In addition, environmental theories were popular. Finally, we have seen that the methods and assumptions of the twin method were criticized by environmentalists and geneticists alike.

Don Jackson's Critique of Schizophrenia Twin Research

In 1960, a chapter containing a powerful critique of genetic studies of schizophrenia appeared in *The Etiology of Schizophrenia*. Its author was the psychiatrist Don Jackson, then Director of the Mental Research Institute in Palo Alto, California. Jackson's chapter was entitled, "A Critique of the Literature on the Genetics of Schizophrenia."[225] Jackson was a well-known pioneer of family systems theory and had been a co-author of a widely discussed article outlining the double-bind theory of schizophrenia.[226] At the time of Jackson's chapter there were five published schizophrenia twin studies, with Kallmann's 1946 report being the largest and most influential. Although Jackson's chapter was about schizophrenia, it was clearly relevant to twin studies of other psychiatric conditions and behavioral traits and, to a lesser extent, IQ. Jackson made eleven major points in his critique, and although all did not originate with him, together they made a strong case that the twin method contained serious and invalidating flaws:

---

223. Neel & Schull, 1954, p. 280.
224. *Ibid.*, pp. 280-281.
225. Jackson, 1960.
226. Bateson et al., 1956.

1) In discussing family (consanguinity) studies, it was noted that conditions can run in families for environmental reasons.

2) There were no genetic studies of schizophrenia in which researchers made diagnoses blindly. The results of these studies were therefore susceptible to the researchers' bias.

3) There were other sources of bias in the diagnostic process, such as the unreliability of schizophrenia diagnoses and the finding that people had a better chance of being diagnosed with schizophrenia the longer they stayed in the hospital. A sampling bias was introduced by the methods used to obtain twin subjects, which could lead to inflated concordance rates.

4) Contrary to genetic expectations, fraternal twins were more concordant than non-twin siblings, even though both sets have the same genetic relationship to each other.

5) Contrary to genetic expectations, female identical pairs were more concordant than male identical pairs.

6) Contrary to genetic expectations, female fraternal pairs were more concordant than male fraternal pairs.

7) Contrary to genetic expectations, same-sex fraternal pairs were more concordant than opposite-sex fraternals.

8) Individual case histories of reared-apart identical twins concordant for schizophrenia do not provide important evidence for genetic factors because they were few in number (two), and because the pairs grew up in similar environments and had an interactive relationship with each other.

9) Identical twins grow up in a more similar environment and are treated more similarly than fraternal pairs. Therefore, greater resemblance of identical pairs for schizophrenia could be explained by the more similar environment they experience.

10) The unique psychological bond or "ego fusion" of identical twins contributes to their higher concordance rate for schizophrenia on the basis of association and identification. Furthermore, the nature of the identical twinship itself might create conditions leading to the identity problems often experienced by people diagnosed with schizophrenia.

11) There is a striking similarity between reports of folie à deux (shared psychotic disorder) and the case histories of identical twins concordant for schizophrenia.

Jackson's chapter had an important impact on the debate over the causes of schizophrenia and raised further doubt that the twin method measured anything more than the greater environmental similarity and "ego fusion" experienced by identical versus fraternal twins.[227] In the period following its publication, twin researchers carefully considered Jackson's observations and attempted to improve their methodology. But most failed to understand that Jackson implied that the twin method was a fundamentally flawed instrument.

As we will see in Chapter 6, most schizophrenia twin researchers recognized that environmental factors influence schizophrenia concordance rate differences to some degree. But how did they know that environmental influences weren't *entirely* responsible for these differences? The answer is that they didn't — they only *hoped* that their studies had measured genetic influences. In 1966, Gottesman and Shields were willing to concede that the greater psychological identification of identical twins could affect concordance rates "provided that the same proportion of potential schizophrenics are held back from overt illness by identifying with a normal twin as those who become ill by identifying with an abnormal one."[228] They provided no evidence in support of this rhetorical attempt to balance the ledgers of the twin method. On what grounds, we might ask, did Gottesman and Shields insist on a one-to-one correspondence between those twins who became concordant for reasons of identification, and those who stayed "well" for the same reason? Why couldn't we assume that, because of identification, *five* pairs become concordant for every *one* that doesn't? Gottesman and Shields's reasoning constituted little more than wishful thinking in the service of keeping a theory intact.

Today, Jackson's critique is a forgotten document in the sense that the twin method is more popular than ever, even though *none of Jackson's 11 major points has ever been refuted.*[229] Twin researchers failed to answer Jackson's most important criticisms, and many agreed with him on several points. Those arguing against him typically focused on Jackson's theory (Point #10) that the unique psychological bond of identical twins not only explains their elevated schizophrenia concordance rates, but is an additional *cause* of "madness." When

---

227. Jackson did, however, believe that "it seems likely that hereditary factors do play a part in at least some of the schizophrenias . . ." (1960, p. 80).
228. Gottesman & Shields, 1966a, p. 55.
229. For a detailed discussion of this point, see Joseph, 2001b. In this article I listed 12 points. I have dropped one here because it is redundant.

defenders of the twin method directly address Jackson's ideas, this is the point they usually focus on. Let's examine the citations in question. According to Jackson:

> The identity problems of the schizophrenic, most often stressed by psycho-dynamically oriented writers, could find no better nidus than in the inter-twining of twin identities, in the ego fusion that in one sense doubles the ego (because the other is felt as part of the self) and in another sense halves it (because the self is felt as part of the other).[230]

> If the psychodynamic thesis is correct, if ego fusion in a particular family environment can be expected to lead to joint madness, then a plausible hypothesis — contrary to the genetical hypothesis — would be that, according to the degree of likeness in siblings, we will find an increased concordance for schizophrenia, without concern for genetic similarity.[231]

> Psychosis by association apparently requires the nidus of social isolation for its hatching . . . These characteristics increase the separation of the twin-pair from the rest of the world and foster a joint ego fusion. . . . The attempt of one to be like the other is not dissimilar to the ego fusion of some twins.[232]

Here, Jackson discusses how the intimate bond and "ego fusion" experienced by identical twins might weaken or blur the psychological boundaries between them. His ego fusion theory describes a psychological process that could cause the co-twins of people diagnosed with schizophrenia to receive the same diagnosis much more frequently than their non-twin siblings. As an alternative "plausible hypothesis" to genetic interpretations of the twin method's results, Jackson theorized that "According to the degree of likeness in siblings, we will find an increased concordance for schizophrenia, without concern for genetic similarity." Slightly modifying Jackson's position, we could say that according to the degree of *environmental similarity* among siblings, we would expect greater behavioral similarity, without concern for genetic relationship. As Jackson argued, this is a plausible interpretation of identical-fraternal twin comparisons (see Chapter 3).

During this period, Rosenthal published a series of critical articles on the twin method, which I view as damage control in response to Jackson's critique.[233] Rosenthal outlined several problems in schizophrenia twin research,

---

230. Jackson, 1960, p. 66.
231. *Ibid.*, p. 67.
232. *Ibid.*, p. 69.

many of which had been discussed by Jackson, and offered suggestions of how twin research could be improved. Like Jackson, Rosenthal pointed to higher concordance rates among female versus male twins, and among same-sex fraternal versus opposite-sex fraternal pairs. In response to Jackson's Point #10, Rosenthal argued that identical twins were not diagnosed with schizophrenia more often than members of the single-born population. However, subsequent data have shown that this question remains open.[234]

Rosenthal concluded his series of articles in 1962 by writing that "the total weight of the evidence" from the schizophrenia twin studies "still strongly favors [the genetic] hypothesis, despite their weaknesses."[235] However, towards the end of his career Rosenthal concluded that both family and the twin studies are "confounded," and that "one can draw conclusions about them only at considerable risk."[236]

Twin researchers have repeatedly addressed Jackson's "ego fusion" theory (Point #10) in their attempts to defeat Jackson's entire argument. Thus, it has become their "straw man" in the case against Jackson. Kringlen's position is typical:

> The findings furthermore cast doubt on Jackson's (1960) "confusion of ego identity theory" of schizophrenia. . . . According to Jackson, the monozygotic twinship involves problems of "isolation, association and identification." This hypothesis stimulated Rosenthal (1960) to investigate if schizophrenia in fact was more frequent in twins. . . . Rosenthal analyzed the twin investigations published by Luxenburger . . . and Essen-Möller and could not support this theory of Jackson.[237]

However, Kringlen and others failed to realize that their argument does not detract from Jackson's basic position, because his theory does not require identical twins to develop schizophrenia at rates higher than the single-born population. Jackson argued that the confusion of ego identity, experienced most strongly by identical twins, could explain why the co-twin of a person diagnosed with schizophrenia is more likely to receive this diagnosis than another family member with whom the twin does not share comparable levels of ego fusion. His

---

233. See Rosenthal, 1960, 1961, 1962a, 1962b.
234. Joseph, 2001b.
235. Rosenthal, 1962b, p. 132.
236. Rosenthal, 1979, p. 25.
237. Kringlen, 1967a, p. 61.

theory centered on why the *second* twin is diagnosed with schizophrenia, not the first.

Although twin researchers have failed to refute any of Jackson's 11 major points, subsequent reviewers have argued that Jackson's theories do not hold up because, they claim, twins are no more susceptible to schizophrenia than are members of the single-born population. However, Jackson's most telling point was that — among pairs with the same genetic relationship to each other — those pairs experiencing a more similar environment and closer emotional bond were consistently more concordant for schizophrenia. For Jackson, this suggested that the identical-fraternal concordance rate difference (already inflated by methodological bias) could be explained on environmental grounds. Using today's behavior genetic terminology, Jackson argued that the theoretical basis of the twin method — the identical-fraternal "equal environment assumption" — is false.

Nadir

At the outset of the 1960s, twin researchers were as isolated as they had ever been. Although Jackson's critique had dealt it a severe blow, the twin method was not quite at the point of being discarded as a pseudoscience. Twin research was able to withstand another blow in 1963, when Pekka Tienari reported a 0% identical twin concordance in his Finnish schizophrenia study.[238] This contrasted dramatically with Kallmann's reported 86% age-corrected rate in his 1946 study. On a more positive note for twin research, the reared-apart twin studies of Shields in 1962 and Juel-Nielsen in 1965 were published during this period. Both investigators concluded in favor of important genetic influences on personality and IQ. In 1966, British psychologist Cyril Burt published the final paper of his subsequently discredited IQ study of 53 pairs of reared-apart twins (see Chapter 4).

Leading twin researchers held a meeting in the fall of 1965 under the auspices of the World Health Organization (WHO). These researchers upheld the validity of twin research while calling attention to some of its methodological shortcomings. They recognized the critics' argument that identical twins experience a greater environmental similarity than fraternals, but concluded that "most shared post-natal experiences of MZ twins are probably not qualitatively different from those shared by DZ partners or even by sibs."[239] On the other

---

238. Tienari, 1963.

hand, they conceded that it "is almost impossible to refute" the critics' argument that higher identical twin concordance could be explained on environmental grounds.[240] The WHO report mentioned the possible benefits of using the co-twin control method, while cautioning that "one would not, of course, risk exposing randomly selected partners in a series of MZ twins to a serious disease," as Mengele had in Auschwitz.[241]

Gottesman and Shields published their British schizophrenia twin study in 1966. They presented their study in a more detailed format in 1972, including case history information as well as diagnoses made by a panel of well-known schizophrenia experts.[242] Gottesman and Shields's stated purpose for performing their study was "to find out what the effects would be on the results when care was taken to avoid, or to make provision for, the alleged sources of error and bias in the earlier 'classical' studies conducted before 1953."[243] Ironically, Gottesman and Shields controlled for most of these sources of error and bias except for the most important and controversial, that is, the argument that the greater resemblance of identical versus fraternal twins can be explained by the former's greater environmental similarity.

Twin Research Makes a Comeback

In the mid-1960s the tide began to turn in the direction of a greater acceptance of important genetic influences on psychiatric disorders. For the most part this was not, as some have put it, due to the work of "ingenious" twin researchers.[244] More important were the schizophrenia adoption studies performed in Oregon and Denmark. In Chapter 7, I review these studies in great detail. Suffice it to say here that, despite their glaring weaknesses, these adoption studies had a tremendous impact. Schizophrenia (and by implication other psychiatric disorders) was increasingly seen as having an important genetic component. Given this shift in the genetic direction, it was natural that twin studies would be seen in a new light and would not have to carry the burden of supporting the genetic position alone. Schizophrenia twin studies could now be seen as having measured the important genetic component

239. World Health Organization, 1966, p. 115.
240. *Ibid.*, p. 115.
241. *Ibid.*, p. 116.
242. Gottesman & Shields, 1972.
243. *Ibid.*, p. xvi.
244. Neale & Oltmanns, 1980, p. 174.

believed to have been established (albeit incorrectly, as we will see in Chapter 7) by adoption research. Gottesman and Shields could defend the validity of the twin method in 1972 by pointing to schizophrenia adoption research:

> The difference in identical vs. fraternal twin concordance rates is not due to aspects of the within-family environment that are more similar for MZ than DZ twins, although there are many such aspects. Studies of MZ twins reared apart as well as adoption and fostering studies show a markedly raised incidence of schizophrenia among relatives even when they were brought up in a different home by nonrelatives.[245]

And four years later they would write, "The old objection that elements of the shared environment 'caused' the higher concordance rates in MZ twins should have been put to rest with the results of the adoption studies" and individual pairs of reared-apart twins.[246]

Thus, adoption research helped legitimize twin research. Family studies could not play this role because they were widely seen as even more vulnerable to environmental confounds than the twin method. At least when assessing behavior and psychopathology, the twin method received a new lease on life in the late 1960s and has been on the ascendancy ever since.

TWIN RESEARCH IN THE 1970S

The 1970s were years of growth and controversy in twin research. Studying twins became a more respectable enterprise during this period, although the old objections remained in many quarters. In 1969, University of California, Berkeley psychologist Arthur Jensen cited twin studies in his controversial article on racial differences in IQ. In the context of this controversy Leon Kamin, in his 1974 work *The Science and Politics of IQ*, addressed problems in IQ twin research at length. Kamin observed that identical-fraternal comparisons are based on the assumption that "environmentally produced within-family variance is equal for the MZ and DZ samples,"[247] but argued that "there is much evidence to demonstrate that in fact the environmental differences to which MZs and DZs are exposed is not equal."[248]

---

245. Gottesman & Shields, 1972, p. 318.
246. Gottesman & Shields, 1976a, p. 371.
247. Kamin, 1974, p. 96.
248. *Ibid.*, p. 97.

61

On the behavioral side, the rising popularity of genetic explanations helped legitimize twin research, and many of Jackson's arguments were ignored or forgotten (Jackson died in 1968). By 1972, Gottesman and Shields were prepared to relegate Jackson's critique to the status of a historical footnote:

> [In 1960] came the book edited by Jackson . . . *The Etiology of Schizophrenia*, containing a critique by him of the "literature of the genetics of schizophrenia." Even today, a decade later, the chapter is considered in many quarters to be the definitive rebuttal to genetic thinking and an excuse to ignore any developments generated by genetic hypotheses; the critique dealt especially harshly with Kallmann's (1938, 1946) findings and conclusions. . . . Although a few of Jackson's points were well taken, most have since been shown to have no force.[249]

Ten years later, Gottesman and Shields saw Jackson as little more than a "vocal critic of genetic interpretations."[250]

Although adoption studies helped breathe new life into twin research, there was a rivalry between schizophrenia twin and adoption researchers during this period. In their first schizophrenia adoption study publication, Seymour Kety and associates wrote that family and twin results were "inconclusive," and that they "fail to remove the influence of certain environmental factors."[251] By the late 1970s, Kety (and Rosenthal, as we have seen) concluded that the twin method was unable to separate potential genetic and environmental factors:

> Twin studies are a more compelling form of genetic data [than family studies], but even twin studies depend on the assumption that the only thing that differentiates monozygotic from dizygotic twins is their genetic relatedness, and that environmental factors are somehow canceled out or randomized. But that is not the case. Monozygotic twins share much of their environment as well as their genetic endowment. They live together; they sleep together; they are dressed alike by parents; they are paraded in a double parambulator [sic] as infants; their friends cannot distinguish one from the other. In short, they develop a certain ego identification with each other that is very hard to dissociate from the purely genetic identity with which they were born.[252]

---

249. Gottesman & Shields, 1972, p. 4.
250. Gottesman & Shields, 1982, p. 73.
251. Kety et al., 1968, p. 345.
252. Kety, 1978, p. 48.

Prominent twin researchers had mixed feelings about schizophrenia adoption research. While generally supportive of them and quite willing to bask in the greater acceptance of genetic theories that adoption studies helped bring about, many were unwilling to accept the twin method's diminished position. Norwegian twin researcher Einar Kringlen lamented in 1976 that "One has the feeling that results from [schizophrenia] twin research have been completely ignored."[253] In 1978, Gottesman wrote, "To state that adoption studies permit the disentangling of heredity from environment may be too pat and misleading."[254] And Gottesman and Shields wrote in 1982, "The strongest evidence implicating genetic factors in the etiology of schizophrenia comes from twin studies."[255]

Still, this was a period of growth for twin research as well as for the fledgling behavior genetics field, which had weathered the Jensen controversy. The first issue of *Behavior Genetics* appeared in 1970. Twin research has been referred to as the "Rosetta Stone" of this field.[256] Psychiatric genetics also received a tremendous boost from schizophrenia adoption research. The stage was now set for the further growth of twin research in the 1980s and 90s.

TWIN RESEARCH FROM 1980 TO THE PRESENT

Twin research and behavior genetics carried their momentum into the 1980s. This decade saw the publication of reared-apart twin studies performed in the United States (Minnesota) and Scandinavia. In particular, the Minnesota studies received a great deal of attention (see Chapter 4), and helped strengthen behavior genetics and twin research in general.

Just as it had in the 1920s, the relatively conservative political climate in the United States and elsewhere provided fertile ground for genetic theories and research. The election of Ronald Reagan in 1980, and the subsequent cutbacks in social programs, were aided by a belief that social problems were the result of biology, as opposed to unequal income distribution, racism, unemployment, etc. Genetic theories were used, as they historically have been used, to provide "scientific" justification for the inequalities of the capitalist economic system.

---

253. Kringlen, 1976, p. 429.
254. Gottesman, 1978, p. 63.
255. Gottesman & Shields, 1982, p. 71.
256. Bouchard, 1999, p. ix.

Genetic and biological theories were also promoted by pharmaceutical companies, who were poised to earn enormous profits from the sale of psychotropic drugs. In addition, memories of the Nazis' use of genetic theories in the service of genocide were fading. American twin researcher David Lykken, for example, bemoaned the fact that Nazi atrocities and the "cruel experiments conducted by Dr. Mengele" at Auschwitz prevented "an entire generation of civilized people" from looking at the role of inherited factors in crime.[257]

Kenneth Kendler and his colleagues began to publish studies based on the Virginia Twin Register. At the beginning of this work in 1983, Kendler published an article in defense of the twin method and the equal environment assumption, arguing that the twin method is not biased by the greater environmental similarity of identical twins.[258] Concluding that the twin method is a valid instrument for the detection of genetic influences on psychiatric disorders, Kendler became the first twin researcher to make a serious attempt to provide theoretical justification for the equal environment assumption, although his analysis was curiously ahistorical and omitted many important facts. For example, Kendler virtually ignored Jackson's critique and failed to answer Jackson's most important arguments. Although Kendler is a leading schizophrenia researcher, he has failed to explain the fact that same-sex fraternal twins have more than double the pooled concordance rate for schizophrenia as opposite sex fraternals, a finding difficult to explain on genetic grounds (see Chapter 6). Nevertheless, Kendler has been one of the world's most famous and prolific twin researchers, and his group's studies have been published in leading medical and psychiatric journals.

The 1980s and 1990s saw an increase in the number of behaviors studied by twin researchers. Predictably, most twin studies found significantly greater identical versus fraternal twin resemblance, leading their authors to conclude in favor of genetic influences on the condition or trait under study.

The greater acceptance of twin research continued into the 1990s. The beginning of the decade saw the publication of the Minnesota reared-apart twin results in *Science*, and twin studies were found more frequently in the pages of leading psychological and psychiatric journals. Popular accounts of behavior genetic research appeared in newspapers, magazines, and books. Concurrently, the Human Genome Project (HGP) was initiated with great fanfare in 1990 and

257. Lykken, 1995, p. 71.
258. Kendler, 1983.

continued its work throughout the decade. In 2001, Kendler identified "five areas of development for twin studies [that] are likely to be particularly fruitful in the coming years." These included longitudinal twin studies, studies of twins and their offspring, the use of biological markers ("endophenotypes") to identify psychiatric conditions, improved statistical methods, and "increased interactions between twin and gene-finding approaches."[259] Human genetic investigators inferred the existence of genes from the results of previous studies of twins and adoptees, and many were by now personally involved in molecular genetic research. Twin research, initiated by Galton five generations earlier, entered the new millennium as strong and accepted as it ever had been.

CONCLUSIONS

I have covered the most important aspects of the history of twin research; at times touching on disturbing times and events. According to William Wright, a journalist generally supportive of twin research, "It is certainly true that the history of twin research is one of the most appalling chapters in science, having been born in Galton's aristocratic notions of the natural worthiness of the English upper class, taken to its evil extreme by eugenicists, and too readily used by American scientists to rationalize racial injustice."[260] Unfortunately, three generations of twin researchers have failed to adequately document the "appalling" history of their discipline.

German racial hygienists and American eugenicists invented the classical twin method in order to convince the socially and politically powerful of the need to eliminate "hereditary tainting" from the population, by any means necessary. They already believed strongly in the importance of genetics and eugenics. But as Galton wrote in 1883, and Siemens noted in the first chapter of *Die Zwillingspathologie*, others remained unconvinced.

My goal in this chapter has been to tell a story that twin researchers have failed to tell. In the following two chapters, I will assess the methodological and theoretical underpinnings of twin research. Only then will we be able to determine its usefulness and validity. In the process we will see more clearly that, beginning with Galton, five generations of twin researchers have failed to

---

259. Kendler, 2001, p. 1012.
260. L. Wright, 1997, p. 33.

understand (1) that environmental factors are the most plausible explanation for the greater behavioral resemblance of identical versus fraternal twins, and (2) that methodological flaws, bias, and the failure to recognize environmental con-founds call into question their claims about studies of twins reared apart.

CHAPTER 3. THE TWIN METHOD: AN ENVIRONMENTALLY CONFOUNDED RESEARCH TECHNIQUE

> If identical twins have the same trait people are inclined to think that this proves that the trait is hereditary. The fallacy is most easily seen if we take an imaginary example. Suppose that a student of human heredity should hail from another planet and that he should be required to use the twin method to find out whether or not people's clothes were a direct consequence of heredity. He would find that identical twins were very often dressed alike, down to the small details, and that this was uncommon with fraternal twins. He would confidently conclude that the choice of clothes was almost an exclusively hereditary trait and might even suppose them to be part of the natural skin of the human animal by the exercise of superficial reasoning.
>
> — *Geneticist Lionel Penrose in 1973.*[261]

The twin method's comparison of the resemblance of reared-together identical versus reared-together same-sex fraternal twins (see Figure 2.1) has been the main way that twins have been used in genetic research. In the previous chapter, I discussed the critics' objections that the twin method is biased because identical twins share more similar environments than fraternals (a violation of Assumption #5, Figure 2.1), implying that some or all of the greater observed resemblance of identical twins is explained by their greater environmental similarity. This question is central to the validity of the twin method.

Critics also have pointed to several other methodological problems with studies using the twin method, such as the lack of an adequate and consistent definition of the trait under study, the use of non-blinded diagnoses, diagnoses that were made on the basis of sketchy information, inadequate or biased methods of zygosity determination, the unnecessary use of age-correction formulas, the use of non-representative sample populations, and the lack of adequate descriptions of the methods used in some of the studies. Moreover, other twin method assumptions outlined in Figure 2.1 are open to question in many cases. Although these problems are important, I will focus on the main theoretical assumption of the twin method for the following reason: if all methodological problems associated with the twin method were resolved, identical

---

261. Penrose, 1973, p. 90.

twins would still resemble each other more than same-sex fraternal twins for most traits (although differences would not be as great as those reported in methodologically flawed studies). Twin researchers are therefore correct that identical-fraternal concordance or correlational differences are real. We will see, however, that there is little reason to accept that identical-fraternal comparisons measure much more — or anything more — than the greater environmental similarity experienced by identical versus fraternal twins.[262]

THE EQUAL ENVIRONMENT ASSUMPTION (EEA)

Two Definitions

The validity of drawing genetic inferences from identical-fraternal comparisons is dependent on the "equal environment assumption" (EEA). The authors of every scientific article and popular work citing these comparisons as evidence in favor of genetics assume (explicitly or implicitly) that the EEA is valid — otherwise, they would conclude that the results are uncertain, or that they suggest the importance of environmental factors.

To begin this discussion, I will outline the two main ways that twin researchers have defined the equal environment assumption.

The traditional EEA definition. The traditional definition was used by most twin researchers prior to 1972 and continues to be used by some in the present day. The EEA is defined as the straightforward assumption that identical and fraternal twins experience similar environments and treatment, and is exemplified by the following quotation from a leading twin researcher: "The basic underlying assumption of the twin method is, of course, that environmental conditions of monozygotic twins do not differ from those of dizygotic twins."[263] Proponents of the traditional definition at times acknowledge that identical twins experience more similar environments than fraternals, and that the twin method might "overestimate heritability" in these cases.

The "equal trait-relevant environment assumption"[264] (referred to here as the "trait-relevant EEA"). An early example of this new definition is seen in a

---

262. In this chapter, I draw upon the ideas expressed in articles I have published on this subject, specifically, Joseph, 1998b, 2000c, 2000e, 2001b, 2002b.
263. Kringlen, 1967a, p. 20.
264. Carey & DiLalla, 1994.

1966 publication by Gottesman and Shields, who wrote that critics must show that the "environments of MZ twins are systematically more alike than those of DZ twins in features which can be *shown* to be of etiological significance in schizophrenia."[265] More recently, Kendler and associates defined the EEA as follows:

> The traditional twin method, as well as more recent biometrical models for twin analysis, are predicated on the equal-environment assumption (EEA) — that monozygotic (MZ) and dizygotic (DZ) twins are equally correlated for their exposure to environmental influences that are of etiologic relevance to the trait under study.[266]

Proponents of the trait-relevant EEA recognize that identical twins are treated more alike and experience more similar environments than fraternals, but claim (a) that the evidence shows that greater environmental similarity does not lead to greater psychological trait resemblance or to higher concordance for psychiatric diagnoses, and (b) that in order to invalidate the finding of genetic factors, critics must identify the trait-relevant environmental factors for which identical and fraternal twins experience dissimilar environments.

Today, Kendler and most other twin researchers recognize that the environments of identical twins are more similar than fraternal twins,[267] and there is plenty of empirical evidence in support of this position. Identical twins spend more time together than fraternals,[268] and more often dress alike, study together, have the same close friends, and attend social events together.[269] James Shields, in his 1954 study of normal twin school children, found that 47% of identical twins, but only 15% of fraternals, had a "very close attachment."[270] In 1967 schizophrenia twin researcher Einar Kringlen published a survey of 117 twin pairs (75 identical, 42 same-sex fraternal; where one or both were diagnosed with schizophrenia) which stands as one of the few post-1960 assessments of differences in twin association and "ego fusion," which in Chapter 2 we saw was central to Jackson's theory. According to Kringlen's survey, 91% of identical twins experienced "identity confusion in childhood," which was true for only

---

265. Gottesman & Shields, 1966a, pp. 4-5. Emphasis in original.
266. Kendler et al., 1993, p. 21.
267. e.g., Bouchard, 1993b; Gottesman & Shields, 1972; Kendler, Neale, et al., 1994; Morris-Yates et al., 1990; Rose, 1991; Scarr, 1968; Scarr & Carter-Saltzman, 1979.
268. Wilson, 1934.
269. Smith, 1965.
270. Shields, 1954, p. 234.

10% of the fraternal twins.[271] Kringlen also found that identical twins were more likely to have been "considered as alike as two drops of water" (76% vs. 0%), "brought up as a unit" (72% vs. 19%), and "inseparable as children" (73% vs. 19%). Kringlen's "global evaluation of twin closeness" showed that 65% of identical twins experienced an "extremely strong" level of closeness, which was true for only 19% of the fraternal pairs. Kringlen's findings illustrate the striking contrast between the environments, both psychological and physical, of identical and fraternal twins.

In the face of this type of evidence, as well as the existing widespread public belief that identical twins are treated more alike and experience more similar environments than fraternals, twin researchers began to adopt the trait-relevant EEA in the 1960s. Today, it predominates. I will look into the validity of the traditional and trait relevant definitions of the EEA a bit later, but first I would like to discuss how twin researchers reconcile their understanding that identical twins experience a greater environmental similarity than fraternals, with their continuing defense of the EEA.

Do Identical and Fraternal Twins Create Differing Environments?

We saw in Chapter 2 that early objections to the twin method by Holmes, Stocks, Wilson and others centered on the assumption that identical and fraternal twins share similar environments. These critics argued not so much that identical twins are treated more similarly because of societal traditions or, as Jackson stressed, that they share a closer psychological and emotional bond. Rather, they argued that identicals experience more similar environments *because of* their more similar genetic makeup. Although these early critics believed in the importance of genetics *in general*, they questioned whether the twin method was able to detect genetic influences on *specific* traits. Schizophrenia researcher Manfred Bleuler (son of Eugen Bleuler, the founder of the schizophrenia concept) made a similar argument in the 1940s, which he later summarized:

> According to the older view, it is axiomatic that identical twins are identical in their inherited predisposition (*Anlage*), while the environment in which the identical twin-partners live is on the average as different as the environment of the fraternal twin-partners. In reality, this last assumption is wrong. Identical twins, on the basis of their common inherited predisposition, create for themselves on the average a much more similar environ-

271. Kringlen, 1967a, p. 115.

70

ment than is the case with fraternal twins. . . . the old idea that the environment of twins varies without regard to their innate predisposition (*Anlage*) is quite erroneous; hence all further deductions which used to be based on this premise must be wrong.[272]

And in the 1970s, he wrote,

> For a long time, high concordance figures for schizophrenic twins were regarded as the "absolute scientific truth" for the great importance of heredity. . . . the line of reasoning of such rigorous precision is based on a fallacy. . . . The fact that two monozygotic twins might live in a more similar environment than two dizygotic twins was forgotten. . . . When monozygotics reveal high concordance figures for psychic disturbances, it could be caused just as easily by the great similarity in hereditary factors as by the great similarity in their human environment.[273]

Beginning in the 1950s, some twin researchers used the argument that identical twins create more similar environments (and that fraternals create more dissimilar environments) *in defense* of the validity of the twin method. For example, here is James Shields in 1954:

> In so far as binovular [fraternal] twins are treated differently from one another and more differently than uniovular [identical] twins, this is likely to be due, not so much to causes outside the twins as to innate differences in the needs of the binovular [fraternal] twins themselves, manifested by different patterns of behavior.[274]

Thus, Shields took the main objection of the critics and stood it on its head: the validity of drawing genetic inferences from identical-fraternal comparisons was now *supported* by the more dissimilar environments "created" by fraternal twins. Shields was attempting to shift the debate from *whether* identical twins experience more similar environments to *why* their environments were more similar. If identical twins' greater environmental similarity was caused by their greater genetic similarity, Shields seemed to argue, the twin method is valid.

This position was taken up by Kallmann in 1958:

---

272. M. Bleuler, 1955, p. 13.
273. M. Bleuler, 1978, p. 432.
274. Shields, 1954, pp. 239-40.

The popular notion that the behavior patterns of one-egg [identical] twins are alike chiefly because of unusual similarity in their early environments has yet to be substantiated. If confirmed, *the argument would only strengthen rather than weaken any correctly formulated genetic theory.* Psychodynamic concepts, too, are built on the premise that man is selective in respect to important aspects of his life experiences an so can be thought of as "creating his own environment" [emphasis added].[275]

Kallmann was arguing that if identical twins' more similar environments caused them to behave more similarly, this "would only strengthen" the genetic position! As dissident psychiatrist R. D. Laing commented, "With this two-headed penny it is not clear how Kallmann can lose."[276] As we will soon discover, most contemporary twin researchers use the very same "heads I win, tails you lose" argument in defense of the twin method's validity.

This argument lay dormant for several years until twin researchers rediscovered it in the 1960s, as they looked for new ways to legitimize their methods. For example, in 1968 twin researcher Sandra Scarr wrote, "Differences in the parental treatment that twins receive are much more a function of the degree of the twins' genetic relatedness than of parental beliefs about 'identicalness' and 'fraternalness.'"[277] The twin method's validity, according to twin researchers, was now based on determining why — not whether — identical twins experience more similar environments than fraternals.

The "Twins create their environment" theory. The EEA was originally defined as the assumption of equal environments between the two types of twin pairs. When empirical studies confirmed what common sense already knew — that identical twins experience more similar environments than fraternal twins — twin researchers retreated to the position that, as Kendler put it, identical twins "might *create* for themselves more similar environments" on the basis of an inherited "similarity in behavior."[278] We shall soon see, however, that the question of *why* identical twins' environments are more alike is irrelevant to the question of whether the twin method measures genetic influences. In addition this concept, which I call the "twins create their environment theory," is problematic.[279] The theory holds that children are born with a genetic predisposition to manifest certain personality and behavior types. Therefore, the behavior of

---

275. Kallmann, 1958, p. 543.
276. Laing, 1981, p. 143.
277. Scarr, 1968, p. 40.
278. Kendler, 1987, p. 706. For a reaffirmation of this position, see Kendler, 2000.

identical twins is more similar due to their identical genetic makeup. Because identical twins exhibit a genetically predisposed similarity of behavior, their parents, and others in their social environment, are induced to treat them more similarly. Fraternal twins, since they are more genetically dissimilar, inherit less similar personalities and therefore elicit less similar treatment. The theory concludes that twins create similar environments to the same degree as their genetic relatedness. But what is conveniently forgotten is this: according to this theory, twins' parents must also be genetically predisposed to manifest genetically determined behavior and response modes.

Proponents of the "twins create their environment theory" implicitly argue that the genetically predisposed behavior of children is able to greatly impact the (predisposed) personalities and response modes of their parents. Identical and fraternal twins are portrayed as genetically programmed to act in rough proportion to the number of genes they share in common, but parents are portrayed as readily able to change their behavior and treatment on the basis of their twins' personalities. Children are characterized by their inborn propensity to display inherited behavior, whereas parents are characterized by their plasticity in reacting to this behavior. If parents can change their behavior on the basis of environmental influences, as this theory maintains, it should follow that children (including twins) would also be able to adjust their behavior in response to environmental stimuli. Genetic influences or not, however, we would expect adults' behavior to be far less malleable than that of five-year-old children. Yet children are portrayed as having a greater ability to change the (genetically predisposed) behavior of adults than adults have to create, through their treatment, similar behavior in identical twins.

This argument was put forward in another form in 1972 by German psychiatric geneticist Edith Zerbin-Rüdin, the daughter of Ernst Rüdin:

> Twin studies, in particular, suggest the presence of a hereditary factor. The large difference between the concordance figures for MZ and DZ twins cannot be explained exclusively by the more similar environment of MZ twins. If MZ twins create a similar environment through their greater similarity, they do so because of the greater inherited similarity in their appearance and response modes. Thus, in a roundabout way, we still come back to the importance of heredity.[280]

279. This idea is referred to by behavior geneticists as the "reactive genotype-environment correlation." See Plomin et al., 1977.
280. Zerbin-Rüdin, 1972, p. 48.

Zerbin-Rüdin *asserts* that response modes are genetic in order to *prove* that behavioral differences are genetic. But her argument goes beyond mere logical error; Zerbin-Rüdin's reference to the striking physical similarity of identical twins must be addressed. Her assertion that "the greater inherited similarity in their appearance" would lead to more similar treatment of identical twins is, in fact, in accordance with the *environmentalist* position, which argues that greater physical resemblance contributes to more similar treatment by the social environment.

Furthermore, identical twins' physical resemblance is the result of the splitting of a fertilized egg; *it is not an inherited characteristic.* Rather, it is a biological phenomenon occurring in a zygote. For readers who view this claim with skepticism, I should emphasize that I am discussing the similarity of identical twins' appearance *to each other*, not to their parents. A randomly selected fertilized egg, taken from a Petri dish and inserted into the uterus of a mother-to-be would, should it split, produce identical twins as identical in appearance as those produced by a split zygote conceived within the biological mother's own body. On the other hand, the degree of physical similarity between fraternal twins *is* influenced by heredity, because in this case we are dealing with two separately fertilized eggs, which require common genes in order to produce physical similarity. But Zerbin-Rüdin missed this point entirely.

Although not discussed by Kendler, Zerbin-Rüdin and others, even if the "twins create their environment" theory were true, the twin method could still be measuring nothing more than environmental factors. Suppose for the moment (1) that the "twins create their environment theory" is correct, and (2) that schizophrenia is caused by brain damage due to the ingestion of mercury. Even if it were true that identical twins' environments were more alike because of their more similar inherited behavior, because identical twins spend more time together, eat more similar foods, etc., they would show higher concordance for exposure to mercury than fraternals, who spend less time together. It is therefore likely that identical twins would show higher concordance for schizophrenia than fraternals, yet schizophrenia would still be completely environmental in origin.

Theoretically, twins' behavior could be largely under genetic control, yet the more similar behavior of identical twins could still lead to higher levels of exposure to pathological environmental conditions for *specific* traits. Therefore, whether identical twins are more alike because they are treated more alike and share a more intimate bond (conventional wisdom), or because they create more

similar environments on the basis of more similar inherited behavioral tendencies (Shields/Scarr/Kendler), the simple fact that identical twins spend more
time together means that the equal environment assumption is false.

An example of twin researchers' theoretical confusion as they attempt to
explain the tortuous logic of the twin method is seen in the following passage by
Kendler:

> Twin studies are based on the assumptions that (a) monozygotic (MZ) and
> dizygotic (DZ) twins share their environment to approximately the same
> degree, but (b) MZ twins are genetically identical whereas DZ twins, like
> normal siblings, have on average only half of their genes identical by
> descent. Although the second of these assumptions is beyond question, the
> first, or "equal environment" assumption, has been a focus of considerable
> controversy. Several studies have shown that measures of the social envi-
> ronment (e.g., sharing the same friends, attitudes of parents and teachers,
> etc.) are more highly correlated among young MZ than among young same-
> sex DZ twins (Kendler, 1983). These results first appear to suggest that the
> equal environment assumption is false. However, reflections suggest
> another possible interpretation. Although the similarity in environment
> might make MZ twins more similar, the similarity in behavior of MZ twins
> might *create* for themselves more similar environments. As recently
> reviewed (Kendler, 1983), these two alternative hypotheses have been sub-
> ject to empirical test in at least nine different studies. Consistently, these
> studies suggest that the environmental similarity of MZ twins is the *result*
> and not the cause of their behavioral similarity. Whereas most of these
> studies examined such traits as intelligence and personality, one specifi-
> cally examined schizophrenia and found no evidence that concordance for
> schizophrenia was produced in MZ twins as a result of the similarity of
> their treatment by the environment (Kendler, 1983). Current evidence sup-
> ports the general validity of the equal environment assumption of twin
> studies.[281]

Kendler began by defining the EEA in the traditional sense, writing that
"twins share their environment to approximately the same degree." He then correctly pointed out that the equal environment assumption has been beset by
controversy, although as usual, he failed to discuss the names, dates, and ideas
associated with this controversy. He then conceded that the traditional EEA is
false, after which, upon "reflection," he offered a "possible" genetically derived
interpretation for identical twins' greater environmental similarity. Kendler

---

281. Kendler, 1987, p. 706.

cited nine separate studies in support of his hypothesis, including one examining schizophrenia. Although Kendler cited his 1983 "Overview" article as the source of the schizophrenia study, this article did not mention a study that "specifically examined schizophrenia and found no evidence that concordance for schizophrenia was produced in identical twins as a result of the similarity of their treatment by the environment." Kendler was probably referring to the summary of his and Robinette's unpublished 1982 study on the relationship between physical similarity and concordance for schizophrenia. The conclusion of this study, as stated in Kendler's 1983 article, was that "no correlation between degree of physical similarity and concordance rate for schizophrenia" was found.[282] Kendler and Robinette, then, only inferred environmental treatment differences on the basis of physical appearance.

Finally, while conceding that identical twins are treated more alike than fraternals, Kendler concluded that the current evidence supports the validity of the equal environment assumption, without changing his definition of the EEA. According to Kendler:

(a) The equal environment assumption holds that identical and fraternal twins "share their environment approximately to the same degree."

(b) Identical twins' environments are more similar than the environments of fraternal twins.

(c) The equal environment assumption is valid.

Logic dictates that Kendler would have to change his EEA definition for conclusion C to follow from premises A and B, but he does not do this. Thus we are supposed to believe something is true after he has just finished telling us that, as he defined it, it is false.

Kendler's Theoretical Retreat

Although Kendler seemed to have staked everything on the theory that twins create their own environments, by 1994 he realized that for many parents, their differential treatment of identical and fraternal twins was an "approach," not simply a reaction to their twins' behavior:[283]

> The tendency for parents of MZ twins to treat their offspring more similarly than parents of DZ twins is, therefore, unlikely to result entirely from a greater similarity in behavior as children of MZ v. DZ twins.[284]

282. Kendler, 1983, p. 1415.
283. Kendler, Neale, et al., 1994, p. 588.

Thus, Kendler recognized that parents' rearing *approaches* might influence the greater behavioral similarity of identical twins. Kendler and colleagues might have closed up shop at this point, but they argued that while "these results may seem like prima facie evidence for rejecting the twin method, such a step would be premature":

> Differential parental treatment of MZ and DZ twins would invalidate twin studies of psychiatric disorders *only if* the type of parental treatment for which MZ twins were more similarly exposed than DZ twins influenced the risk for the psychiatric disorders under examination.[285]

For Kendler, therefore, to invalidate the twin method it must be shown that identical twins' more similar treatment could increase their concordance for psychiatric disorders. According to Kendler, the EEA predicts that identical twins "would develop similar phenotypes regardless of the similarity of their social environment."[286] But a simple example casts doubt on Kendler's prediction. Suppose that a pair of identical twins is separated at birth and raised in different social environments. One twin is shifted from one foster home to another, is beaten, unloved, neglected, and abandoned several times during childhood. The other twin is raised in an exceptionally caring home in which he or she is loved and treasured, respected, and is well taken care of. For Kendler's theory to hold, he must conclude that both twins would be at equal risk for childhood or adult psychiatric diagnosis. However, it is unlikely that the twins in this example would be at equal risk for psychiatric disorders, as there is considerable research evidence indicating that these experiences contribute to the causes of, or prevent the appearance of, these disorders.[287]

A similar hypothetical scenario has been discussed by Kendler himself:

> If we selected 100 genetically diverse individuals and took them on a "love-boat" cruise in which all of their needs (e.g. entertainment, food, and love) were met, we would find, I predict, both low levels of anxiety and depression and very little variability. Almost everybody would be happy. Suppose, however, that we then expose these same 100 individuals to a highly stressful experience, such as combat. The average level of anxiety and depression

---

284. *Ibid.*, p. 588.
285. *Ibid.*, p. 588.
286. Kendler, 1983, p. 1414.
287. Bentall, 2003; Read et al., 2004.

would increase but, more important, so would variability. Some individuals who are good at coping would deal effectively with the adversity and demonstrate few psychiatric symptoms. Others, with poorer coping abilities, would become very symptomatic.[288]

Here, Kendler recognized that a given individual could be "happy" or "depressed" on the basis of a good or bad environment. There is no disputing this. But Kendler's EEA theory cannot allow that an identical twin sent into combat after taking the cruise would be more likely to become depressed or anxious than his or her genetically identical co-twin who remains on the "love-boat," since as we have seen, Kendler insists that identical twins would develop the same phenotype regardless of the similarity of their environments. From the genetic standpoint, however, there is no difference between *an individual* placed in different environments, and *identical twins* placed in different environments.

To the extent that Kendler's example as analogous to a twin pair's rearing environment, because virtually everyone (including Kendler) agrees that identical twins share more common experiences and are treated more alike than fraternals, they would be much more likely, for example, to go on a cruise or into combat *together*. Conversely, the more variable environments of fraternal twins (or even, as Kendler would argue, their more dissimilar genetic relationship) would make it much more likely that one twin would go into combat, while the other would choose to stay on the cruise. Because Kendler would consider the twin method "invalid" if such environmental differences "influenced the risk for the psychiatric disorders under examination," his position that the same individual could be depressed or happy — depending upon which environment he or she experienced — contradicts the main premise of his defense of the twin method. Nevertheless, in 2001 Kendler continued to regard the twin method as "a vibrant part of the field of psychiatric genetics . . ."[289]

A CRITIQUE OF THE TRAIT-RELEVANT EQUAL ENVIRONMENT ASSUMPTION

Family Studies and the Trait-Relevant EEA

Having subtly redefined the basic theoretical assumption of the twin method by adding the crucial trait-relevant condition, twin researchers charged

---

288. Kendler, 1995, p. 896.
289. Kendler, 2001, p. 1005.

critics with the responsibility of identifying trait-relevant environmental factors. According to twin research Thomas J. Bouchard, Jr.,

> The equal environment assumption [of the twin method] is required only for trait relevant features of the environment; features of the environment that have causal status. Causal status must be demonstrated, not assumed. . . . It is absolutely mandatory that Hoffman demonstrate that the differential treatments she cites have a causal influence on the traits whose similarity she is trying to explain. This is a very difficult task.[290]

However, if identical twins are treated more alike, spend more time together, have more common friends, and experience greater levels of identity confusion, they are more likely to be similarly exposed to "trait-relevant" environmental factors — known or unknown. Contrary to the wishes of the trait-relevant EEA proponents, *twin researchers* bear the burden of proof for showing that identical and fraternal twins are not differentially exposed to potentially relevant environmental factors. A basic tenet of science is that the burden of proof falls on the claimant, not on critics.[291]

The trait-relevant EEA becomes even more implausible when we realize that, as we saw in Chapter 1, most genetic researchers now recognize that the results of *family* studies are plausibly explained by environmental factors. However, genetic researchers do not make the trait relevant argument for family studies. On the contrary, they recognize that, because family members share a common environment, genetic conclusions based on family data are confounded by environmental factors. Similarly, the twin method is confounded by the greater environmental similarity of identical twins versus fraternal twins, even if (like family studies) the specific trait-relevant environmental factors are unknown. An example of genetic researchers' failure to apply their observations about family studies to the twin method is found in an article by twin researchers Michael Lyons, Kendler, and their colleagues. Although these investigators recognized that family studies are "confounded" because "offspring are likely to share exposure to toxic or infectious agents that could lead to similar outcomes for the siblings,"[292] they failed to recognize that *the twin method is similarly confounded by the fact that identical twins share exposure to potential toxins to a greater*

---

290. Bouchard, 1993b, p. 33.
291. Lilienfeld et al., 2003.
292. Lyons et al., 1991, p. 124.

degree than fraternal twins. Lyons and colleagues ask for, in effect, a sort of "special exemption" for the twin method in the face of the overwhelming evidence that, in addition to sharing a greater genetic similarity, identical twins experience more similar environments than fraternals.

Without realizing it, trait-relevant EEA proponents have redefined the assumption in a way that ultimately must lead to the obsolescence of the twin method. If we extend their logic to family studies, it follows that the familiality of a condition demonstrates its genetic basis, unless it can be shown that the affected families were exposed to "trait-relevant" environmental influences. Because family studies would thereby prove the existence of genetic factors, there would be little reason to conduct twin studies at all. Therefore, the now definitive nature of the new "trait-relevant" family studies will eventually drive the twin method out of existence as a research method. *The trait-relevant equal environment assumption has transformed the twin method into little more than a special type of family study.*

In family studies, diagnosed individuals share a more similar environment with members of their family than with members of the general population. In twin studies, identical twins share a more common environment and a greater psychological association than fraternal twins. Although most twin researchers do not dispute this, they argue that the correlation between common genes and common environment confounds family studies but *does not* confound the twin method. Their argument, however, could just as easily be used in support of drawing genetic inferences from family studies. They could argue, for example, that although family members share a more common environment than shared by a group of randomly selected members of the population, critics (of drawing genetic conclusions from family studies) incorrectly conclude that family members are more likely to be exposed to trait-relevant environmental factors.

But twin researchers do not make this argument. Instead, they reserve the trait-relevant EEA *for the twin method* while correctly understanding that environmental factors confound genetic inferences from family studies. In doing so they commit the logical fallacy of "special pleading" — that is, they apply standards to one research method but refuse to apply them to another method equally worthy of these standards. In fact, their decision to place "the burden of proof" on critics of drawing genetic inferences from the twin method — but not on critics of drawing genetic inferences from family studies — is entirely arbitrary and contrary to basic scientific principles. A broad overview of the family study method has led twin researchers, correctly, to the understanding that genetic and environmental influences cannot be disentangled using this method. For this reason, genetic researchers do not test for family members' "trait-relevant" exposure to etiologically relevant factors. Twin researchers, however, have failed to explain why the trait-relevant condition is applicable to the twin method, but not to family studies.

What Leading Twin Researchers Say About the EEA

Let's examine some of the ways in which leading twin researchers have discussed and defended the validity of the EEA, which will be followed by my commentary. I have chosen certain passages in order to make particular points, but I do not mean to imply that these investigators' views differ greatly. They are actually quite similar.

Plomin, DeFries, McClearn, and Rutter. Plomin, DeFries, McClearn, and Rutter are the authors of a 1997 textbook on behavior genetics, where they defined the EEA as follows:

> The *equal environments assumption* of the twin method assumes that environmental similarity is roughly the same for both types of twins reared in the same family.... The equal environments assumption has been tested in several ways and appears reasonable for most traits.[293]

They argued that identical twins share a more common environment in part because "some experiences may be driven genetically," concluding that "Such differences between identical and fraternal twins in experience are not a violation of the equal environments assumption because the differences are not caused environmentally."[294] This is a problematic position, as seen in the following example. Suppose a twin study is performed to investigate possible genetic influences on lung cancer. For the purpose of this example, we will assume that there is a genetic basis for the desire to smoke tobacco, meaning that identical twins are more likely to be concordant for cigarette smoking than fraternals. According to Plomin and associates' logic, since *tobacco smoking* is "driven genetically," the higher identical twin *lung cancer* concordance rate is evidence of

---

293. Plomin, DeFries, et al., 1997, p. 73. Earlier, Plomin & DeFries (1985, p. 17) held a position very similar to the tradition EEA definition: "In general, identical twins do not appear to be treated much more similarly than fraternal twins. Even though identical twins are slightly more similar than fraternal twins on a few environmental measures, these experiential differences do not make a difference in their behavior."
294. *Ibid.*, p. 73.

a genetic predisposition for lung cancer — a clearly erroneous conclusion because lung cancer could still be completely explained by exposure to tobacco smoke, which is an environmental occurrence. As another example if, as some maintain, there is a gene for "novelty seeking," it is more likely that both members of an identical twin pair versus a fraternal pair would attempt to climb Mt. Everest.[295] Could we then conclude in favor of a genetic predisposition for frostbite?

Rowe. In his 1994 book *The Limits of Family Influence*, behavior geneticist David Rowe wrote that "the question is not whether MZ twins receive more similar treatments (they do, and to claim otherwise would be foolish), but whether these treatments influence a particular trait."[296] For Rowe, while the few remaining proponents of the traditional EEA are "foolish," the equal environment assumption remains tenable because (according to research) greater intrapair environmental similarity does not predict twin resemblance for psychological traits.

Rowe addressed the environmentalist argument that societal treatment of identical twins on the basis of their greater physical similarity might cause them to behave more similarly, claiming that "matching people in physical appearance should have little effect on similarity in their psychological traits . . ."[297] Rowe then cited an example:

> How alike in personality or musical talent is even the best Elvis Presley lookalike to the King of Rock and Roll? If we gathered 10 Elvis lookalikes together, would they be alike in personality at all? The remarkable similarity of MZ cotwins is attributable to genes' creating matching neurons. Personality and temperament reside in the brain, not a face.[298]

But would Elvis have been the "King of Rock and Roll" had he looked like Winston Churchill, or if his facial pigmentation, like Chuck Berry's, had been dark? One could argue that Elvis, while certainly talented, would not have been a major figure in the history of music were it not for his Caucasian ancestry plus his appeal to some for his attractiveness, and to others for his "cool." Presley did not write his own music and was dependent on skilled songwriters to generate hits; they could have sold their music to other performers had Presley not gained

295. Ebstein et al., 1995.
296. Rowe, 1994, p. 45.
297. *Ibid.*, p. 48.
298. *Ibid.*, p. 48.

a mass following. Would an identical twin brother of Elvis have had the same chance of being the King if he had sustained serious burns and had disfiguring scars on his face? Would movie studios have cast Elvis in several feature films, had he been unattractive?

According to Rowe, "personality and temperament reside in the brain, not a face." Thus, Rowe seemed to discount the relevance of environmental factors in shaping psychological traits, in addition to exhibiting questionable logic. Would he argue that education is unimportant because "learning takes place in the brain, not in a book," or that post-traumatic stress reactions are unrelated to trauma because "personality and temperament reside in the brain, not on a battlefield"? In fact, the logic of Rowe's argument demands that we not stop at the face, and that we discount all environmental and sensory stimuli as affecting personality and temperament, since these also do not "reside in the brain."

Faraone, Tsuang, and Tsuang. According to psychiatric geneticists Faraone, Tsuang, and Tsuang,

> The validity of twin studies . . . requires us to assume that differences between MZ and DZ twins are due to genes; we must assume that environmental determinants of similarity are identical for the two types of twin-pairs. But if the similarity of MZ twin environments is greater than that of DZ twin environments, then the twin method will overestimate the importance of genetic influences.[299]

Faraone and colleagues pointed to several studies showing that identicals experience more similar social environments, "Thus, twin studies may overestimate heritability if these differences in environmental similarity are etiologically relevant to the disorder under study."[300] The claim that etiologically relevant environmental factors might "overestimate heritability" is common, and deserves a brief comment.

The question centers on whether environmental factors influencing identical-fraternal differences point to genetic influences being overestimated, or whether they indicate that the twin method is unable to disentangle these potential influences. Because identical twins experience more similar environments, one could legitimately transform the twin method into a measure of the differing *environments* of the identical and fraternal twinships, based on the assumption that genetic factors are not involved. We might call this the "GEA,"

---

299. Faraone, Tsuang, & Tsuang, 1999, p. 39.
300. *Ibid.*, p. 39.

or "genes excluded assumption." Theoretically, any twin study designed to assess the genetic component of a trait could be transformed into a twin study utilizing the same trait as a measure of the differing environments of the identical and fraternal twinships. All that is required is exchanging one assumption for another, in this case the GEA for the EEA. We could pick a psychiatric disorder with an unlikely genetic component as a way of testing the plausibility of the EEA or GEA. An example is anorexia. Studies have shown that identical twins are about six times more likely than fraternal twins to be concordant for anorexia (45% vs. 7% pairwise concordance).[301]

Bouchard. Although more well known for his studies of twins reared apart (see Chapter 4), Bouchard has been a resolute defender of the twin method:

> We . . . assume that common family environment influences are the same for both types of twins (equal environment assumption). This assumption has been constantly challenged by critics of the twin method. It has, however, been studied by twin researchers for many years and the general consensus is that it is well supported.[302]

But twin researchers are hardly impartial testers of a method that is, in many cases, their life's work. The methods they use for evaluating data, the data they choose to collect, and the conclusions they reach, are influenced by their pre-existing belief in the twin method's validity. In attempts to test (or more accurately, uphold) their method, contemporary twin researchers — without exception — overlook the fact that twin studies suffer from most of the problems which they acknowledge invalidate genetic inferences drawn from family studies.

Segal. In her 1999 book *Entwined Lives*, behavior geneticist Nancy Segal wrote that twin researchers "assume that environmental influences (e.g., parental discipline or teacher expectations) affecting traits are the same for identical or fraternal twins. This is called the *equal environment assumption (EEA)*."[303] According to Segal, the critics

---

301. Treasure & Holland, 1995.
302. Bouchard, 1997a, p. 279.
303. Segal, 1999, p. 318.

conclude that *similar treatment* not *similar genes* causes identical twins' behavioral resemblance. Conversely, they reason that because fraternal twins look different people treat them differently; thus fraternal twins show less similarity due to environmental effects, not genetic ones. Critics, therefore, believe that identical and fraternal twins' unequal treatment violates the EEA and invalidates findings.[304]

Segal revisited Rowe's 1994 argument about facial appearance, where he "reminds us that faces and brains are not one, so similar treatment will not affect biological functions underlying personality."[305] Another way of saying this is: Treatment does not affect personality, which is biological and genetic. The argument in favor of the trait-relevant EEA thus boils down to the circular claim that the twin method suggests genetic influences on personality and behavioral differences because . . . personality and behavioral differences are genetic.

Segal mentioned Rowe's 1994 rhetorical question asking whether we would expect 10 Elvis lookalikes to be similar in personality at all, implying (like Rowe) that we wouldn't. However, we *should* expect personality similarities among these impersonators! Far from constituting 10 randomly selected members of the world's population, they have several factors in common, which include: (1) most of them are Caucasian; (2) most of them are male; (3) they are willing to perform publicly, which suggests certain styles such as extroversion, narcissism, etc.; (4) they possess musical and singing ability; (5) they are more likely to have been reared in the culture of the southern United States than randomly selected people; and (6) they are more similar in age than randomly selected people, and therefore are more likely to have been raised in the same national birth cohort. Thus, we would expect these biologically unrelated lookalikes to correlate significantly above zero on personality measures, since sex and age effects alone can have a "substantial" effect on psychological trait correlations.[306] This underscores the basic problem of Bouchard, Segal, and associates' conclusions from their reared-apart twin studies, which is the subject of Chapter 4.

Lyons. Twin researcher Michael Lyons wrote in 2001 that if twin pairs (combining identical and fraternal) resemble each other at rates exceeding chance expectation, "we can conclude that there is a family influence on the characteristic we are studying."[307] Because most psychiatric twin studies find a

---

304. *Ibid.*, p. 318.
305. *Ibid.*, p. 319.
306. McGue & Bouchard, 1984, p. 325.

much greater than chance resemblance among both identical and fraternal twins, Lyons implicitly recognized that "family influences" affect concordance rates for the various disorders. However, Lyons went on write, "Because the family environment promotes resemblance equally in MZ and DZ pairs, if we observe significantly greater similarity among MZ pairs than among DZ pairs, we can conclude that the characteristic we are observing is influenced by genetic factors."[308] Thus, Lyons recognized that family influences are indeed "trait-relevant" for most psychiatric disorders, while at the same time arguing that "family environment promotes resemblance equally" among identical and fraternal twins.

According to most twin researchers, although identical twins experience a more similar family environment than fraternal twins, no one has shown that these differing environments are trait-relevant. Conversely for Lyons, family environments are trait-relevant, but identical and fraternal twins' family environments are equal. The formulation that reflects reality, however, borrows from both of these positions: Family environments are trait-relevant, and identical twins experience far different family environments than fraternal twins. The only valid conclusion is that the twin method is hopelessly contaminated by environmental influences.

***

My aim has been to illustrate the way that leading contemporary twin researchers write about the EEA. What their writings show is that intellectually gifted scholars and writers who have dedicated their lives to human genetic research and the twin method become inarticulate, contradictory, and confusing when attempting to uphold the validity of a thoroughly indefensible theoretical assumption.

ATTEMPTS TO TEST THE VALIDITY OF THE EQUAL ENVIRONMENT ASSUMPTION

As we have seen, the broad consensus among contemporary twin researchers includes the following points:

● The observation, by twin method critics, that identical twins experience more similar environments than fraternals do, is correct.

307. Lyons & Bar, 2001, p. 318.
308. *Ibid.*, p. 318.

- Although identical twins experience more similar environments, the evidence suggests that this does not seriously bias the results of twin studies. The twin method remains valid because identical twins' greater environmental similarity is the result, not the cause, of their greater genetic similarity.

- Identical and fraternal twin environments must be shown to differ on environmental factors etiologically relevant to the trait under study. Furthermore, the burden of proof for demonstrating differential trait-relevant environments falls on critics.

- On the basis of the "trait-relevant" definition of the equal environment assumption, the twin method remains a viable instrument for the detection of genetic influences on psychological trait differences and psychiatric disorders.

A common theme of Kendler's articles of the 1980s and 1990s was that the EEA has been tested "in five different ways."[309] Kendler and other twin researchers concluded that the results of these tests uphold the validity of the twin method. Plowing through this body of research is a tedious affair, and I will not subject readers to a critique of this literature here. Critical analyses have appeared elsewhere, where the general conclusion was that the studies comprising the five ways of testing the EEA don't provide much evidence in its favor.[310] In the words of Pam and colleagues, in their 1996 review of the "EEA test" literature, "much of the research supporting the EEA does not meet scientific standards, is often sophistical in the way inferences are drawn from the data, and slanted in terms of objectivity."[311] Here, I will limit the discussion to the three most frequently cited studies in support of the EEA and the twin method.

Loehlin and Nichols, 1976

Behavior geneticists John Loehlin and Robert Nichols's 1976 twin study is one of the most frequently cited investigations in the behavior genetics literature. Its results are discussed in reference to several theoretical pillars of behavior genetic methods and findings, including the alleged importance of "non-shared" environmental influences on psychological trait differences.[312]

---

309. For example, see Kendler, 1983, 1993.
310. See Joseph, 1998a, 1998b; Pam, 1995; Pam et al., 1996; K. Richardson, 1998.
311. Pam et al., 1996, p. 359.

Loehlin and Nichols analyzed data from 850 pairs of reared-together twins and their parents. Their sample was obtained from high school juniors who took the National Merit Scholarship Qualifying Test (NMSQT) in the spring of 1962. Approximately 600,000 students took this examination, and examinees were asked to indicate if they had a twin sibling. The researchers obtained the names of 1,507 same-sex twin pairs, who were sent questionnaires and were asked to participate in the study. The final sample of 850 twins was based on pairs supplying "reasonably complete" data, plus a completed parent questionnaire.[313]

Loehlin and Nichols found that identical twins correlated significantly higher than fraternals on personality and cognitive ability measures. This difference, of course, would be expected by "hereditarians" and "environmentalists" alike. Behavior geneticists view this study as groundbreaking due to the finding of a "weak overall relationship between differences in childhood treatment and differences in personality . . ."[314] According to Loehlin and Nichols,

> Differences in the childhood treatment of twins are not very predictive of adolescent personality differences. Identical twin differences — though necessarily environmental — are not better predicted from treatment differences than are fraternal-twin differences. Stronger attempts at parental influence do not result in more similar pairs of twins. Attempts to treat twins alike do not lead to greater similarity between them.[315]

They argued that identical twin correlations are not predicted by more similar parental treatment, which was measured by the treatment-related responses on the parent questionnaire. This implied that the twin method is not biased by the more similar parental treatment experienced by identical twins.

Since the late 1970s, the results of this study have been cited repeatedly by behavior geneticists as supporting the validity of the EEA. Scarr and Carter-Saltzman's claims are typical:

> If the usually observed behavioral differences between MZ cotwins are smaller than those between DZ cotwins, then critics assume that the experiential differences between DZs are a sufficient explanation for their greater behavioral differences. If that were true, then MZs who are treated more differently by their parents than other MZs would be less similar than MZs who are treated more alike. This was the logic of the study of Loehlin

312. See Plomin & Daniels, 1987.
313. Loehlin & Nichols, 1976, p. 7.
314. *Ibid.*, p. 60.
315. *Ibid.*, p. 92.

and Nichols (1976), who calculated the correlations between environmental differences of MZ cotwins on those variables that differentiated MZ and DZ twins and personality and intellectual differences between MZ cotwins. They found little relationship between differences in parental treatment and experiences and test score differences. Thus the differences in treatment between MZs who were treated similarly and those who were treated differently could not account for the magnitude of the differences between MZs and DZs.[316]

More recently, in 2001, Bouchard and Loehlin wrote, "Loehlin and Nichols . . . long ago showed that similarity in treatment could not account for twins' similarity in personality."[317]

A closer look at Loehlin and Nichols's study reveals problems which include (1) a sample that was heavily biased in the direction of more similar or "twinlike" pairs; (2) a reliance on questionnaires and tests that were completed at home, thereby outside of the investigators' control, and were subject to twins collaborating on their responses; and (3) the selective analysis of data.[318] The most serious problem, however, is Loehlin and Nichols's reliance on responses from retrospective questionnaires, filled out by the twins' parents, to determine treatment similarity. Several researchers have found that parental recall is of questionable reliability and is influenced by parents' views at the time of questioning.[319]

In a large 1970 study assessing the accuracy of parental recall, Yarrow and associates compared the retrospective reports of 224 mothers and 190 children to ratings, observations, and reports obtained contemporaneously on the children, which included parental reports made during the child's early years. Yarrow et al. found only a modest correlation between their baseline information and mothers' later accounts, finding that mothers "painted a 'better' or 'more satisfactory' picture than yielded by the baseline."[320] Consistent with Robbins's 1963 findings, the researchers "found mothers' reports conforming to child care values current at the time of reporting."[321] Yarrow and associates concluded that the use of retrospective data — especially in areas with a "high potential for ego involvement" — is an unsatisfactory substitute for a longitudinal study, and that

---

316. Scarr & Carter-Saltzman, 1979, p. 528.
317. Bouchard & Loehlin, 2001, p. 251.
318. This last point is detailed by Kamin, in Eysenck vs. Kamin, 1981.
319. e.g., Bradburn et al., 1987; Halverson, 1988; Hardt & Rutter, 2004; Holmberg & Holmes, 1994; Reuband, 1994; Robbins, 1963; Yarrow et al., 1970.
320. Yarrow et al., 1970, p. 67.

if used, researchers must be aware of the data's "serious limitations."[322] They concluded with "a plea for caution . . . the research practitioner [should] not be too hasty in the application of the retrospective method and the consumer of the outcome of such research [should] attend to the signs saying *caveat emptor*."[323]

Loehlin and Nichols's parent questionnaire contained 56 treatment-related items, and parents reported treating their twins alike, both identical and fraternal, on the vast majority of experiences. Possible responses to the treatment related items were: (1) "Twin one only," (2) "Twin two only," (3) "Both twins," and (4) "Neither twin." Answers #1 and #2 were counted as one point on scales measuring the frequency of reported differential treatment of the twins by their parents. According to the results, it appears that parents treated their twins remarkably alike. Based on 56 questions, parents recalled that they had treated (1) their male identical twins the same on 99.14% of the questions, (2) their female identical twins the same on 98.98% of the questions, (3) their male fraternal twins the same on 97.44% of the questions, and (4) their female fraternal twins the same on 97.7% of the questions.[324] (These figures are in conflict with the "twins create their environment theory," which predicts that parents would treat their fraternal twins far more differently than identicals.)

---

321. *Ibid.*, p. 72. Robbins (1963) studied parental recall of occurrences in the lives of their three-year-old children. These parents had been participants in a longitudinal study since the birth of their child, and their written descriptions of experiences with their children while they were happening were known to the researchers. When parents were asked to recall events that had occurred in their child's life just two years earlier,

> The parents were quite inaccurate in their memory of details about child rearing practices and early developmental progress, in spite of the frequent rehearsal of these data due to their participation in the longitudinal study, and in spite of the relatively young age of the children. . . . Inaccuracies tended to be in the direction of the recommendations of experts in child rearing, especially on the part of the mothers. (Robbins, 1963, p. 261)

We should keep in mind that some of the events recalled by the NMSQT parents went back more than 16 years. Robbins found that parents tended to remember inaccurately in the direction of the recommendations of child rearing experts (such as Dr. Spock). This meant that parents, probably without consciously lying, had a tendency to idealize their role. They answered questions about their child rearing practices in the direction of how they, and society, believed that they should have raised their children, and there is reason to believe that the NMSQT parents answered questions about the treatment of their twins in much the same way.

322. Yarrow et al., 1970, p. 69.
323. *Ibid.*, p. 74.
324. Loehlin & Nichols, 1976, p. 55, Table 5-8.

Combining all responses, parents answered "neither twin" or "both twins" *98.3% of the time.* According to Loehlin and Nichols, "it was exceptional for a parent to mark any treatment item as different for the twins; for the identicals it was, in fact, positively rare."[325] Rare indeed. For identical twins ages 0-12, parents reported treating them differently once every 175 questions.[326] These questionnaire responses are therefore of doubtful reliability in assessing the relationship between twin correlations and parental treatment.[327] Clearly, parents answered questions on the basis of an idealized image of their child-rearing practices.

Ironically, although most people recognize that identical twins are treated far more alike than fraternals, there appears to be one exception to this near unanimous opinion: the NMSQT twins' parents. What better evidence that these parents represented an extremely atypical population, or, what is much more likely, that the retrospective questionnaire method is a flawed instrument for assessing parents' treatment of their children?[328] As Yarrow and associates concluded, retrospective questionnaires are an inadequate and potentially misleading substitute for a longitudinal study.

The numerous commentators citing Loehlin and Nichols's study rarely mention its various methodological problems, many of which were discussed by the investigators themselves. For example, Loehlin and Nichols recognized that their twin sample was biased because it (1) contained an excess of identical twins, (2) contained an excess of female twins, (3) consisted only of twins who were willing to respond to a mailed questionnaire, and (4) consisted of high-achieving Merit Scholarship test takers.[329] They also cautioned that their sample was probably selected (biased) for similarity of experience,[330] and that in general, interpretations of their data "do not altogether exclude a completely environmentalist position."[331] Finally, although Loehlin and Nichols were well

---

325. *Ibid.*, p. 56.

326. By my calculation.

327. Parent responses to the treatment-related questions could have been checked by looking at comparable twin questionnaire items, were it not for the fact that not one of the 1,092 questions put to the twins dealt directly with how they were treated by their parents. The closest that any twin questionnaire items came to addressing parental treatment were questions #1069 and #1070, which asked the respondent to state which twin "Is more liked by your mother," and "Is more liked by your father" respectively. The possible answers were "I am," "Both the same," or "My twin is" (Loehlin & Nichols, 1976, Student Questionnaire Appendix, p. 23). Male identical twins' answers correlated at only .21 for the question about their mothers, and .26 about their fathers (Appendix B, p. 26), even though their parents had answered that they treated both male identical twins the same 99.14% of the time.

aware of the "fallibility of parents' retrospective accounts of the early history of their children," they believed that "fallible data interpreted with some caution are clearly better than no data at all . . ."[332] On the contrary, 25 years of secondary sources' caution-free and uncritical acceptance of Loehlin and Nichols's unwarranted conclusions have provided support for the EEA and the twin method. In this instance, faulty conclusions based on "fallible data" have been far *worse* than no data at all.

Scarr and Carter-Saltzman

The twin studies by American behavior geneticist Sandra Scarr in 1968, and Scarr and Carter-Saltzman in 1979, are the most frequently cited examples of the "reverse zygosity" test of the twin method's validity.[333] This method looks at twins whose zygosity status has been misidentified by twins or their parents. For example, some genetically identical pairs believe themselves, or are believed to be, fraternal. Test score correlations of these mislabeled twins are compared to correctly identified twins in order to test the effects of "true" zygosity (genetic effect) versus "perceived" zygosity (environmental effect). Scarr saw this comparison as an important test of the EEA's validity:

---

328. Loehlin & Nichols's parent questionnaire item #288 (Appendix A, p. 12) asked parents whether they treated their twins differently or alike, and provided five different response choices. For all twin pairs, parents answered "We tried to treat them exactly the same" eight times more frequently than they answered "We tried to treat them differently" (Loehlin & Nichols, 1976, p. 87). These results most likely reflected the wishful thinking that parents displayed in their responses to this question — for how many parents would be willing to admit to others (or even to themselves) that they treated their twins differently? Question #288 was also unclear about what is meant by "different treatment." Did this mean that twins were treated as separate and different human beings, though equally, or as it probably appeared to most parents, that they played favorites among the twins? One could argue that the question only asked if parents "tried" to treat their twins equally, but common sense should tell us that it is unlikely that so many parents actually did — for how can anyone treat two people "exactly the same?"

329. Loehlin & Nichols, 1976, pp. 8-9.
330. *Ibid.*, p. 87.
331. *Ibid.*, p. 94.
332. *Ibid.*, p. 47.
333. Scarr, 1968; Scarr & Carter-Saltzman, 1979. The reverse zygosity test was proposed by Kety as early as 1959 (Kety, 1959).

Critics of twins studies have assumed that differential treatment of MZ and DZ pairs constitutes *prima facie* evidence of bias. It does not. The direction of effect in the correlation between zygosity and environmental similarity is not at all clear. It is possible that the greater genetic similarity of MZ twins leads to more similarity in their environments. Parents and others may respond to the behavioral similarity of MZ pairs with more similar expectations for them, and identical twins themselves may select and attend to more similar aspects of their environments.[334]

Thus, Scarr invoked the "twins create their environment theory" of Shields and Kallmann in defense of the twin method's validity.

Despite being widely cited in support of the twin method, Scarr's studies have problems, which include: (1) the sample sizes of misidentified twins were small. In the 1968 study, there were only 11 misidentified pairs. In the 1979 study there were 41 pairs (out of a total of 342) where both members of the pair were wrong about their true zygosity status. (2) The results, as least as they pertain to personality, were mixed. In 1968 Scarr concluded, "Data from the Vineland Social Maturity Scale and the Adjective Check List suggest that beliefs about zygosity also have an effect on MZ pairs . . . ,"[335] and in 1979 Scarr and Carter-Saltzman found that perceived zygosity had an effect on fraternal twin resemblance for personality measures. (3) According to Kamin, who had access to the 1979 raw data, a closer look at these figures reveal environmental effects on cognitive ability measures not shown in the "blunted" environmental scales reported in the published paper.[336] (4) After carefully examining Scarr and Carter-Saltzman's IQ data, developmental psychologist Ken Richardson concluded that one of the key results "does *not* confirm the equal environments assumption in the way that is frequently claimed."[337]

According to Scarr and Carter-Saltzman, "perceived similarity is not an important bias in IQ tests," although it might have more of an effect on personality. They ended with an endorsement of the twin method:

> The critical assumption of equal environmental variance for MZ and DZ twins is tenable. Although MZ twins generally experience more similar environments, this fact seems to result from their genetic similarities and not to be a cause of exaggerated phenotypic resemblance.[338]

334. Scarr & Carter-Saltzman, 1979, p. 528.
335. Scarr, 1968, p. 40.
336. Kamin, in Eysenck vs. Kamin, 1981, pp. 130-131.
337. K. Richardson, 1998, p. 147.

The problem with this argument is that — even if it were true — the EEA is not supported. The twin method is used for a wide range of traits and disorders for which the *reason* that identical twins experience more similar environments is completely irrelevant. As a critic correctly observed, "a greater similarity in environment induced by greater similarity in genetics still means that monozygotic twins have more similar, or shared, environments than dizygotic twins."[339]

In addition, the argument that identical twins' genetically determined similarity in personality causes their parents and others to treat them more similarly overlooks an important point which I discussed earlier: aren't parents and others, according to this theory, reacting to twins on the basis of *their own* genetically shaped behavior? How can twins' behavior be genetically determined, but parents' behavior be ever-changing and adaptive to their twins' needs? In the childhood struggle to create similar environments, Scarr's postulated genetically-driven twins' behavior must confront a formidable foe — the genetically-driven and culturally-shaped behavior of their parents.

This discussion should, one hopes, put to rest the idea that the twin method is valid or "tenable" if twins' environments are more similar because, as Kendler and others have speculated, identical "twins might *create* for themselves more similar environments."[340] Because identical twins experience more similar environments than fraternals — regardless of how genetic factors might influence this — the twin method is unable to disentangle potential genetic and environmental influences.

CONCLUSIONS

In this chapter I have focused on the assumption that identical and fraternal twins experience similar environments, known as the equal environment assumption. Until the early 1970s, most twin researchers defined the EEA as the assumption that identical and fraternal twins experience similar environments, or that, if the environments were different, the EEA remained valid because identical twins created more similar environments by virtue of their greater genetic similarity. Most twin researchers now recognize that identical twins experience

338. Scarr & Carter-Saltzman, 1979, p. 541.
339. Hoffman, 1991, p. 188.
340. Kendler, 1987, p. 706.

94

more similar environments than fraternals but, contrary to scientific principles, require *critics* to show that environments differ on "trait-relevant" factors in order to invalidate the twin method. As Bouchard has written, "It is certainly true that MZ twins experience more similar environments than do DZ twins," while adding that "no one has been able to show that such imposed similarities in treatment are trait-relevant."[341] But it is simply not necessary for critics to identify trait-relevant factors in order to invalidate the twin method.

For several decades, twin researchers defined the EEA in the traditional sense. When the evidence showed overwhelmingly that the assumption was false, they did not discard the twin method as they should have done. Instead, they subtly redefined the EEA in ways that allowed them to go on with their work, and the new unspoken definition became the "unequal-environments-don't-matter assumption."[342] Twin researchers make the "trait-relevant" argument in support of the twin method, but fail to apply it to family studies. Their error lies in the fact that once identical-fraternal environments are acknowledged to be different, the twin method is transformed into little more than a special type of family study because, in both family and twin studies, potentially causative environmental and genetic factors cannot be disentangled. Twin researchers have led the twin method into a theoretical blind ally because they have unwittingly linked its fate to the family study method.

As three generations of critics have argued, the twin method is an environmentally confounded research technique unable to disentangle possible genetic and environmental factors. Thus, any conclusions in favor of genetic influences on psychiatric disorders or psychological trait differences derived from the twin method should be disregarded. Like family studies, the twin method offers only an "initial hint" that trait differences might have a genetic component. Twin researchers' attempts to test the EEA have yielded unconvincing results, and it is doubtful whether any test could establish the validity of the twin method in the face of the greater environmental similarity experienced by identical versus fraternal twins. The most reasonable conclusion we can draw from identical-fraternal comparisons is that environmental factors cause higher concordance or correlations among identical versus fraternal twins. Researchers wishing to assess possible genetic factors must find other methods to do so.

---

341. Bouchard, 1997b, p. 134.
342. Pam et al., 1996, p. 354.

CHAPTER 4: GENETIC STUDIES OF TWINS REARED APART: A CRITICAL REVIEW

> Many of the criticisms of behavior genetic studies that we offer may seem to place unusual demands for proof upon the researchers. But, given the potential social misuse of conclusions in this field, we believe that the utmost care is required.
>
> — Critics Paul R. Billings, Jonathan Beckwith, & Joseph S. Alper, in 1992.[343]

> Another area for consideration is illustrated by a case study from my own research . . . that brought out some startling similarities for two women who were both Baptists and who both considered volleyball and tennis their favorite sports. They preferred math and English in school (and listed shorthand as their least favorite subject). Both were studying nursing, and both preferred their vacations in historical places. Perhaps most astonishing — these two women were *not* identical twins and were not related in any way.
>
> — Psychologist W. Joseph Wyatt in 1993.[344]

In recent years a great deal of attention has been paid to studies of reared-apart twins. Many people, unconvinced by studies of twins reared together, have been convinced by these studies. Although reared-apart identical twins might appear to be able to unlock the mysteries of the nature-nurture debate, in this chapter we will see that all studies published to date have been unable to achieve this.

There are two streams of evidence coming from the twins reared apart (TRA) model: the "folklore" stream and the "scientific" stream. The folklore stream consists of stories released by twin researchers or reported by journalists about separated twins who, upon being reunited, are said to share an amazing set of common features, traits, preferences, etc. The scientific evidence consists of data presented in books and peer-reviewed academic journals, where the psychological trait resemblance of reared-apart twins is compared with that of reared-together twins. The researchers usually conclude that their results support the existence of important genetic influences on psychological trait differences, while adding (in recent studies) that "shared family environment" con-

---

343. Billings et al., 1992, p. 236.
344. Wyatt, 1993, p. 1295.

tributes little. I use the term "MZA" (monozygotic twins reared apart) as a shortcut for reared-apart identical twins, and "DZA" (dizygotic twins reared apart) for reared-apart fraternals. I do so only to distinguish them from reared-together identical twins (MZTs) and reared-together fraternals (DZTs), since as we will see, few of these twins were truly reared-apart from birth in uncorrelated environments.

PART I: IDENTICAL TWINS REARED APART AS FOLKLORE

Suppose we pick up the newspaper and read about MZAs who, it is said, share remarkable similarities. They are said to share the same taste in food, have given their children similar names, work at the same type of job, have similar quirky habits, and so on. What should we conclude? Certainly, this story is newsworthy and interesting. But does it also provide important information about genetic influences on human behavioral differences? At first glance it appears that it would; after all, these are two people who grew up apart yet share a remarkable set of similarities. What else besides the genes they share could account for these similarities?

After getting past these initial reactions, critically-minded readers might ask some questions: how were these twins discovered? Who discovered them? Did the people who reported them have information about dissimilar pairs, which they were not reporting? Had the twins met before being studied? Did the twins have personal or financial motives for exaggerating their similarities and degree of separation? Did the people reporting their story have ulterior motives for exaggerating their similarities and downplaying their differences? Are the twins available for independent verification of their stories? And most importantly, did factors other than, or instead of, genetic identity contribute to their similarities? I will examine these questions as we assess the stories of reared-apart identical twins.

To begin, most pairs come to the attention of researchers and journalists *because of* their similarities. It's similar to the old "dog bites person versus person bites dog" rule in journalism. A dog biting a person isn't newsworthy because it's a common occurrence, whereas a person biting a dog is news because it is an *unusual* occurrence. If we were to read several articles describing a person biting a dog, it would be mistaken to conclude that people bite dogs more frequently than dogs bite people. The same is true for twins. Stories of similar twins are news because they are interesting and compelling; stories about dissimilar twins are not.

The next question is whether the pair is representative of MZAs as a whole. Because many of the pairs we have heard about since the early 1980s have been identified by the Minnesota researchers, let's start there. As of 2002, the Minnesota investigators reported having given personality tests to 74 pairs of MZAs.[345] Unfortunately, they have released information on only a few selected pairs whose stories, in most cases, support the genetically-oriented views of the researchers.

Many popular accounts of MZA studies have discussed Jim Lewis and Jim Springer, commonly referred to as the "Jim Twins." These identical twins were separated at birth and were reunited at age 39. The Jim Twins were said to share an uncanny set of similarities, such as the names of their wives and children, career choices, and preferences for particular brands of beer and cigarettes. They were invited to the University of Minnesota by Bouchard, and became the first pair of separated twins in the Minnesota Study of Twins Reared Apart (MISTRA).

The most celebrated case was the twin pair Oskar Stöhr and Jack Yufe. These identical twins were born in 1933 in Trinidad to a Jewish father and a German mother. A few months after they were born, their parents divorced. Jack remained in Trinidad with his father, while Oskar went to live in Germany with his mother. Oskar was raised a Catholic Nazi in Germany, whereas Jack was raised as a Jew in Trinidad and spent time on an Israeli kibbutz.[346] Although the twins grew up to be what their caregivers had raised them to be, this was not the way the case was presented. The two men were described as leading "markedly different lives,"[347] and a supposedly amazing set of similarities was discovered after they arrived at the airport:

> Both were wearing wire-rimmed glasses and mustaches, both sported two-pocket shirts with epaulets. They shared idiosyncrasies galore: they like spicy foods and sweet liqueurs, are absentminded, have a habit of falling asleep in front of the television, think it's funny to sneeze in a crowd of strangers, flush the toilet before using it, store rubber bands on their wrists . . .[348]

---

345. Markon et al., 2003.
346. Holden, 1980.
347. *Ibid.*, p. 1324.
348. *Ibid.*, p. 1324.

However, Jack and Oskar had met previously and had been in postal contact for more than 25 years.[349] After their cases were reported in the press, they sold their life stories to a Los Angeles film producer.[350]

Jack and Oskar had a personal interest in exaggerating (or inventing) their similarities and underreporting previous contact. It is possible that other pairs exaggerated their similarities in order to please the investigators, or in an effort to attain celebrity status. The Jim Twins hired an agent at one point.[351] According to a description of Daphne and Barbara (the "Giggle Twins"), another celebrated pair of MISTRA MZAs,

> At the end of their week in Minneapolis, as they settled down to a spectac-
> ular lunch to celebrate the end of their week-long testing, one final coinci-
> dence emerged. Both had told Bouchard the same lie. "We both said we
> wanted to be opera singers and neither of us can sing a note," Barbara con-
> fessed. And they both broke into peals of laughter, yet again.[352]

As Kamin and Goldberger observed, "This anecdote, intended to emphasize the similar behavior of MZAs, makes it clear that twins could and did lie about themselves to the investigators."[353]

Often overlooked is that similarities can be found between any two people, if this is what one is looking for. As Bouchard conceded, "Some of those similar-ities are surely coincidental — complete strangers at cocktail parties routinely discover 'astonishing' occurrences in their lives; imagine what they might find after fifty hours of filling out questionnaires."[354] Psychologist W. J. Wyatt and his colleagues assessed the similarities of 25 biologically *unrelated* pairs of college students matched on age and sex, and compared them with a group of reared-together identical twins. One pair of unrelated individuals had a lot in common: "Both are Baptist; volleyball and tennis are their favorite sports; their favorite subjects in school were English and math (and both listed shorthand as their least favorite); both are studying nursing; and both prefer vacations at historical places."[355]

349. See Horgan, 1993; L. Wright, 1997.
350. Horgan, 1993.
351. Donald Dale Jackson, 1980.
352. Watson, 1981, p. 43.
353. Kamin & Goldberger, 2002, p. 86.
354. Bouchard, 1997c, p. 53.
355. Wyatt et al., 1984, p. 64.

Another celebrated pair, Jerry Levey and Mark Newman, two New Jersey firefighters, were brought together on the basis of being identified by another firefighter:

> Capt. Jim Tedesco looked across a crowded room at a firefighters convention and was startled at what he saw. There was Mark Newman, and that was impossible. He knew that Newman was back home in Paramus, N. J. But how many bald, 6-foot-6, 250-pound-plus New Jersey volunteer firemen are there who wear droopy mustaches, aviator-style sunglasses and a key ring on the belt on the right side? [356]

In other words, Jerry and Mark were discovered because they were similar; had they been dissimilar, they might never have been reunited. It is not, as journalists often write, "spooky" or "eerie" that these twins were both firefighters, wore droopy mustaches, aviator glasses, and a key ring on the right side. On the contrary, they were discovered *because* they were both firefighters, wore droopy mustaches, aviator glasses, and a key ring on the right side. Had Jerry been a florist instead of a firefighter, Captain Tedesco would not have made his discovery. I should reemphasize that Jerry and Mark's story is compelling from the human interest perspective. Personally, I'd want to read all about it. Nevertheless, it tells us little about genetic influences on behavioral and personality differences.

Like other publicized pairs, we are told about what Mark and Jerry have in common, but little about what sets them apart. For example, it was said that both men are bachelors, are attracted to similar women, and enjoy hunting, fishing, going to the beach, John Wayne movies, pro wrestling, and Chinese food. However, because journalists and "scientists" highlight the similarities, we could reasonably assume that they are *dissimilar* for other, unmentioned comparisons. What were Jerry and Mark's taste in music, magazines, sports, newspapers, or shoes? What were their favorite high school subjects, planets, animals, baseball players, or musical instruments? What were their choices of toilet paper, deodorant, toothpaste, motor oil, paint, or insecticide? The list is endless. But because readers are presented with the similarities, they are led to conclude that the twins are more alike than they really are. "When any two biographies are avidly compared, at least some overlap is likely to be found." The bleating of a nit-picking critic, perhaps? Hardly. This sentence can be found in a 1992 article by Minnesota twin researchers Lykken, McGue, Tellegen, and Bouchard.[357]

---

356. Lang, 1987, p. 63.

Photographs are also used to create the impression of similarity. In Minnesota twin researcher Nancy Segal's *Entwined Lives*, she presented a photograph of Jerry and Mark each holding a can of beer in his right hand, each curling his little finger under the can. Segal commented, "Note that the twins' little fingers support their can of beer."[358] There are photographs in other publications, however, in which only *one* twin is seen curling his little finger under his beer.[359] Why did Segal choose one photo and not the other? Like the journalists, Segal ignored non-genetic explanations for why these twins may have held their beers the same way in her photo. Some possibilities include:

- That Jerry and Mark took cues from each other before the photo was taken in order to appear more "twinlike."
- That many people hold cans this way, which would mean that their similarity in this regard would not be so astonishing.
- That men hold cans in this manner more often than women.
- That members of the working class are more likely to hold cans this way than randomly selected people.
- That New Jersey firefighters traditionally hold beer cans this way.
- That Jerry and Mark were asked by the photographer or the journalist to hold the cans this way.
- That people growing up in their area and culture (Jerry and Mark lived only 65 miles apart[360]) tend to hold cans this way.

For a journalist to cite this photo as suggesting a genetic basis for pinkie curling is understandable. For a "scientist," it is inexcusable.

Nor is this the first time that twin researchers have used photographs of reared-apart twins to create the illusion of similarity. In his classic review of the older TRA studies in *The Science and Politics of I.Q.*, Kamin commented on a photo of twins "Edwin and Fred" found in the Newman and associates 1937 study,

---

357. Lykken et al., 1992, p. 1566.
358. Segal, 1999, p. 144.
359. For example, see Lang, 1987, p. 63.
360. Lang, 1987.

The case study includes a photograph of the twins side-by-side "at the time of their first meeting." The photograph was "reproduced by permission of *The American Weekly*." The twins are remarkably alike in appearance. They are wearing identical pin-striped suits, and identical striped ties. These, of course, might have been bought "at the time of their first meeting." Perhaps it is relevant to note that Fred was unemployed at the time of the [Newman, Freeman, & Holzinger] study.[361]

For the first seven or eight years of the Minnesota study, stories about twins such as I have described were the only information available to the public. A commentator observed that these stories about twins were similar to those offered in support of astrology or extrasensory perception (ESP),

> Striking coincidences are reported as supposed grounds for belief in the phenomenon itself. In literature about astrology and E.S.P., cases where forecasts came true or where thought of a friend was immediately followed by a phone call from that friend are offered as evidence. The cases where forecasts failed or where a thought of someone is not followed by a phone call from that person are forgotten or left unmentioned.[362]

The problem with placing faith in testimonials was described by Robert T. Carroll in *The Skeptic's Dictionary*. Carroll wrote that "Testimonials and vivid anecdotes" are often used to support beliefs in the "transcendent, paranormal and pseudoscientific." "Nevertheless," he continued, "testimonials and anecdotes in such matters are of near zero value in establishing the probability of the claims they are put forth to support. . . ." Carroll wrote that accounts of people's encounters with angels, aliens, ghosts, miraculous dowsers, a Bigfoot, or psychic surgeons are "of little empirical value in establishing the reasonableness of believing in such matters." The reason is that these accounts are unreliable and biased: "They are of no more value than the televised accounts of satisfied customers of the latest weight loss program. . . ."[363] While it might sound strange to compare stories about twins to astrology and ghost sightings, there are parallels. These include the selective presentation of information by parties attempting to persuade, the frequent inability to verify information, and the desire on the part of the audience to want to believe. Carroll's reference to television commercials is relevant to this discussion. When we see people celebrating the fact that

---

361. Kamin, 1974, p. 54.
362. Dusek, 1987, p. 21.
363. Robert T. Carroll, "The Skeptic's Dictionary" website: http://skepdic.com/testimon.html.

Detergent X removed the stains from their shirts, few people, if they stopped to think about it, would say that the commercial depicted a random sample of people washing their clothes with Detergent X. They would realize that the makers of Detergent X consciously chose to show only people happy with their product. Why then are people willing to believe in the power of genetics on the basis of a few pairs of selected twins? Possibly because they do not believe that scientists (and to a lesser degree journalists) would try to "sell" them something.

Clearly, the selected information on selected pairs released by the Minnesota investigators or reported by journalists proves nothing about the importance of genetics, but it does show the potentially misleading nature of these stories. The US Pentagon used the Minnesota strategy successfully during the 1991 Gulf War. They showed photographs of missiles hitting their targets to the public; they did not show those missing their targets, or those hitting schools, hospitals, bus stations, and so on.

Judith Harris concluded in *The Nurture Assumption* that "These true stories of reared-apart identical twins are a testimony to the power of genes,"[364] but it could just as easily be said that the testimonials of people who believe in alchemy or astrology suggest the power of these pseudosciences.

Influence of the Cohort

We have seen that there is little reason to regard individual cases of reared-apart twins as anything more than interesting stories. This brings us to a fundamental question to which I have already alluded and which bears directly on the upcoming discussion of the scientific studies of reared-apart twins: are there important environmental factors shared by both reared-together and reared-apart identical twins that would lead them to resemble each more than two randomly selected members of the world's population? The answer is a resounding yes. These include that they are exactly the same age; that they are the same sex; that they are almost always the same ethnicity; that their appearance is strikingly similar, which probably will elicit similar treatment; that they usually are raised in the same socioeconomic class; that they usually are raised in the same culture; that they shared the same prenatal environment; and that they typically spent a certain amount of time together in the same family environment, were aware of each other's existence when studied, and often had regular contact over a long period of time. All of these factors work towards

364. Harris, 1998, p. 33.

increasing the resemblance of reared-apart twins, yet are rarely discussed in popular accounts of individual pairs.

The mere fact that two people are born on the same day can have an important impact on their subsequent behavior and beliefs, because people with the same date of birth are members of a national birth cohort. Examples of how often we make statements recognizing the influence of this cohort include, "she grew up during the Depression and so she doesn't trust banks," "he smoked marijuana because everyone was doing it in the 60s," "young people today aren't interested in politics," "their disagreements are due to the generation gap," and so on. A 50-year-old woman loves The Beatles for the same reason that her parents love Benny Goodman and her children love Britney Spears — because it's the music she grew up with.

We also recognize that men and women in Western cultures tend to think and act somewhat differently. Thus, we would expect two women or two men to be more alike than a man and a woman. In addition, people raised in the same subculture are more similar because of cultural influences; hence statements about New Yorkers being one way, Californians acting another way, Southern hospitality, country folks, city slickers, and so on. The same could be said about cultural differences between various ethnic groups. As for physical appearance, we have ideas about how attractiveness or lack of attractiveness can create or block opportunities, influence personality, etc. If the "Elephant Man" had had an identical twin brother from whom he had been separated at birth, it would be reasonable to expect *both* to have suffered anxiety in social situations. Social class also has an important influence on beliefs, preferences, tastes, and values. Had Jerry and Mark been raised by French aristocrats, we might have discussed the reasons why they curled their little fingers under a glass of Dom Perignon instead of a can of beer.

In fact, most pairs of reared-apart identical twins share all of the common factors I have just outlined and, as I discuss more fully in the next section, often had frequent contact during the years they were "separated." Reared apart identical twins, as one author observed, are "not so much similar to each other as they are similar to people of their eras and SES [socioeconomic statuses]."[365] Even personality trait theorists such as Gordon Allport recognized that the impact of culture on personality is "indisputable," and that "Everyone admits

---

365. Farber, 1981, p. 77. Farber was speaking of twins' dental records in this quotation, but her observation applies to most aspects of twin resemblance.

that culture is vastly important in shaping personality."[366] In most cases, reared-apart and reared-together identical twins share at least seven different cultural influences: national, regional, ethnic, religious, economic class, birth cohort, and gender cohort.

Another example of the failure to recognize cultural and cohort influences on twin resemblance is found in journalist Kay Cassill's description of reared-apart identical twins Keith Heitzman and Jack Hellback, who grew up in Louisiana:

> Although the mighty Mississippi divided these two physically, it could not separate their parallel lives. The welder from one side and the pump mechanic from the other found that they are both allergic to ragweed and dust. Both had done poorly in school. Both disliked sports and had cut their gym classes whenever they could. They are both addicted to candy. Their similarity of dress includes a penchant for wearing cowboy hats, which matches their parallel interest in guns and hunting.[367]

As a critic commented, "Even if 'the mighty Mississippi divided' the twins, the fact that they both wear cowboy hats and like hunting is not that unusual for two [white] working-class men in the same region of Louisiana."[368] The same point can be made about the celebrated "Jim Twins," two working-class white men who grew up in the same region of Ohio at the same time.

Returning to Oskar Stöhr's childhood in Nazi Germany, Bouchard has been quoted as saying that the "twin who was reared in Germany belonged to the Hitler Youth, but virtually every young man in this community had to belong to it."[369] Bouchard thereby attested to the influence of the cohort. Millions of young Germans joined the Hitler Youth not only because they were required to, but because they were caught up in the spirit of the region and era in which they found themselves. As an extreme example of the confounding nature of cohort effects, suppose a 20-year-old pair of male German MZAs, separated at birth and brought up in different parts of the country, had been reunited in 1942. A researcher or journalist would have noticed several striking similarities. Both might have been wearing swastika armbands, have had short-cropped hair, have given stiff arm salutes, and had railed against "the enemies of the Reich." The values, beliefs, behaviors, and even wardrobes of these twins might have been

366. Allport, 1961, p. 165.
367. Cassill, 1982, p. 183.
368. Dusek, 1987, p. 21.
369. Quoted in Cassill, 1982, p. 193.

106

remarkably similar, yet it would have been fallacious to conclude that these similarities were caused by common genes.[370]

* * *

The stories of individual pairs of reared-apart identical twins tell us little
to nothing about genetic influences on human psychological trait differences. As
behavior geneticist Richard Rose commented, they make "good show biz but
uncertain science."[371] Judith Harris has written that "there are too many of these
stories for them all to be coincidences,"[372] and it is true — they are not coincidences; they are selectively reported "show biz" combined with a stunning
failure to recognize the environmental factors influencing these twins' similar
behaviors.

PART II: SCIENTIFIC STUDIES OF TWINS REARED APART

In Part II, I focus the systematic studies of reared-apart twins, which typically assess twins through the use of personality and IQ test score correlations.
Although one might assume that these scientific studies are vastly different from
the single cases reported in the media, the investigators have made many of the
same errors made by the journalists.

Until 1979, there were only three systematic studies of reared-apart identical twins: Newman, Freeman, and Holzinger in 1937, who studied 19 pairs (and

---

370. As an experiment, I call on readers to identify a biologically-unrelated person of the
same sex who is close in age and was reared in the same class and culture. Sit down
with this person and list all of the things you have in common — likes, dislikes,
hobbies, etc. Look hard for similarities, since journalists and genetically-oriented
researchers do the same with their twins. When you are finished, you might have a
list as impressive as any you have read about separated twins. Of course, it is also
possible that you will have little in common with this person, just as there are probably several dissimilar pairs of reared-apart twins for every one reported. If you and
your partner discover many similarities, consider contacting the media to see if they
are interested in reporting your story. If you can't find similarities, you probably
won't bother contacting the media because it is doubtful that they would be interested. Also be aware that between you, similarities can be invented. The inherent bias
in the stories of reared-apart twins will become more apparent after completing this
experiment. As another exercise, imagine that you have just been reunited with your
reared-apart identical twin sibling who was (somehow) born 200 years before you
and grew up in another part of the world. How much do you think that you would
have in common with this person?
371. Rose, 1982, p. 960.
372. Harris, 1998, p. 293.

107

used MZTs and DZTs as controls); Shields in 1962, who studied 44 pairs (and used MZTs and DZTs as controls); and Juel-Nielsen's 1965 study, where 12 MZA pairs were investigated (no controls were used; Juel-Nielsen updated his study in 1980). Newman et al. found an MZA Binet IQ correlation of .67, and an MZT correlation of .91. The MZA correlation for the Woodworth-Mathews personality test was .58; for the MZTs it was .56.[373] Shields reported IQ correlations of .77 for MZAs, and .76 for MZTs.[374] For personality, he found MZA and MZT "Neuroticism" correlations of .53 and .38 respectively, and .61 and .42 for "Extraversion."[375] Juel-Nielsen reported an MZA Wechsler-Bellevue IQ test correlation of .62, but did not calculate total sample personality test correlations.[376]

The Minnesota Study of Twins Reared Apart (MISTRA) was initiated in 1979 by Bouchard and his colleagues at the University of Minnesota, who have included David Lykken, Nancy Segal, Matt McGue, Auke Tellegen, Leonard Heston, and Elke Eckert. The study continued until 2000, with some of the participants returning for a second assessment several years after their first one.[377] Since 1988 the MISTRA group has published papers reporting test score correlations for MZAs, DZAs, and other types of twins.[378] Two other TRA (twins reared-apart) studies were published in the 1980s — a study from Finland, and the Swedish Adoption/Twin Study on Aging (SATSA).

The controversy surrounding British psychologist Cyril Burt's claim to have given IQ tests to 53 MZA pairs has been discussed in detail elsewhere, and I will therefore touch only on the main points. Irregularities and invariant correlations were first reported by Leon Kamin in 1973 and were published in his 1974 book, *The Science and Politics of I.Q.* Two years later, London *Sunday Times* reporter Oliver Gillie wrote an article in which he raised doubts that Burt's research assistants "Miss Conway" and "Miss Howard" had ever existed.[379] British psychologist Leslie Hearnshaw published a biography of Burt in 1979 in which he argued convincingly that Burt had invented much of his twin data. In part because he was well respected by even Burt's strongest supporters, Hearnshaw's

373. Newman et al., 1937, p. 347.
374. Shields, 1962, p. 61.
375. *Ibid.*, pp. 69-70.
376. Juel-Nielsen, 1980, p. 107.
377. Johnson et al., 2004.
378. The first MISTRA paper making these comparisons was published in 1988 by Tellegen et al.
379. Gillie, 1976.

conclusions were widely accepted. Although some have subsequently argued that the charges of fraud leveled at Burt are unsubstantiated, it is widely agreed that, fraud or not, Burt's twin data cannot be used.[380]

Kamin also produced the first in-depth critical analysis of problems with the other TRA studies. He demonstrated that many of the twins in the Newman, Shields, and Juel-Nielsen studies had frequent contact and often were reared in correlated environments. Kamin also questioned the validity of some of the measures used to test twins, and observed that twins were often tested by the same investigator, which could lead to bias. Kamin argued that IQ correlations were subject to the confounding influence of sex and age effects, and that Newman's study recruited twins on the basis of similarity and their pre-existing knowledge of each other, which meant that his sample was biased toward similarity because twins who were unaware of each other were excluded.

Kamin's chapter on TRA research was written in the context of the larger debate over the importance of genetic influences on IQ scores, which had been reignited by Jensen's 1969 article, "How Much Can We Boost IQ and Scholastic Achievement?" There, Jensen argued that IQ was "highly heritable" and that tests score differences between American whites and blacks were plausibly explained, at least in part, by genetic factors (that is, were due to the alleged genetic inferiority of African-Americans). Jensen cited Burt's study, which in turn motivated Kamin and others to look more closely at this and other reared-apart twin studies.[381]

Susan Farber published her *Identical Twins Reared Apart: A Reanalysis* in 1981, which remains the most in-depth work ever published on the subject. Farber's book contained a wealth of thoughtful discussion, theorizing, and detailed statistical analysis. Drawing from Kamin's critique and adding her own observations and analysis, this book is a devastating indictment of the logic and

---

380. See Hearnshaw, 1979. For the arguments of Burt's defenders, see Fletcher, 1991; Joynson, 1989. Although Fletcher and Joynson attempted to exonerate Burt of the charges of scientific fraud, they did not call for the rehabilitation of Burt's data. According to Bouchard & Pedersen (1999, p. 73), "there is agreement that Burt's data should not be included in any summaries of kinship correlations of the IQ literature."

381. Contemporary popularizations of behavior genetic research usually present Kamin as, at best, a defender of ideas that research has shown to be untrue. At worst, he is presented as a discredited, politically-motivated villain seeking to impede the work of "true scientists" (see, for example, William Wright's *Born That Way*). In fact, Kamin has provided a valuable service to society. He deserves praise for his willingness to publicly expose bad research, as well as some of the unsavory history of his own profession.

methods of these studies. Farber observed that, like the Newman study, Shields's sample was biased in favor of similarity because he had recruited twins on the basis of a media appeal.[382] Looking at all pairs reported in the literature (including single-case studies), Farber pointed out that due to ascertainment bias, "approximately 90 percent of the known cases of separated MZ twins have been studied precisely because they were so alike," and that conclusions about their similarity were therefore based on "circular reasoning."[383] Farber viewed Juel-Nielsen's study as being less affected by similarity bias because he located his 12 pairs through the use of registers and other records, "thus ensuring that their knowledge of each other was not the prerequisite for selection."[384] According to Farber, the twins recruited to these studies, whose similarities were so eagerly generalized to the non-twin population, were not even representative of the population of reared-apart twins.

Like Kamin, Farber questioned the twins' degree of separation. Of the 121 reported pairs in studies and single case reports, she found only 3 cases in which the twins were separated during the first year of life, were reared with no knowledge that they had a twin, and were seen at the time of their first meeting. "Of the 121 cases reported in the last fifty years," wrote Farber, "only three are 'twins reared apart' in the classical sense."[385] The "most accurate description of this sample," she concluded, is "MZ twins partially reared apart."[386] Farber also observed that about 90% of MZAs were born into poor families, and that "anywhere from 50 to 75 percent of the twins were reared in clearly deprived homes. . . . Only two or three individuals were adopted into professional families."[387] She said the collected data were "not good and do not meet even elementary design requirements."[388] A well-designed study, according to Farber, would use randomly selected and assigned MZAs who were "reared as single individuals with no contamination from twinning," and having no "organic but nongenetic features within pairs (such as prematurity, birth trauma) to artificially inflate or deflate heritability estimates."[389] Finally, Farber called for the inclusion of sub-

---

382. This point was also made by Taylor in 1980, and by Peter Mittler in his 1971 book on twin research.
383. Farber, 1981, p. 36.
384. *Ibid.*, p. 17.
385. *Ibid.*, p. 60.
386. *Ibid.*, p. 273.
387. Farber, 1981, p. 62.
388. *Ibid.*, p. 22.
389. *Ibid.*, p. 30.

jective impressions in addition to test results, as well as the creation of a central registry so that others would have access to the data. For Farber, a registry would act as a safeguard "against the disreputable claims and use of data that have occurred in the past and undoubtedly will occur in the future."[390] One objective of this chapter is to show that Farber's criticisms are valid, and that the MISTRA investigators ignored most of her admonitions.[391]

The Myth of the Separated Identical Twins

I have borrowed this phrase — "The myth of the separated identical twins" — from a chapter in Howard Taylor's *The I.Q. Game*. The reason is simple: Most twins in these studies were *not* separated at birth and raised in separate environments. Rather, the studies suffer from problems which include:

- Twin pairs sometimes were separated only after having been reared together for several years.
- Pairs were often raised by different members of the same family, and usually were placed into families correlated for socioeconomic status.
- Many pairs were aware of each other's existence and had frequent contact during much of their lives.
- Most pairs were brought to the attention of researchers on the basis of their similarity or knowledge of each other's existence.
- The material used to evaluate the similarities or correlations of twins was collected by the same researchers.
- Personality and environmental resemblance was assessed by non-blinded raters.

The study of separated twins is both a twin study *and* an adoption study, because one or both twins have been removed from their biological mother and placed into an adoptive or foster family, or with a relative. A critical assumption of all adoption studies is that children are randomly placed into available adoptive homes (see Chapter 7). In the case of separated twin studies, it must be

---

390. *Ibid.*, pp. 273-274.
391. The individual reviews by behavior geneticists Gottesman, Heston, Loehlin, and Rose recommended Farber's book as an interesting, if flawed, analysis (see Gottesman, 1982; Heston, 1981; Loehlin, 1981; Rose, 1982). On the other hand, Bouchard called it a "pseudo-analysis" in his 1982 review. By 1999 Bouchard & Pedersen were reporting that Farber's book was "severely criticized" by Heston, Loehlin, Rose, and Bouchard (Bouchard & Pedersen, 1999, p. 73).

shown that MZA trait resemblance is not the result of common factors in the twins' environments, or if it is, that researchers properly control for these factors. Genetic inferences from studies of twins reared together depend on the identical-fraternal equal environment assumption (Chapter 3). Genetic inferences from studies of *reared-apart* twins rest on the assumption that twins' environments were no more alike than those of a group of randomly selected biologically-unrelated paired individuals.

Examples of contact and late separation. In striking contrast to the newer studies, the authors of the older investigations included hundreds of pages of case history material. Looking through these descriptions, we have numerous examples of twins for whom the designation "reared-apart" is very misleading.

In Newman and colleagues' 1937 study of 19 MZAs, the age at separation ranged from 2 weeks to 6 years, and most pairs had considerable contact before being tested. For example, Pair I "had lived together for a year before tests were given," Pair II "had lived together 7 years when tests were given," Pair IV had "been acquainted and have visited back and forth at intervals all their lives," Pair VI was "living together when examined at the age of 58 years," Pair XIII had "visited at long intervals during the last 4 years," and Pair XIX was separated at age 6 and "spent 6 months studying nursing together when they were 17 years old."[392]

In Shields's 1962 study, we find the following descriptions. Each depicts a different pair of "reared-apart" identical twins.

- 14-year-old male twins who "did not meet or know they were twins till they were 5 or 6. From the age of 9 they met once a week when Kenneth's family came to live in the town"[393]
- 17-year-old male twins whose "mother died the day after the twins were born. The paternal aunts decided to take one twin each and they have brought them up amicably, living next door to one another. . . . They are constantly in and out of each other's houses."[394]
- 38-year-old male twins who, at age 11, lived "in cottages next door to one other and attended the same school."[395]

392. Newman et al., 1937, pp. 144-145.
393. Shields, 1962, p. 163.
394. *Ibid.*, p. 164.
395. *Ibid.*, p. 176.

Chapter 4: Genetic Studies of Twins Reared Apart: A Critical Review

- 39-year-old male twins who were separated "at 4 years. . . . The twins came home again to mother at 14. Both twins say they thought a lot about one another as children."[396]

- 8-year-old female twins who were "brought up within a few hundred yards of one another, but no contact between the families. Attracted to each other at the age of 2. . . . gravitated to one another at school at the age of 5. In order to separate them Mrs. E. removed Jessie to another school but they continued to meet in the park."[397]

- 23-year-old female twins "separated at 6 months. . . . They were reunited from 5 to 7 years, then parted again. They meet every weekend now."[398]

- 30-year-old female twins "separated at 13 months. . . . Though attending different schools, the twins met about twice a week and sometimes spent holidays together."[399]

- 32-year-old female twins "separated at 5 years. . . . brought up in the same town, not always same school; met regularly."[400]

- 38-year-old female twins "separated at 8 years. . . . Until then the twins had done everything together; thereafter they continued to attend the same school in the village where both families lived . . ."[401]

- 38-year-old female twins who "were brought up together till the age of 7. . . . The twins meet about once a fortnight when they visit their brother in hospital."[402]

- 39-year-old female twins "separated from birth till 11 years and again at 16." Viola was taken by the grandparents at birth, "while Olga remained with the parents, living nearby in the same industrial area. . . . There were occasions when [Viola] was taken back to mother, but until 11 she was always returned to the grandmother before the day was out. At that age she went to live with the parents and was forbidden to see her grandmother."[403]

- 40-year-old female twins separated at birth, but "probably together at various times before school age. . . . They were reunited most of the time from 5 to 15, generally with the aunt, but sometimes with the mother."[404]

396. *Ibid.*, p. 180.
397. *Ibid.*, p. 189.
398. *Ibid.*, p. 192.
399. *Ibid.*, p. 192.
400. *Ibid.*, pp. 193-194.
401. *Ibid.*, p. 206.
402. *Ibid.*, pp. 208-209.
403. *Ibid.*, p. 210.
404. *Ibid.*, p. 214.

- 42-year-old female twins "separated soon after birth. . . . The twins generally spent summer holidays together at each other's homes."[405]
- 43-year-old female twins "separated at 4 years (not met since)." Twins began corresponding at age 36. "Though they have never met they have developed an intense affection for one another and derive great emotional satisfaction from their correspondence, which supplies something in their lives which they sometimes feel to be lacking. . . . When Berta sends a letter with a lipstick kiss she presses it warmly against her lips. She longs for her twin so much that it hurts." [406]
- 45-year-old female twins "separated at birth. . . . From just before their ninth birthday they lived as close neighbors in a coastal town, attending the same school until they were 15. They were closely attached and went about a lot together . . ."[407]

There are many more examples in Shields's study, but from these descriptions it is clear that few of his pairs were truly "reared-apart." As Shields himself acknowledged, his twins' environments "have not as a rule been extremely different."[408] In fact, 30 of his 44 MZA pairs were raised in different branches of the same family, and 33 pairs differed in family socioeconomic status "only to a modest extent." According to Shields's criteria, twins were considered to be reared-apart if they had been raised in different homes for at least five years during childhood.[409] This did not, however, prevent behavior geneticists Jinks and Fulker, in their influential 1970 paper on biometrical analysis, from writing that there is "no evidence of correlated environments" in Shields's MZA sample.[410]

Let's turn to Juel-Nielsen's 1965 study of 12 MZA pairs. Although these twins were more separated than those of Newman or Shields, the designation "reared-apart" is questionable for several pairs. Case III twins Maren and Jesnine were separated at 6 weeks because of the death of their mother, but "were put into the care of two of their paternal aunts, who were both married to farmers and lived in the twins' home county, about 10 kilometers apart from each other."[411] Ingegerd and Monika (Case IV) were separated at 12 months but were reunited from ages 7-14. Martha and Marie (Case VI) were reared together until

405. *Ibid.*, p. 219.
406. *Ibid.*, pp. 221, 223.
407. *Ibid.*, p. 225.
408. *Ibid.*, p. 20.
409. Shields, 1962, p. 27.
410. Jinks & Fulker, 1970, p. 337.
411. Juel-Nielsen, 1980, Part II, p. 68.

age 3 1/2 and later attended the same school. 50-year-old twins Kamma and Ella (Case VII) were separated after one day of life but grew up in rural districts 40 miles from each other: "Although the twins never came into very close contact with each other, they had both had a feeling of strong solidarity since they met at the age of 12 . . ."[412] Signe and Hanne (Case VIII) were 54-years-old when interviewed. Although separated at 3 weeks, they discovered each other at age 14 and "corresponded with each other during the next few years . . ."[413] Astrid and Edith (Case XI) were separated at 3 1/2 years. Viggo and Oluf (Case XII) were separated at 5 3/4 years.

In the Swedish study (SATSA), which provided only quantitative data for its subjects, the investigators considered pairs to be reared apart if they had been separated before age 11 [414] (average age at separation = 2.8 years[415]), and about 75% had some degree of contact after separation. How much contact was not reported by the SATSA investigators apart from a complicated and inadequate mathematical formula. Nevertheless, according to the formula the SATSA twins (average age: 65.6 years) were "separated" for an average of only 10.9 years at the time of testing.[416] Twins supplied information by mail, and many were not investigated personally. In the Finnish study, 12 of the 30 MZA pairs were separated after they turned five-years-old, and the degree of post-separation contact is unclear.[417]

Environmental similarity. It is common for separated twins to be raised by members of the same family, often with the biological parent(s) caring for one of the twins. Like individual pairs such as the Jim Twins and Jerry Levey and Mark Newman, they often share many cultural influences as well as a similar appearance. For these reasons, and quite apart from possible genetic influences, we would expect even twins reared apart from birth who were unaware of each other's existence to share many similarities. The investigators' failure to recognize common environmental factors was a major theoretical blind spot in the earlier studies. For example, Juel-Nielsen noted his twins' "outstanding similarities" despite experiencing "markedly different upbringings and later lives," and concluded, "These similarities must be related to the genotypical identity of the

---

412. *Ibid.*, p. 176.
413. *Ibid.*, p. 201.
414. Pedersen, Plomin, et al., 1992, p. 347.
415. Pedersen, McClearn, et al., 1992.
416. Pedersen, Plomin, et al., 1992, p. 347.
417. Langinvainio, Koskenvuo, et al., 1984.

twins, and must be taken as an expression of the importance of genetic factors for the normal development of personality."[418]

Kamin observed that Newman and colleagues recruited their reared-apart identical pairs on the basis of similarity, and that they excluded several pairs showing marked behavioral differences. They excluded potential participants if they answered the following question in the negative: "Do you yourselves believe that you are far more alike than any pair of brothers or sisters you know of?"[419] The investigators were attempting to screen out potential DZA pairs because of the great expense of transporting them to and lodging them in Chicago. Kamin noted that the offer of an all-expense-paid trip to Chicago in the middle of the Depression may have induced twins to exaggerate the extent of their separation. According to Newman et al., "Pair after pair, who had previously been unmoved by our appeals to the effect that they owed it to science and to society to permit us to study them, could not resist the offer of a free, all-expenses-paid trip to the Chicago Fair."[420] Several pairs were excluded because they lived far apart from each other and would have created too much difficulty and expense for the researchers. According to Taylor, "such twins might have been culturally or environmentally separated as well."[421]

Newman and associates actually were aware of the possibility that their MZA group was biased in favor of similarity, writing, "When it comes to the separated cases. . . . It seems possible that our group is more heavily weighted with extremely similar pairs than with identical twins of less striking similarity."[422] In an attempt to gather new cases after the study's publication, Gardner and Newman corresponded with a pair of separated male twins in Pennsylvania. Although one of the twins was "anxious to have their case studied," his brother "is, we fear, somewhat of a hoodlum and refuses to submit to examination." Gardner and Newman commented, "This is unfortunate in view of the fact that the two brothers now seem to be so different in their personality traits."[423] This case illustrates a source of bias in reared-apart twin studies: Gardner and Newman were unable to study these twins *because of* their personality differences.

---

418. Juel-Nielsen, 1980, p. 77.
419. Newman et al., 1937, p. 135.
420. *Ibid.*, p. 134.
421. Taylor, 1980, p. 82.
422. Newman et al., 1937, p. 31.
423. Gardner & Newman, 1940, p. 119.

In spite of problems such as similarity bias and the questionable "separation" of twins, the early investigators recognized that MZAs differed in many respects, and that environmental differences have an impact. As Newman wrote three years after the publication of his study, "if one thing is clear in our results it is that fairly large environmental differences do modify physical, mental and temperamental traits and produce proportionately large differences even between hereditarily identical individuals."[424] Juel-Nielsen described the differences he found among his twins as follows:

> In all 12 pairs there were marked intra-pair *differences* in that part of the personality governing immediate psychological interaction and ordinary human intercourse. . . . the twins behaved, on the whole, very differently, especially in their cooperation, and in their form of and need for contact. Corresponding with these observations, the twins gave, as a rule, expression to very different attitudes to life, and very divergent views on general culture, religion and social problems. Their fields of interest, too, were very different. . . . Those twins who had children treated, on the whole, their children differently, and their ideas on upbringing were, as often as not, diametrically opposed. Characterologically, the twins presented differences in their ambitions and in their employment of an aggressive behavior. Emotionally, there was a deep-going dissimilarity with regard to the appearance of spontaneous emotional reactions or to the control of affective outbursts. Various traits of personality found their expression in differences in taste, mode of dress, hair style, use of cosmetics, the wearing of beard or of glasses.[425]

Thus, Juel-Nielsen found "marked differences" in personality, emotion, and character among the genetically identical MZAs he studied. One of his MZA pairs, "Kaj and Robert," met for the first time at age 40. According to Juel-Nielsen, Robert was "revolted by the glimpses he got of Kaj's way of living," and in Robert's words, Kaj was "the most unpleasant person I have ever come across."[426] Clearly, if the Jim Twins were similar because of their identical genes, all reared-apart twins should be equally similar.

Contemporary genetic researchers, when they discuss these studies, tend to overlook twins' environmental similarity and the dubious claims of separation. The evidence is usually reported by critics. After Jensen calculated an

424. *Ibid.*, p. 126.
425. Juel-Nielsen, 1980, p. 75.
426. *Ibid.*, Case Histories, p. 132.

overall intraclass IQ correlation of .824 for the Newman, Juel-Nielsen, Shields, and Burt studies, while downplaying the twins' degree of contact, he wrote that this correlation "may be interpreted as an upper-bound estimate of the heritability of IQ in the English, Danish, and North American Caucasian populations sampled in these studies."[427] More recently, behavior geneticist David Rowe conceded only that "MZ twins reared apart are sometimes not perfect separations." In another gross understatement, Rowe spoke of "the occasional social contact of 'separated' twins," which "may not introduce strong biases."[428]

We have come to the end of a brief discussion of the earlier reared-apart twin studies. Readers interested in further analyses of these investigations should consult previous publications.[429] I will now focus on the Minnesota study, which has been the subject of much attention since the early 1980s.

THE MINNESOTA STUDY OF TWINS REARED APART

The Minnesota studies helped tilt the nature-nurture debate in the direction of nature. Only "the most extreme skeptic," argued Bouchard and Loehlin in 2001, would not agree that "virtually all human psychological traits are influenced by genetic factors to a significant degree."[430] The MISTRA studies have been the subject of numerous articles and several books popularizing behavior genetic research, where sweeping and dramatic statements have been made about its findings. As Lawrence Wright wrote in *Twins*, "The field of psychology has been shaken by separated-twin studies,"[431] and "The science of behavioral genetics, largely through twin studies, has made a persuasive case that much of our identity is stamped on us from conception."[432]

The results I now present are taken from a 1990 MISTRA article published in the prestigious journal *Science*, which compared MZA and MZT correlations for physical variables, IQ, and personality test scores. The full scale WAIS IQ scores showed an MZA intraclass correlation of .69 (48 pairs), while the MZT correlation was .88 (40 pairs). For personality measures, MZA and MZT correla-

---

427. Jensen, 1970, p. 146.
428. Rowe, 1994, p. 39.
429. For further analysis of the pre-MISTRA reared-apart twin studies, see Farber, 1981; Joseph, 2001d; Kamin, 1974; Taylor, 1980.
430. Bouchard & Loehlin, 2001, p. 243.
431. L. Wright, 1997, p. 8.
432. *Ibid.*, p. 143.

tions were each about .50. Because the MZA and MZT correlations were similar, the investigators concluded that psychological traits are strongly influenced by genetic factors, and that "shared family environments" have little influence. The authors of subsequent MISTRA papers published in the 1990s reached similar conclusions for a host of psychological traits, interests, and tendencies. (The SATSA and Finnish studies produced lower intraclass personality correlations, ranging from .20 to .40.[433] The SATSA first principal component cognitive ability intraclass correlation was reported as .78, although the subtest correlations ranged from .15 to .65.[434])

What are we to make of these findings? Do they indeed justify the conclusions of both the investigators and the popularizers of the MISTRA studies? The answer to these questions will occupy the remaining pages of this chapter.

Secret Data, Secret Lives

In the aftermath of Kamin's and Farber's analyses of the older studies, one might have expected reared-apart twin researchers to change the way that they study twins, and in some respects they did. The Minnesota investigators were careful to test twins separately and to have different people administer the tests. This reduced the possibility of unconscious rater bias, as discussed by Kamin. They also attempted to minimize ascertainment bias by vigorously recruiting all pairs they became aware of regardless of whether they were MZA or DZA. In many ways, however, the Minnesota studies are methodologically *inferior* to the older investigations.

A major problem has been the investigators' refusal to make life history and test score data available for independent analysis. The authors of the older studies published hundreds of pages of biographical and test score information. Unfortunately, the MISTRA researchers have not published this information, and have refused to make it available to others. According to Bouchard, each twin was given a Life History Interview, Clinical Interview, Sexual Life History Interview, Life Stress Interview, Child Rearing/Schooling Interview, and Briggs Life History Questionnaire.[435] However, the investigators have reported only minimal quantitative data on rearing environments and have repeatedly denied

---

433. The SATSA personality results are found in Pedersen et al., 1988; the Finnish results are from Langinvainio, Kaprio, et al., 1984.
434. Pedersen, Plomin, et al., 1992.
435. Bouchard, 1984.

critically-minded reviewers access to the interview data.[436] Bouchard refused to make MZA data available to a prominent critic (Leon Kamin), even under conditions where pairs are identified by code numbers and where information about age is omitted to guarantee non-identification.[437]

The Minnesota researchers have claimed that they are forbidden by federal law,[438] or by the University of Minnesota Human Subjects Committee, from releasing data on their twin subjects. According to journalist William Wright, "On agreeing to take part in the study, the twins signed an informed-consent agreement in which the Minnesota scientists promised not to divulge information that could be traced to specific twin volunteers."[439] However, as just one example, this does not explain how the investigators were able to discuss the developmental and sexual histories of six MZA pairs in a 1986 study of homosexuality.[440] The investigators discussed cases while changing details "in order to protect the twins [*sic*] identities," and mentioned no legal or ethical restrictions on releasing or publishing this information.[441] In 1991, Bouchard and colleagues were challenged by Beckwith et al. in the pages of *Science* to publish case history information.[442] Although William Wright would subsequently claim that "because of the informed-consent agreement, which promised the twins anonymity, the [1990 *Science*] article provided no biographical information . . . ,"[443] Bouchard and colleagues did not mention any such restriction in their reply to Beckwith.[444]

During the 1980s, Bouchard and associates supplied information on their twins to journalists, which led to the publication of their stories in leading US magazines and newspapers. In addition, they released selected information on twins who signed the informed consent agreement, even though they had requested anonymity. Peter Watson devoted an entire chapter of his 1981 book about twins to MISTRA-supplied stories, which included at least two case histories of anonymity-seeking twin pairs:

---

436. Horgan, 1993; L. Wright, 1997.
437. Leon J. Kamin, personal communication of 4/8/2001.
438. See W. Wright, 1998.
439. *Ibid.*, p. 215.
440. The homosexuality study is by Eckert et al., 1986.
441. Eckert et al., 1986, p. 422.
442. Beckwith et al., 1991.
443. W. Wright, 1998, p. 56.
444. Bouchard et al., 1991.

> This completes the number of identical twins raised separately whom Bouchard has seen and whose names he is willing to release. For a variety of reasons the other five sets of twins he has seen *do not want any personal publicity*, and two I now mention are therefore identified only by a "case number" [emphasis added].[445]

Watson described the two most similar pairs; he chose not to discuss the other three. Apparently, it was acceptable to supply information to a journalist on pairs who "did not want any personal publicity," if it helped support the genetic bias of the MISTRA researchers.

The Minnesota investigators have kept important information about the twins they have studied secret. In doing so, they ignored Farber's timely call for making this type of data available to others for independent review. They also ignored Arthur Jensen, who in many ways is Bouchard's intellectual inspiration.[446] According to Jensen, MZA data should be made available to "anyone who wishes,"

> Especially rare data, such as those of monozygotic twins reared apart, siblings from cousin matings, double first cousins, and the offspring of two mated pairs of monozygotic twins . . . should be published in full, along with complete descriptions of the tests or measurements and procedures. Perhaps this should be a general requirement for the publication of studies based on such valuable data, so that quantitative analytical techniques other than those used by the original author can be applied to the data by anyone who wishes.[447]

Unfortunately, Bouchard did not follow Jensen's 1974 recommendation, making it very difficult for independent researchers to analyze the data and offer alternative interpretations, as they had with the earlier studies. With access to the raw data, a reviewer could determine the similarity of twins' childhood environments and could assess their degree of contact, interaction, and mutual influence. A reviewer could also examine the quantitative data in order to see if there were important trends overlooked by the original investigators. Apart from an early article discussing the first 15 pairs, Bouchard and colleagues have failed to provide a table containing information such as sex, age, age of separation, and so forth for the individual pairs. Only mean figures for the various twin types have been provided. In 1978, Shields wrote that one of the "obvious limitations"

---

445. Watson, 1981, p. 61.
446. Bouchard, 2004; L. Wright, 1997.
447. Jensen, 1974, pp. 26-27.

of Burt's study was "the lack of information provided about the environments and life histories of the twins."[448] Shields continued, "Burt provided very scanty information in his 1966 paper — he did not even give their ages."[449] In fact, Shields could just as easily have been criticizing the MISTRA studies.

How "separated" were the MISTRA twins? An important task for any reanalysis of the MISTRA data would be to determine the degree of the twins' separation. Like the Newman and Shields studies, MISTRA twins were recruited on the basis of similarity and their pre-existing knowledge of each other. We cannot determine how "separated" these twins really were, however, because the investigators have not published life history information. Instead, they developed the wholly inadequate "contact time" formula. Time spent before and after separation is included in the total and reflects the amount of time (measured in months) that twins spent together before being studied. Bouchard and McGue wrote that their MZAs had "minimal contact" before entering the study, implying that a pair of separated twins could influence each other only when together.[450] However, intimate relationships are based on an ongoing association between people, which is not necessarily limited to the time they are in physical proximity. According to the MISTRA contact time formula, "Twins who met for a week at Christmas and for a week in the summer each year over a 10-year period are credited with 20 weeks of contact."[451] Thus, a MISTRA table would show such a pair to have had as much contact as twins who had spent only the first 20 weeks of life together. Clearly, these two pairs cannot be regarded as having had the same influence on each others' development. Instead of regarding twins who spent 14 days a year together over a 10-year span as having accrued 20 weeks of contact time, we must understand that these "reared-apart" twins *had a 10-year relationship*. And according to Bouchard and colleagues' 1990 *Science* article, the total contact time range was 1-1233 weeks.[452] This means that the members of at least one pair had over 23 years of contact with each other.[453] In addition, as Kamin noted, volunteer twins seeking inducements or scientific approval, on whose verbal accounts contact time has been calculated, might be tempted to underreport the amount of contact they had.

448. Shields, 1978, p. 79.
449. *Ibid.*, p. 80.
450. Bouchard & McGue, 1990, p. 267.
451. *Ibid.*, pp. 266-267.
452. Bouchard et al., 1990, p. 224.
453. This point was made by K. Richardson (1998).

The MISTRA contact time formula does not capture the essence of a relationship between two people, and its usefulness rests on the assumption that environmental similarity and mutual influence increase proportionately with each hour twins spend together. Unfortunately, the investigators presented no evidence supporting this assumption. In her discussion of the older studies, Farber observed that the investigators often assumed that "a brief meeting leaves insignificant impact" because they, like Bouchard, viewed contact in terms of quantity as opposed to quality. Although it may be possible to assess the contact, environmental similarity, and mutual influence of twins on the basis of an examination of their life histories, the impact of these factors cannot, as Bouchard and colleagues imagine, be measured *as a function of time.*

Farber pointed out that losing one's virginity may occupy only a moment in time, and asked, "Does this mean that the event is forgotten or had no psychological impact?"[454] The concept of post-traumatic stress disorder (PTSD) is based on the recognition that a brief yet traumatic experience can severely impact a person's subsequent psychological functioning, yet the trauma (or "contact") itself may consume only a few seconds of a person's life. Obviously, these examples are more relevant to behavioral than to cognitive ability (IQ) similarity. In any case, the MISTRA contact time formula is no substitute for the publication and availability of the twins' life history information.

How the researchers' assumptions influenced interpretations. When the investigators found large differences between the members of MZA pairs, they often devised explanations that minimized the importance of the social and family environment. Here, I offer three examples.

Lawrence Wright discussed a male MISTRA MZA pair who had been reared in very different environments. One of the twins had been adopted by illiterate parents; the other had been raised by a "better educated" family. The difference in their IQ scores was 29 points, with the better-educated twin scoring higher. According to Wright, Lykken believed that "the lower-IQ twin may have suffered some kind of brain damage at birth."[455] But why should we accept such unsupported speculation?

The second pair, described by Leonard Heston, consisted of two 55-year-old female MZAs who were separated at four months and were adopted into different homes.[456] "Twin A" appeared "overweight and slovenly" and expressed

454. Farber, 1981, p. 19.
455. L. Wright, 1997, p. 71.

"chronic dissatisfaction with her life." She had been treated for depression for eight years before being studied, had experienced anxiety all her life, and had several unexplained medical complaints. She had completed high school at age 17 and worked as a clerk for the next 30 years. Twin A described her marriage at age 38 as "miserable." Her MZ sister, "Twin B," weighed 60 pounds less than Twin A and was described as an "engaging, even vivacious woman." Twin B had gone to college and had worked as a head technician in a large hospital laboratory. She was happily married and described her life as fulfilling and happy. According to Heston, his MISTRA colleagues "searched exhaustively through [the twins'] histories for some explanation of the striking differences between these women." A clear bias is evident in this description, since the researchers did not exhaustively investigate the histories of *similar* twins for the purpose of discovering an environmental explanation for their resemblance. It turned out that Twin B had been taking thyroid medication since age 18, but Twin A had not. The MISTRA laboratories determined that Twin A's thyroid function was in the "low normal range." Heston offered Twin B's thyroid medication as a "tenuous explanation" for the twins' differences.

As a third example, in *Entwined Lives* Segal briefly discussed two MISTRA MZA pairs showing "marked differences on the MMPI [Minnesota Multiphasic Personality Inventory, a personality test purporting to measure psychological disturbance]." Although she might have attributed the results to environmental influences, Segal instead concluded that MMPI score differences were "traceable to the development of psychiatric disorder in one case, and probable brain damage due to head injury in the other case."[457] Unfortunately, neither Segal, Heston, nor Lykken indicated whether these pairs and their test scores were retained in the study, leading a reviewer to wonder whether they were removed on the basis of a suspicion of organic problems.

The researchers' beliefs and assumptions have influenced this study at every step: The investigators recruited the Jim Twins to the study because they were so similar, and then offered them as evidence of the similarity of MZAs. On the other hand, they assumed that differing pairs became that way due to nongenetic factors. It is remarkable that Bouchard et al. included the Jim Twins in their study and figured their test scores into all of their subsequent statistical formulations. Like most previous studies, the Jim Twins came to the attention of

456. See Heston, 1988, pp. 211-212.
457. Segal, 1999, p. 134.

researchers *because of* their similarity. (Bouchard recruited them after reading about their reunion in the newspaper.) The Jim Twins epitomize the circular reasoning discussed by Farber.

Instead of publishing case history material, the MISTRA investigators have provided only mean scores from a questionnaire, the Family Environment Scale (FES). One might ask why this scale would be necessary in light of the fact that twins underwent extensive interviews covering most aspects of their life histories, or how it could be considered a substitute for an evaluation of family environment by blinded raters. The FES, in fact, does not measure socioeconomic status, but only social environment within the family.[458] Bouchard and McGue claimed that FES correlations show that only "modest" selective placement occurred in their MZA sample.[459] But why must we take their word for it? Until proven otherwise, we can safely assume that the MISTRA twins' "separation" is as dubious as it was in the studies of Newman et al., Shields, and Juel-Nielsen. Until the data is made available to "anyone who wishes" (or at least to any credentialed reviewer who wishes), *we must reject all conclusions about the role of genetics.*

This position is unrelated to whether or not the researchers are legally or ethically required to keep their data secret; it is more a question of the social and political importance of the study's conclusions and implications. What Bouchard and colleagues ask of us is that we radically alter our views on the importance of heredity, with a possible dramatic shift in social policy, on the basis of how *they* interpret the data on a few hundred individuals. Essentially, they ask us to believe on faith. However, science demands that claims be subject to replication. When this is extremely difficult due to the nature of the subjects under study (such as reared-apart twins), it becomes mandatory that independent analysts be permitted to analyze the data and offer alternative interpretations.

Environmental Factors Contributing to MZA Resemblance

We have seen that there are many environmental influences shared by MZTs and even the most perfectly separated MZAs. Not only did the early investigators fail to control for these influences, they did not even think of them. They simply did not understand that because two people are the same sex and the same age, and grow up in similar cultures, they will display a certain degree

---

458. See Moos & Moos, 1986.
459. Bouchard & McGue, 1990, p. 278.

of psychological trait resemblance. Today, these influences remain misunderstood by the authors of popular works and even by some researchers. According to the SATSA investigators, for example, "the phenotypic resemblance of identical twins reared apart in uncorrelated environments can be attributed to genetic influences."[460] And according to psychiatric geneticists Faraone and Tsuang, "Since MZ twins reared apart do not share a common environment, any phenotypic similarity must be due to genetic factors. We cannot invoke shared environment as a cause of phenotypic concordance."[461] Even Bouchard has written that "When identical twins are reared apart, their personality correlations must be an effect of genetics"[462] (although as we will see, Bouchard and associates used an age and sex correction formula).

Let's review the main environmental influences causing MZAs to resemble each other more than pairs of randomly selected members of the world's population:

- They are the same age.
- They are the same sex.
- They are almost always the same ethnicity.
- Their appearance is strikingly similar, which probably will elicit similar treatment.
- They usually are raised in the same socioeconomic class.
- They usually are raised in the same culture or subculture.
- They frequently share the same religious beliefs.
- They are both volunteers. Studies have shown that research volunteers share several psychological and behavioral characteristics.[463]
- They shared the same prenatal environment.
- They typically spent a certain amount of time together in the same family environment, were aware of each other's existence when studied, and often had regular contact over a long period of time.

---

460. Pedersen et al., 1992, p. 256.
461. Faraone & Tsuang, 1995, p. 91.
462. Bouchard, 1997c, p. 54.
463. Rosenthal & Rosnow, 1975.

While the early investigators failed to recognize any of these factors, Bouchard and associates attempted to control for two of them. In 1984, McGue and Bouchard published a paper outlining their strategy to deal with sex and age confounds.[464] They recognized that age and sex effects for most psychological variables are "substantial,"[465] and wrote that researchers can either ignore these effects or can use information on age-sex effects obtained from within the twin sample. Thus, they recognized that the previous investigators "ignored" environmental effects providing a "substantial" contribution to MZA resemblance. McGue and Bouchard devised a formula, based on the twin sample itself, which they claimed adjusts twin correlations to account for age and sex effects. They used this formula to adjust all subsequent MISTRA twin correlations, and it was also used by the SATSA investigators. Unfortunately, MISTRA publications have not reported the original raw correlations alongside those that were adjusted, which would allow readers to see how much impact the adjustment had on the data.[466]

It is unlikely that the magnitude of sex and age effects can be determined through the use of twin data, as McGue and Bouchard attempted to do in their complicated formula. In any case, their formula purports to adjust for only two influences on MZA similarity, which is completely inadequate. However, there is a method that may be able to assess the effects of common age and sex, but which has been overlooked by all reared-apart twin researchers, and which relates to the central thesis of this chapter: Genetic influences can be assessed only by comparing MZA resemblance to the resemblance of biologically unrelated pairs of age- and sex-matched strangers.[467]

MISTRA MZA personality and IQ test score correlations. The Minnesota samples consist of twins whose life histories, degree of separation, quality of contact, similarity of socioeconomic status (SES), and cultural background have been kept secret by the investigators. Apart from this important problem, MISTRA results are based on the assumption that, using psychometric tests, "intelligence" and "personality" can be quantified and accurately measured. I will discuss the merits of this assumption as it relates to IQ in Chapter 9. Here, the discussion is limited to personality.

---

464. McGue & Bouchard, 1984.

465. *Ibid.*, p. 325.

466. The McGue & Bouchard 1989 MISTRA IQ publication listed corrected and uncorrected correlations for environmental measures, but not for the IQ correlations themselves.

There are many ideas about what constitutes "personality." For some, personality is an underlying set of traits that are stable over time and can predict the behavior of a person in a given situation. Personality in this sense is a construct inferred from behavior and from the way that people fill out personality inventories. Trait theorists such as Gordon Allport began by identifying thousands of words in the English language that describe distinctive forms of personal behavior. Researchers grouped these terms into categories and claimed the existence of traits such as introversion, neuroticism, authoritarianism and so on. For others, behavior is best understood in the context of the specific situation in which people find themselves, and the existence of stable traits is rejected.[468] Personality theory is one of the most contended areas in psychology. The existence, nature, causes, and endurance of traits are still debated, and there no accepted unifying personality theory in psychology. For this reason, introductory personality textbooks often consist of a series of chapters discussing various theories and their proponents.

For traits theorists such as Allport, personality can be studied in the same way as stars and planets are studied:

> Does an astronomer who studies Arcturus think of the star as a construct tied together by a name? Hardly; to him it is a celestial body, truly existing, and possessing a composition and structure that he will try to scientifically comprehend. When a biologist dissects a plant he does not believe that the plant's structure and physiology reside only in his manipulations. Personality is even more difficult to study than stars or plants, but the situation is the same.[469]

---

467. In 1931, Sims obtained test scores of 203 sibling-pairs raised in the same home and recorded an intraclass correlation of .40. He then tested 203 unrelated pairs matched on the basis of age, school attended, and "home background." The unrelated group correlated at .29, which was not significantly different from the sibling pairs. Sims then formed a group of 203 unrelated pairs matched only on age and school attended. He found that this group correlated at .04. Sims concluded that "intelligence, as far as we are today able to measure it, is greatly influenced by environment" (p. 65). In discussing this report, Kamin (1974, p. 81) noted that "This sophisticated and elegant study has simply disappeared from contemporary reference lists." In a subsequent paper, Bouchard (1983) wrote that Sims had obtained a correlation of .04 for the group of unrelated students matched on age (and school attended), but Bouchard neglected to cite the .29 figure for the group matched on these factors plus home background.
468. See Mischel, 1968.
469. Allport, 1961, p. 28.

Thus Allport argued that "personality" has as much physical reality as a plant or a star.

For proponents of the theory of stable underlying human traits, the next step was to claim that traits can be *measured and quantified*. The Minnesota results depend on the assumption that traits both exist and can be measured. People can feel happy or sad, but can we measure and quantify how much happiness or sadness we possess? Can "extroversion" be quantified?

My purpose is to illustrate the questionable nature of personality as a quantifiable entity, as an example of one of the many MISTRA assumptions that is rarely if ever mentioned. Twin research, as we have seen, has a long history of failing to adequately discuss and provide evidence in support of its many assumptions.[470] In many cases, conclusions in support of genetic factors depend on the validity of many assumptions, all of which must be true in order to sustain the conclusion. Typically, twin researchers test their assumptions only after critics point to the lack of evidence in their favor. If the MISTRA researchers listed all of the assumptions that must be true in order to accept their conclusions, far fewer people would accept them. The implicit MISTRA assumptions about personality form only a small subset of the assumptions of the entire project. All of the following must be true before we can even begin to consider the possibility that the MISTRA investigators' conclusions are valid:

- Humans exhibit underlying stable personality traits which can be identified through the use of statistical methods.
- Personality traits can be identified through the use of words describing behavior.
- Personality traits can be measured and quantified on the basis of people's responses to a questionnaire.
- Personality traits are normally distributed in the population (in a bell-shaped curve).

We have seen that MISTRA personality studies found MZA correlations of about .50. Because MZT correlations are also in this range, the investigators concluded that genes and non-shared family environments are the main influences on personality. Leaving all other problems aside, this conclusion is faulty because it fails to account for the environmental factors shared by MZAs and

---

470. See Taylor, 1980 for a detailed discussion of this point.

MZTs, and there are ways of checking this. A search through the psychological literature finds a few studies that looked at age and sex affects on test scores of genetically unrelated people. One study was performed by Martin and colleagues in 1981.[471] The subjects (N = 179, 118 female, 61 male, mean age 28.4 years) completed the California Psychological Inventory (CPI), which was used in the MISTRA studies and which purports to measure "normal" psychological traits or "folk concepts." Each CPI scale was correlated with the age of the respondents. Remarkably, the mean correlation between age and scale scores was .28 across all 18 scales, with 10 scales showing a correlation of .35 or higher. Martin and colleagues' sample was drawn from a population of college students who were probably similar in socioeconomic and educational status, and therefore had more in common than age alone. Nevertheless, if these figures roughly reflect age effects in the general population, the influence of common age, which is only one of many environmental similarities shared by MZAs, is enough to account for more than half the value found in MISTRA MZA personality correlations.

We can also compare MISTRA personality results with those of other types of studies less vulnerable to environmental confounds. In a 1998 study coming out of the Colorado Adoption Project, the investigators found that the mean personality scale correlation between birthparents and their adopted-away biological offspring (N = 245) — a relationship that they considered "the most powerful adoption design for estimating genetic influence"[472] — was zero (.01 to be precise).[473] The results of this longitudinal adoption study by leading behavior geneticists are strikingly different from the MISTRA results, and we could reasonably conclude that at least one of these studies contained invalidating methodological flaws.

What About the DZA Correlations?

The MISTRA was the first TRA study to collect a sizable sample of reared-apart fraternal twins (DZAs). Although a greater MZA versus DZA correlation can be explained on environmental grounds (see below), from the genetic standpoint MZAs *must* correlate higher, because of their identical genetic makeup. Interestingly, there is little published evidence that the MISTRA same-sex DZAs correlate very differently than MZAs on personality measures and IQ tests.

---

471. Martin et al., 1981.
472. Plomin et al., 1998, p. 211.
473. *Ibid.*, p. 214.

The evidence suggests that DZAs would correlate lower than MZAs for purely environmental reasons. Parents, families, and adoption agencies probably would attempt to place identical twins into more similar homes because of the greater physical similarity of MZAs and societal expectations that they should be treated more similarly. In the cases of DZAs separated after infancy, differences in their physical appearance would make it more likely that they would be placed into less correlated environments than MZAs, whose similar appearance might increase the desire to place them into more similar environments. Due to their less similar physical resemblance, DZAs would experience more dissimilar appearance-related treatment in their social environment.

In addition, MZAs are more likely to be aware of each other's existence and to have had more pre-study contact than DZAs. Several separated MZAs located their twin siblings, who lived in neighboring communities (Jerry and Mark, for example), because people had confused them. Upon discovering the existence of a separated co-twin, an MZA would be more likely to seek out a twin sibling in the belief that this person would be just like him or her. This idea is supported by the fact that the MISTRA MZA:DZA sample ratio includes far more MZAs than chance would expect, and that MISTRA MZAs had about twice as much pre-study contact time than DZAs.[474] Therefore, several environmental factors suggest that separated identical twins would resemble each other more on psychological traits than separated same-sex fraternals. According to a 1998 paper by Bouchard et al., MZA/DZA mean correlations on the 20 California Personality Inventory (CPI) folk scales were .46 and .27 respectively, with several scales showing no statistically significant differences, and it is likely that the environmental factors described above contributed to the mean scale difference.[475] Moreover, the 53 DZAs in the sample included 16 opposite-sex pairs (30%), whose correlations were not listed separately and whose scores probably reduced the total mean DZA correlation. Bouchard et al. should have excluded opposite-sex DZAs from this comparison because, as we have seen, they recognized that sex (and age) effects on psychological variables can be "substantial"[476] (although elsewhere they claimed that "any bias introduced by combining opposite-sex and same-sex twin pairs is small to negligible"[477]).

---

474. Bouchard & McGue, 1990, p. 267; Bouchard et al., 1998, p. 309.
475. Bouchard et al., 1998, p. 314.
476. McGue & Bouchard, 1984, p. 325.
477. Krueger et al., 2003, p. 815.

Where are the MISTRA DZA IQ correlations? The investigators' failure to adequately report DZA IQ correlations was documented in Kamin and Goldberger's 2002 review of the MISTRA studies,[478] and I will discuss and comment further on their findings here. Kamin and Goldberger noted that as of 2000, Bouchard et al. had failed to publish Wechsler IQ correlations for their entire DZA sample, in direct contrast to their publication of DZA *personality* correlations. According to Bouchard and colleagues, they did not report DZA correlations in their 1990 *Science* article "due to space limitations and the smaller size of the DZA sample (30)," even though they viewed MZAs and DZAs as "a fascinating experiment of nature," who "provide the simplest and most powerful method for disentangling the influence of environmental and genetic factors on human characteristics."[479] Space limitations? One column in Table 4 of their 1990 article would have been sufficient to supply DZA personality and IQ correlational data. Smaller sample size? The researchers considered a sample of 30 DZAs sufficient to provide enough data to view this "most powerful" comparison in a 1990 MISTRA personality paper, where they reported the correlations of 26 DZA pairs.[480] Bouchard and colleagues' decision not to publish the IQ correlations of their 30 DZA pairs in their 1990 *Science* article was completely arbitrary and was made by researchers who possessed the raw data at the time they made this decision.

Let's look at some examples of Bouchard and colleagues' selected and incomplete reporting of their MISTRA DZA IQ data. In a 1989 paper by McGue and Bouchard, the authors reported MZA and DZA correlations on tests of "Special Mental Abilities." For Verbal Reasoning, the correlations were MZA = .44, DZA = .49; for Spatial Abilities, MZA = .52, DZA = .27; for Perceptual Speed MZA = .46, DZA = .30; for Visual Memory, MZA = .29, DZA = .22.[481] As we see, only one of the four variables saw a substantially larger MZA correlation. To the extent that DZA correlations are lower, they could be explained by the environmental factors I have already discussed. Moreover, 16% (4/25) of the DZAs were

---

478. Kamin & Goldberger, 2002.
479. Bouchard et al., 1990, p. 223. Four years earlier, Bouchard & colleagues wrote, "DZA twins allow us to test the two most common competing hypotheses proposed as alternatives to the genetic hypothesis as an explanation of the similarity between MZA twins: placement bias and recruitment bias. We believe placement to be approximately comparable and we will be able to evaluate this assumption" (Bouchard et al., 1986, p. 300).
480. Bouchard & McGue, 1990, pp. 280-281.
481. McGue & Bouchard, 1989.

opposite-sex and therefore should have been removed from the DZA sample. Denise Newman and colleagues published a 1998 MISTRA article on adult ego development which compared subsamples of MZAs and same-sex DZAs. As Kamin and Goldberger noted, this was the first published report of MISTRA full-scale Wechsler IQ DZA correlations. Although the full-scale IQ correlations for the MZAs and DZAs in the subsample were .75 and .47 respectively, the Verbal Reasoning correlations were reported as .46 and .53.[482] A 1999 MISTRA study of authoritarianism, by McCourt and colleagues, reported correlations for "general cognitive ability" as .74 MZA, .53 DZA.[483] The MZAs and DZAs were a subsample of twins who had been tested for "right-wing authoritarianism," and 26% (10/38) of the DZA sample consisted of opposite-sex pairs. Finally, the MISTRA group failed to report DZA IQ correlations in a 2004 article on intelligence.[484]

Clearly, the MISTRA investigators have failed to produce adequate DZA IQ score correlations in spite of having provided correlations for DZA personality scores. An example of a more recent publication providing personality data is Bouchard and colleagues' 1998 paper reporting CPI correlations for 53 DZA pairs.[485] In a paper published the following year, Bouchard and Pedersen reported DZA personality correlations in both the SATSA and the MISTRA, yet failed to publish MISTRA DZA IQ correlations even though they discussed IQ.[486] As we have seen, the reported IQ correlations reveal instances when, contrary to genetic predictions, there were no significant differences between the genetically identical MZAs, versus DZAs, who share only 50% of their genes on average. Moreover, true differences that might exist can be explained on environmental grounds.

---

482. Newman et al., 1998, p. 992.
483. McCourt et al., 1999, p. 1001. In his 1996 Galton Lecture, Bouchard reported the SATSA "first principal component" mental ability correlations (MZA = .80, DZA = .32), but failed to report the MISTRA DZA correlations (see Bouchard, 1996).
484. Johnson et al., 2004.
485. Bouchard et al., 1998.
486. Bouchard & Pedersen, 1999.

MZTs (OR DZAs) AS THE CONTROL GROUP: THE INVALIDATING FLAW OF ALL
TWINS REARED APART STUDIES

In 1981, behavior geneticist Richard Rose reviewed Farber's *Identical Twins
Reared Apart* and speculated that cohort effects have confounded the results of
MZA studies:

> A colleague suggests that we cannot know [the importance of MZA resem-
> blance] without necessary control data on similarities found in pairs of age-
> matched strangers. . . . Were one to capitalize on cohort effects by sampling
> unrelated but age-matched pairs, born, say, over a half-century period, the
> observed similarities in interests, habits, and attitudes might, indeed, be
> "astonishing."[487]

Rose thereby highlighted the invalidating flaw of the studies of Newman,
Shields, Juel-Nielsen, and Bouchard — the investigators mistakenly compared
MZA correlations to the correlations of MZTs or DZAs, when the only valid
control group would consist of biologically unrelated pairs of strangers matched
on the environmental variables shared by MZAs. As its name implies, the
purpose of a control group is to control for factors other than those tested by the
researchers' hypothesis, but MZA-MZT comparisons fail to control for many of
the environmental factors I have outlined in this chapter.

Readers may find it difficult to believe that prominent researchers could
have made such an elementary mistake. Genetic studies, however, have a long
history of these types of errors. For example, over 40 years went by before the
majority of twin researchers recognized that reared-together identical twins
experience a more similar environment than reared-together fraternals. Psychologist Stephen Ceci discussed another example in his book, *On Intelligence*. Ceci
took a second look at IQ pioneer Lewis Terman's influential *Genetic Studies of
Genius*, where high IQ children (who came mostly from the upper classes) were
tracked throughout their lives in order to test whether IQ correlates with
"success." Terman compared the outcomes of his "geniuses" to the general population, and Ceci observed, "But surely, this is the wrong comparison group!
Terman should have compared the graduation rates, college grades, and earnings
of his selected students to those of others from their same family background
social class."[488] Ceci called for an alternative control group: "What is needed is a
sample of men with unremarkable IQs who came from the same family social

---

487. Rose, 1982, p. 960. Rose identified his colleague as G. P. Frommer.
488. Ceci, 1996, p. 60.

background as Terman's high-IQ men . . . and who were subjected to the similar special treatments."[489] Bouchard and his predecessors simply repeated the critical error made by Terman, that is, they used the wrong comparison group.

Behavior geneticists may object to my call for the formation of an alternative control group of biologically-unrelated matched strangers, by arguing that Bouchard and colleagues corrected their correlations for age and sex effects. They might also argue that the finding of greater IQ or personality resemblance for MZAs versus same-sex DZTs is proof of the importance of genetics factors. They might present other data and other comparisons. However, these potential objections do not eliminate the necessity of using the correct control group.

What a Valid MZA Study Might Look Like

Although we can draw no valid conclusions in favor of genetics from any TRA study published through mid-2004, a description of a valid MZA study seems in order. First, a systematic ascertainment of MZAs separated at or near birth would be performed in a way that does not lead to the formation of a similarity-biased sample. Twins having had post-separation contact, or who were aware of each other's existence, could not participate in the study. After an experimental group of MZAs is collected in this manner, twins would be evaluated separately and would not be reunited until after the evaluation process is completed. The resemblance of this group of MZAs would be compared with the resemblance of a control group consisting of *biologically unrelated* pairs of strangers sharing all of the following characteristics: they should be the same age, they should be the same sex, they should be the same ethnicity, the correlation of their rearing environment socioeconomic status should be similar to that of the MZA group, they should be similar in appearance and attractiveness, and the degree of similarity of their cultural backgrounds should be equal to that of the MZA twins. Finally, they should have no contact with each other until after they are evaluated and tested. These controls will constitute the Unrelated group. Of course, it is not possible for unrelated pairs to share a common prenatal environment, which twins do share.

At this point we can compare the resemblance of MZAs versus that of the Unrelated pairs. After the scoring is completed, the code concealing the individual's group and pair status will be broken, and each set of scores will be assigned to the proper MZA or Unrelated pair. The genetic hypothesis predicts that MZAs would correlate higher than the Unrelated pairs on the basis of their

---

489. *Ibid.*, p. 61.

identical genetic makeup. The environmental hypothesis predicts that there would be no differences between the MZA and Unrelated groups, other than those accounted for by prenatal factors.

CONCLUSIONS

Contrary to popular belief, media reports describing the behavioral resemblance of ostensibly reared-apart twins, while entertaining, tell us little if anything about the role of genetic influences on psychological trait differences. The same is true for the systematic investigations of reared-apart twins, which include the Minnesota studies. The problems with these studies include the dubious "separation" of twins, the similarity bias of the samples, the failure to publish or share raw data and life history information for the twins under study (MISTRA), and the researchers' bias in favor of the genetic explanations. Most importantly, there has been a failure to recognize and control for the many environmental similarities shared by reared-apart twins. Therefore, we can draw no valid conclusions in support of genetic influences on psychological trait variation from studies of twins reared apart published to date.

In many respects the "journalistic" and "scientific" approaches to MZA reporting converge, in that the similarities of the twins presented (1) were based on twins who were discovered *because* they were similar, (2) were reported by people who were interested in demonstrating similarity, and (3) are plausibly explained by factors other than the twins' common genes. Bouchard dismissed Farber's 1981 book as a "pseudo-analysis," and proceeded to complete a study whose flaws had just been laid bare by Farber.[490] In one important sense, Farber's book is a manual on how *not to* perform a study of reared-apart twins. An updated edition of Farber's book could use Bouchard's MISTRA as a case study.

In the early part of 2001 the sequence of the human genome was published simultaneously in *Science* and *Nature*. One of the lead investigators, Craig Venter of Celera Genomics, concluded that although the Human Genome Project found surprisingly little evidence that genes influence behavioral differences, we know that genes are important because of the results from reared-apart twin studies. He was simply perplexed by the contrasting results. In this chapter I have tried to provide an explanation for the seemingly contradictory evidence reviewed by Venter and others.

490. Bouchard, 1982.

Chapter 5. The Heritability Concept: A Measure of Inheritance or Inherently Misleading?

Heritability analysis of human behaviour has become the dominant paradigm in academic psychology and now appears prominently in introductory texts, where it is presented to naïve students who have no understanding of the false assumptions inherent in the calculations. The preeminence of heritability analysis is the outcome of a power struggle, not the resolution of a debate among scientists.

— Dissident behavior geneticist Douglas Wahlsten in 1994.[491]

Hereditarians *claim* to have the same interest as anti-hereditarians, to determine how much change is possible so that we may act rationally in social programs. But if that is really the hereditarian agenda, why do they keep studying heritability, which simple logic tells us cannot give the answer to this problem. Why do they not design studies to ask the questions about changeability directly? Because the answer would come out in the wrong direction.

— Biologist Richard C. Lewontin in 1987.[492]

It is ironic that heritability estimates are made most often in behavior-genetic research using human subjects, where the only real application of h2 estimates — selective breeding — is prohibited.

— Psychologist Scott F. Stoltenberg in 1997.[493]

"The nature-nurture debate is over. The bottom line is that everything is heritable . . ."

— Behavior geneticist Eric Turkheimer in 2000.[494]

Heritability is perhaps the most misunderstood and misused concept in the psychological and psychiatric literature, as well as in popular works based on this literature. This admittedly difficult concept, widely used in the behavior genetics field, has helped solidify a belief in the importance of genetic influences on human psychological trait differences. Behavior geneticists frequently claim that both IQ and personality "are highly heritable," although the meaning of this statement is often misunderstood. This chapter attempts to clarify the meaning of heritability and will look closely at two important questions relating to the

---

491. Wahlsten, 1994, p. 254.
492. Lewontin, 1987, p. 32.
493. Stoltenberg, 1997, p. 92.
494. Turkheimer, 2000, p. 160.

concept: (1) What is the purpose of a heritability coefficient?, and (2) What does heritability say about the importance of genetic influences on psychological trait differences?

Heritability, according to Plomin and colleagues, is "the proportion of phenotypic variance that can be accounted for by genetic differences among individuals."[495] In this case "phenotypic variance" refers to the distribution of a trait in a population. Heritability is expressed as a number ranging from 0.0 (no heritability) to 1.0 (complete heritability). Plomin and colleagues made several observations about the concept: (1) Heritability refers to the genetic contribution to trait variation in a particular population; it does not describe the importance of genetic factors as they relate to an individual. (2) Heritability is applicable to a specific population, at a specific time, and in a specific environment. (3) A finding of genetic influences on differences *within* populations does not mean that genetic factors contribute to differences *between* populations. (4) High heritability does not mean that environmental changes or interventions cannot have an important impact. (5) High heritability does not imply that traits are fixed or unchangeable: "finding genetic influence on complex traits does not mean that the environment is unimportant."[496]

These observations are critical if one is to properly understand the meaning of heritability, yet they are often unknown, misunderstood, or ignored by people using or attempting to understand the concept. Before going on, therefore, we will look at each point more closely.

(1) *Heritability refers to the genetic contribution to trait variation in a particular population; it does not describe the importance of genetic factors as they relate to an individual.* Researchers often make this point, yet many people still believe that heritability refers to the portion of an individual's manifestation of a trait attributable to genetic factors. In fact, the concept refers only to the percentage of the population variance that can be attributed to genetic factors, and there must be variation in order to calculate heritability. This leads to a paradoxical situation in which a trait could be 100% inherited, yet have a heritability of zero — human

---

495. Plomin, DeFries, et al., 1997, p. 79.
496. *Ibid.*, p. 85.

beings having two eyes, for example. Because virtually all people with one eye became that way on the basis of an environmental occurrence, heredity explains none of the observed variance of "eyedness," and heritability is therefore zero.[497] Thus, a genetically determined trait could have a heritability of zero. It is also important to understand that heritability is the property of a population and not of a trait itself.[498]

(2) *Heritability is applicable to a specific population at a specific time and in a specific environment.* In Kendler's words, "a heritability estimate is a population- and time-specific 'snapshot.'"[499] Heritability estimates applicable to a particular population in a particular environment could be meaningless in another environment.[500] Suppose we have a population of cows and wish to breed for increased milk production. The fact that we might be able to use a heritability coefficient to successfully breed for milk production in a particular environment (characterized by geography, climate, feed, etc.) does not mean that we would obtain the same result in another environment. For this reason, Plomin and associates wrote that "heritability describes *what is* in a particular population at a particular time rather than *what could be.*"[501]

(3) *A finding of genetic influences on differences within populations does not mean that genetic factors contribute to differences between populations.* Most genetic researchers concede this point, even if their conclusions do not always flow from it. In his controversial 1969 article, Jensen concluded that, due to the allegedly high IQ heritability within the US black and white populations, genetic factors play a role in mean IQ differences between blacks and whites.[502] The erroneous nature of this extrapolation was demonstrated by Richard Lewontin in his well-known "sack of seeds" example.[503] We could take two handfuls of genetically heterogeneous open-pollinated corn seed, wrote Lewontin, and plant one handful in nutrient-rich soil, and the other in nutrient-poor soil. After a few weeks, we measure the height of the plants. Because no environmental variation in either field was allowed, differences in plant height within each field is completely explained by genetic variation (heritability = 1.0). However, the mean difference

---

497. Kamin, in Eysenck vs. Kamin, 1981.
498. Jacquard, 1983.
499. Kendler, 2001, p. 1005.
500. Feldman & Lewontin, 1975.
501. Plomin, DeFries, et al., 1997, p. 84.
502. Jensen, 1972.
503. Lewontin, 1976, pp. 14-15.

in plant height between the two fields is the result of differences in soil quality (heritability = 0.0). Thus, a trait could have a heritability of 1.0 within each of two populations, with the difference between populations being completely and easily explainable on environmental grounds.

(4) *High heritability does not mean that environmental changes or interventions cannot have an important impact.* Researchers usually concede this point as well, yet it is commonly believed that a highly heritable trait is difficult to change through treatment or intervention. A classic example of this fallacy is phenylketonuria (PKU), a genetic disorder of metabolism which, without a specific environmental intervention, causes mental retardation. Although the population variance for susceptibility to PKU is completely explained by genetic factors (heritability = 1.0), the administration of a low phenylalanine diet to the at-risk infant prevents the disorder from appearing. PKU exemplifies Lewontin's observation that a "trait can have a heritability of 1.0 in a population at some time, yet could be completely altered in the future by a simple environmental change."[504] Interestingly, in a society with universal and accurate screening for PKU, the heritability of the disorder would be 0.0, because 100% of the *variation* of PKU in the population would be explained by infants' failure to receive the proper environmental intervention. Thus, a "highly heritable" trait could be drastically altered or eliminated through a simple environmental intervention, and a heritability estimate can change from 1.0 to 0.0 even though susceptibility for the trait continues to be transmitted genetically.

In his 2001 book *The Dependent Gene*, developmental psychologist David Moore argued that while heritability addresses factors influencing a trait's variation, it says nothing about a trait's *causes*. As an analogous example he cited snowflake formation, which requires both high relative humidity and a temperature below 32 degrees Fahrenheit. Because the temperature at both poles is constantly sub-freezing, while relative humidity varies, all snowfall *variation* at the North and South Poles can be explained by differences in relative humidity. This does not mean, however, that temperature is not an important factor in the creation of snowflakes, or even that relative humidity is more important than temperature. Conversely, Moore observed that snowfall variation in a tropical country such as Costa Rica (which receives snow at the highest mountaintops) is accounted for by *temperature* variation alone. Thus, investigators seeking to determine the causes of snowflake formation would find "temperature heritabil-

---

504. Lewontin, 1974, p. 400.

ities" of 0.0 (explaining *none* of the variation) at the poles, and 1.0 (explaining *all* of the variation) in the tropics. Self evidently, these figures tell us nothing about the relative importance of humidity and temperature on snowflake formation; they merely reflect the fact that a key variable has been held constant. Moore concluded, "since heritability estimates account for variation *and do not explain causation*, they tell us *nothing at all* about how genetic and environmental factors influence the development of our traits."[505]

(5) *High heritability does not imply that traits are fixed or unchangeable.* This point is related to point #4 and has also been a common criticism of hereditarian interpretations of the heritability coefficient. The classic example is human height, which has increased steadily in the industrialized nations (most dramatically in Japan). This increase has occurred in spite of the frequent claim that human height is a highly heritable trait.

ORIGINS OF THE HERITABILITY CONCEPT

Although the word "heritability" has been used since the 19th century, the concept (as it is understood today) was described by Jay Lush as early as 1936.[506] Lush, a professor of animal breeding at Iowa State College, was interested in facilitating the breeding of animals for economically desirable traits. Animal breeders needed a quantifiable method for predicting the results of a program of selective breeding for these desired traits. Lush wrote that heritability estimates could be derived from the resemblance of genetically related animals: "All methods of estimating heritability rest on measuring how much more closely animals with similar genotypes resemble each other than less closely related animals do."[507]

Lush produced his clearest statement on heritability in a 1949 paper.[508] Here, he defined heritability as "the fraction of the observed or phenotypic variance which was caused by differences between the genes or the genotypes of the individuals."[509] Lush discussed two types of heritability: broad and narrow.

---

505. Moore, 2001, p. 43. Emphasis in original. A discussion of the origins of the heritability concept can be found in Bell, 1977.
506. See Bell, 1977.
507. Lush, 1945, p. 93.
508. Lush, 1949.
509. *Ibid.*, p. 356.

Broad heritability refers to the entire genotype, which includes genes which interact non-additively. *Narrow heritability* refers only to additive genetic effects, meaning those genes which are of interest to plant and animal breeders. Lush understood that highly heritable traits in animals were nonetheless influenced by the environment:

> Most of the numerical estimates for heritability of economically important characters have been surprisingly high in view of the fact that many of these characters are known to be readily and strongly influenced by changes in nutrition and by other environmental circumstances.[510]

This observation stands in direct contrast to Jensen's 1969 position, where he argued that there is little hope of significantly "boosting" IQ scores because intelligence is "highly heritable" — an argument revisited in Herrnstein and Murray's *The Bell Curve*.[511] Lush understood that trait resemblance among livestock relatives was influenced by environmental factors, writing, "The correlation between relatives is likely to have an environmental component."[512] Lush also viewed comparisons of identical twin *livestock* as vulnerable to environmental influences: "Both members of a pair of identical twins would be subject to many of the same peculiarities of environment. . . . Consequently this method overestimates heritability to the extent that the common environment of identical twins is important."[513]

WHAT IS THE PURPOSE OF A HERITABILITY COEFFICIENT?

We have seen that the heritability concept was created for the purpose of assisting animal and plant breeders to select for economically desirable traits. Because the statistic is now widely used in human populations for IQ and personality, and for psychiatric constructs such as schizophrenia and ADHD, one might ask what relevance heritability has for humans.

For Lush, "To know whether heritability is high or low is important when making efficient breeding plans."[514] A heritability estimate is important, according to Lush, because "The possibilities of improving farm animals by

510. *Ibid.*, p. 372.
511. Herrnstein & Murray, 1994.
512. Lush, 1949, p. 360.
513. *Ibid.*, p. 362.
514. *Ibid.*, p. 359.

suitable genetic methods are real and important."[515] I. M. Lerner discussed heritability in some detail in his 1958 book on genetics and selection, referring to it as a "measure of *accuracy of selection*."[516] Lerner cautioned, however, that a heritability estimate was only valid for the particular population and generation from which data are obtained.

Fuller and Thompson's 1960 *Behavior Genetics* is usually regarded as the founding document of the behavior genetics field. According to its authors,

> The heritability of a trait is of primary importance in behavior genetics, since it provides a measure of the expected rate of gain from selection. . . . For the practical breeder or the psychologist trying to develop a strain characterized by particular behavioral qualities, heritability determinations provide a guide to the most efficient method of selection.[517]

Towards the end of their book Fuller and Thompson wrote that while heritability estimates have "no logical meaning when applied to an individual," with regards to a population, "one may well ask how much of the observed variation in behavior is attributable to genetic differences and how much to environment."[518] Fuller and Thompson appear to have been the first authors to extend the definition of heritability to include the estimation of "how much" of the observed variation is due to genetic factors, although many previous investigators had expressed interest in this question without linking it to heritability estimates.

In a 1978 work, Fuller and Thompson wrote that the heritability concept "was developed by animal and plant breeders who were concerned with predicting the results of selection," while noting that heritability "has been taken over frequently in behavior genetics, sometimes with a less limited sense than is implied in the preceding definition."[519] They wrote that broad heritability was useful "for quantifying the extent of genetic influence on continuously distributed characters," and that "because the term is often misinterpreted, it has been suggested that behavior geneticists avoid the concept, but there is as yet no satisfactory alternative for quantifying the extent of genetic influence on continuously distributed characteristics."[520]

---

515. *Ibid.*, p. 374.
516. Lerner, 1958, p. 63.
517. Fuller & Thompson, 1960, p. 64.
518. *Ibid.*, p. 317.
519. Fuller & Thompson, 1978, p. 60.
520. *Ibid.*, p. 61.

In one of his early publications, behavior geneticist Gerald McClearn wrote about heritability in the context of animal research:

> One of the central concepts of modern genetics is that of *heritability*, which is defined as the ratio of the variance attributable to additive gene effects to the total phenotypic variance. This quantity represents the genetic contribution which is "useful" in the sense that it provides for firm prediction of the outcome of various matings (e.g., in a selection program).[521]

Again, heritability is "useful" as a means of predicting the outcome of matings. Although McClearn made several references to the importance of estimating the extent of genetic versus environmental influences in a population, he did not link the heritability concept to this idea.

In 1969 Jensen wrote, "Heritability is a population statistic, describing the relative magnitude of the genetic component (or set of genetic components) in the population variance of the characteristic in question."[522] Jensen's definition derived from Fuller and Thompson's 1960 extension of the heritability concept from an estimate for use by breeders, to an indicator of the magnitude of genetic influences on the population variance for the trait in question. I shall return to this point later.

In the 1996 edition of his influential *Introduction to Quantitative Genetics*, geneticist D. S. Falconer wrote,

> The heritability of a metric character is one of its most important qualities. It expresses . . . the proportion of the total variance that is attributable to differences of breeding values, and this is what determines the degree of resemblance between relatives. But the most important function of the heritability in genetic study of metric characters is its predictive role, expressing the reliability of the phenotypic value as a guide to the breeding value.[523]

Like most earlier writers, Falconer discussed a heritability estimate as being important in the context of selective breeding.

Returning to Plomin and colleagues' 1997 behavior genetics textbook, we have seen how these authors discussed the limitations (and by implication misuse) of the concept. Given these limitations, how did Plomin and colleagues describe the purpose of a heritability coefficient? According to them, heritability

---

521. McClearn, 1964, p. 234.
522. Jensen, 1972, p. 115.
523. Falconer & Mackay, 1996, p. 160.

is the "statistic that estimates the genetic effect size. . . . Heritability is the proportion of phenotypic variance that can be accounted for by genetic differences among individuals."[524] They defined "effect size" as the "proportion of individual differences for the trait in the population accounted for by a particular factor."[525] Thus according to Plomin et al., "genetic effect size" and "heritability" are synonymous. They identified broad heritability as describing the genetic effect size of a population, whereas narrow heritability is "interesting in the context of selective breeding studies."[526]

Plomin and colleagues defined heritability and discussed its limitations, yet offered no explanation of its practical use in human populations other than to estimate the genetic effect size. In light of their discussion of the concept's limitations, they failed to convincingly demonstrate the heritability statistic's *raison d'être* in human populations.

As we have seen, the two most frequently cited justifications for a heritability estimate are: (1) that it can predict the results of a breeding program, and (2) that it represents the percentage of phenotypic variance in a population attributable to genetic factors (that is, the magnitude of genetic influences). Regarding the first point, heritability estimates were designed for (and continue to be widely and profitably used in) agriculture. They might also be useful in experiments with laboratory animals under conditions where breeding and environment can be controlled. In human populations, a selective breeding program is called eugenics, which few people currently using heritability estimates advocate.

Turning to the second justification, is it true that a heritability coefficient estimates the magnitude of genetic influences on a trait? Several critics have argued that it does not. According to dissident behavior geneticists Terry McGuire and Jerry Hirsch, whereas narrow heritability (h2) is an important statistic in selective breeding programs, broad heritability (H2) has been used by psychologists "in the belief that it is their nature-nurture ratio or index to the causes of a trait."[527] However, H2 does not

---

524. Plomin, DeFries, et al., 1997, p. 79.
525. *Ibid.*, p. 312.
526. *Ibid.*, p. 300.
527. McGuire & Hirsch, 1977, p. 46.

describe the average influence of heredity in determining the level of trait expression in a population. Broad heritability estimates only the contribution of genetic factors (additive, dominance deviation, interaction and genotype-environment covariance) in the total phenotypic variance for one trait in a specified population.

McGuire and Hirsch wrote that while H2 is "of little interest to geneticists," it has been used and "misused" in psychology.[528]

As noted, many critics have correctly pointed out that "high heritability" implies nothing about the likelihood of a particular intervention's success. For example, a finding that children's reading ability is "highly heritable" does not mean that a new teaching method could not dramatically improve reading scores.[529] Thus, a heritability estimate for reading ability has no practical purpose, and, in general, heritability "is nearly equivalent to no information at all for any serious problem of human genetics."[530]

Although heritability estimates have no practical purpose apart from animal and plant breeding, they are frequently misunderstood by people unfamiliar with the concept. It is often believed that the words "heritable" and "inherited" are synonymous. Hirsch, however, has emphasized that "heritability" and "heredity" are "two entirely different concepts that have been hopelessly conflated" in several texts.[531] "Because of their assonance," he wrote, "when we hear one of the two words, automatically we think the other."[532] Stoltenberg proposed that the word "heritability" be replaced by "selectability" so as not confuse the "scientific" and "folk" definitions of heritability. For Stoltenberg, the "folk definition" of heritability is synonymous with "inherited." A change in terms would go a long way in eliminating the confusion between "heritability" and "inherited," although as Stoltenberg pointed out, many investigators would object to the eugenic implications of the word "selectability." "Because of this," wrote Stoltenberg, "researchers would need to carefully consider the utility of calculating such a statistic, and when it is deemed necessary, careful use of language would be required."[533]

---

528. *Ibid.*, p. 46.
529. Hirsch, 1970.
530. Feldman & Lewontin, 1974, p. 1168.
531. Hirsch, 1997, p. 220.
532. *Ibid.*, p. 220.
533. Stoltenberg, 1997, p. 96.

The unfortunate conflation (or confusion) of "heritable" and "inherited" (or "hereditary") is more understandable when we realize that even the creator of the heritability concept confused the terms. In a paper Lush considered "one of my best summaries on the [heritability] topic,"[534] he wrote:

> If a character can be influenced strongly by environmental variation, its *heritability* would be low in a population in which the environment varies widely. But in another population in which the physical control is so rigid that the environmental variations are very slight, the same character would appear highly *hereditary* [emphasis added].[535]

And on the preceding page Lush spoke of "heredity in the broad sense," when he was clearly referring to heritability in the broad sense.[536] Of course, it is unlikely that Lush anticipated that his concept would be used in human populations. In later years he would complain that "it does little good to warn against oversimplifying the idea of heritability. Some feel it necessary to repeat those warnings each time they write. Others ignore the warnings." Lush believed that the time had come for "coining a new word . . . to be presented *precisely* and in a way in which it *cannot be misunderstood*."[537]

An Exchange on the Meaning of Heritability

In a 1990 *Behavioral and Brain Sciences* target article, maverick Canadian behavior geneticist Douglas Wahlsten remarked, "The only practical application of a heritability coefficient is to predict the results of a program of selective breeding."[538] Of the 29 authors commenting on Wahlsten's article (both favorably and unfavorably), only a few offered alternative applications. According to Cheverud, heritability analysis "is commonly used because it can at least provide evidence that genes play a role in the development of a particular phenotype."[539] Crow commented, "one can be interested in the heritability of IQ without advocating a breeding program," and claimed that low heritability "says that changes of exiting variables within the existing range can have a substantial effect on the trait," whereas with high heritability, "environmental manipula-

---

534. Quoted in Bell, 1977, p. 298.
535. Lush, 1949, p. 358.
536. *Ibid.*, p. 357.
537. Quoted in Bell, 1977, p. 299. Emphasis in original.
538. Wahlsten, 1990, p. 119.
539. Cheverud, 1990, p. 124.

tions will have little effect unless they extend beyond the existing range, or bring in new factors."[540] Crusio wrote that heritability figures had been justified in the past in order to convince psychologists and ethologists of the causal role of heredity in psychological trait differences, while "It is by now clear that this approach is basically sterile and that these efforts should be abandoned."[541] Detterman wrote that "Heritabilities are nothing more than a way of representing portions of variance according to a particular model."[542] According to Goodnight, heritability estimates "reflect the potential for evolutionary change in a particular population at a particular time."[543] Finally, McGuffin and Katz wrote that heritability estimates can help researchers choose the correct definition of psychiatric diagnoses such as schizophrenia, and claimed that specific numerical estimates are of little or no interest to researchers.[544]

Let's examine these claims a bit more closely. Cheverud wrote that heritability estimates provide evidence that genes play a role in the development of a trait. However, the findings of genetic kinship studies — if each study is methodologically sound and if its underlying assumptions are valid — are sufficient to demonstrate this. For example, if the all assumptions of the classical twin method are correct (see Figure 2.1), a significantly higher identical versus fraternal twin correlation suggests the role of genetics for the trait in question. A heritability coefficient is merely the product of this finding, rather than evidence in favor of genetic factors.

Crow claimed that environmental interventions for traits with high heritability have little effect "unless they extend beyond the existing range, or bring in new factors." The problem as this applies to humans is that environments are frequently capable of going "beyond the existing range," and "new factors" are always possible; a new drug or teaching method, for example. This is particularly true in our current age of fast-paced technological changes.

Detterman's response to Wahlsten repeated the common position: Heritability estimates help gauge the magnitude of genetic effects for a particular trait using a particular model. However, we have seen that there is little evidence in favor of this position. Goodnight argued that heritability estimates "reflect the potential for evolutionary change" in a population. This is similar to the claim

540. Crow, 1990, p. 127.
541. Crusio, 1990, p. 128.
542. Detterman, 1990, p. 132.
543. Goodnight, 1990, p. 134.
544. McGuffin & Katz, 1990.

that high heritability implies that environmental interventions are unlikely to be successful.

McGuffin and Katz argued that it is not important to quantify precise estimates of the heritability of a trait, a position taken up by other researchers. According to behavior geneticists Rutter, Silberg, and Simonoff in 1993,

> We may well agree with the environmentalists' criticism that knowledge of the level of heritability of any given trait is of very little interest with respect to policy and practice because quite a high heritability does not rule out effective environmental intervention. Nevertheless, it is a serious mistake to suppose that behavioral genetics is mainly involved with quantifying heritability. There is indeed very little interest in calculating the precise level of heritability as such . . .[545]

In fact, behavior geneticists are very interested in calculating heritability coefficients, as seen in current research as well as in various studies published in the period when Rutter and colleagues made this claim. For example, Bouchard and McGue calculated heritability coefficients for personality variables from the Minnesota Study of Twins Reared Apart, Tsuang and Faraone estimated the heritability of several mood disorders, Goodman and Stevenson did the same for inattentiveness and hyperactivity, and in the edited book in which the Rutter et al. chapter appeared, Cardon and Fulker presented exact heritability estimates for various dimensions of cognitive ability.[546]

In summary, when confronted with Wahlsten's contention that a heritability estimate's only practical application is for use in a selective breeding program, several leading genetic investigators were unable to provide a viable alternative explanation. This included Plomin's response, where he did not address the issue.[547] Of note is geneticist Oscar Kempthorne's comment on Wahlsten's article, where he observed that "most of the literature on heritability in species that cannot be experimentally manipulated, for example, in mating, should be ignored."[548]

---

545. Rutter et al., 1993, p. 437.
546. Bouchard & McGue, 1990, p. 277; Tsuang & Faraone, p. 1990, p. 74; Goodman & Stevenson, 1989, p. 697; Cardon & Fulker, p. 1993, p. 112.
547. Plomin, 1990.
548. Kempthorne, 1990, p. 139.

A Heritability Estimate: The Product of Environmentally Confounded Research

Behavior geneticists' calculations of heritability are based on the assumption that genetic makeup and environment do not interact, when it is clear that they do.[549] Obvious examples are height and skin color. In the United States, a person genetically coded to have dark skin pigmentation will usually experience a far different (i.e., vastly inferior) environment than someone born with light pigmentation. In the same way, adult males who are four feet tall experience different physical and social environments than adult males who are six feet tall. Kempthorne argued that because of gene-environment interaction, "we cannot reasonably state that such and such a proportion of variance is 'due to' heredity," and concluded that "'heritability' does not even exist in the human IQ context."[550]

Even if heritability were a valid and important concept for humans, the fact remains that heritability estimates are derived from twin, family, and adoption studies. I have already documented the problems associated with genetic conclusions based on twin research, and as we will see more clearly in later chapters, adoption studies have their own set of problems and potential confounds. In twin research, heritability usually is estimated as $2(rMZ - rDZ)$, where rMZ is the identical twin concordance or correlation, and rDZ is the fraternal twin concordance or correlation.[551] Put more simply, heritability is calculated by doubling the reared-together identical-fraternal correlational difference. For example, if identical twins correlate at .70 for IQ, and fraternal twins correlate at .40, twin researchers would estimate IQ heritability as .60. However, these estimates are based on the validity of the equal environment assumption and are therefore dubious. In studies of twins reared apart, the MZA intraclass correlation is sometimes said to represent a direct estimate of heritability. These studies, as we saw in Chapter 4, contain methodological flaws and questionable assumptions which call into question their role as indicators of genetic influences.

Human genetic researchers often write that the results from family, twin, adoption, studies converge on the conclusion that genes are an important factor influencing human psychological trait variation. In fact, the results from these

549. McGuire & Hirsch, 1977; Wahlsten, 1990.
550. Kempthorne, 1978, p. 19.
551. Falconer, 1965; Smith, 1974.

studies are plausibly explained on the basis of environmental factors, methodological error, and bias. Thus, even if heritability estimates were valid and useful in human populations, it is doubtful that accurate figures could be obtained from these methods.

HERITABILITY AND PSYCHIATRIC DISORDERS

Although the heritability debate has centered mainly on IQ and to a lesser extent personality, heritability estimates are also used in psychiatry. Schizophrenia has been the most studied of all psychiatric disorders and will serve as an example. Kendler calculated heritability coefficients for schizophrenia on the basis of the results from twin studies.[552] He concluded that "the major twin studies of schizophrenia agree in estimating the heritability of liability of schizophrenia at between 0.6 and 0.9," while noting that these figures are based on the multifactorial threshold model "which may or may not be appropriate for schizophrenia."[553] According to Kendler and Diehl, the "heritability of liability. . . . ranges from 0.0 if genetic factors play no role in susceptibility of a disorder to a maximum of 1.0 if genes entirely determine disease risk."[554] A similar formulation is found in Faraone, Tsuang, and Tsuang's 1999 psychiatric genetics textbook.[555] Kendler's description obscures the critical point that a simple environmental intervention could completely eliminate a true genetic disorder, such as PKU. Conversely, we have seen that a heritability estimate of 0.0 for a trait does not necessarily mean that "genetic factors play no role in susceptibility," but only that such factors do not play a role in the population variance of the condition.

The questionable methods Kendler used to calculate twin concordance rates, from which he calculated heritability, are beyond the scope of this chapter, as are the problems with schizophrenia genetic studies in general (see Chapters 6 and 7). What is relevant here is the meaning of Kendler's heritability estimates and how they might be misunderstood by people interested in the causes of schizophrenia. A reader of Kendler who understands heritability in the "folk" meaning of the term would conclude that schizophrenia "is 60–90% inherited."

---

552. Kendler, 1988.
553. *Ibid.*, p. 17.
554. Kendler & Diehl, 1995, p. 945.
555. Faraone, Tsuang, & Tsuang, 1999, p. 32.

A more knowledgeable reader with an accurate yet incomplete understanding of heritability estimates might conclude that genetic factors are more important than environmental factors. As we have seen, both of these conclusions are erroneous. Gottesman and Shields concluded that "70% of the variance in the combined liability to developing schizophrenia is genetically transmitted and 20% is 'culturally' transmitted."[556] Again, many people would mistakenly interpret this statement as meaning that 70% of schizophrenia cases are the result of genetics, and that 20% are due to cultural transmission. Nevertheless, the authors of a 2004 article in *The American Journal of Psychiatry* wrote about the "societal burden" of "complex genetic brain disorders," where they estimated that the "heritable" component of schizophrenia costs the United States nearly $40 billion annually.[557]

In fact, heritability estimates for psychiatric conditions are as misleading as they are for IQ and personality, for even if it were true that a diagnosis such as schizophrenia had an important genetic component, it could still develop that, like PKU or farsightedness, a simple environmental intervention would greatly reduce or eliminate its impact. That this has not occurred with schizophrenia is due to the failure to discover specific environmental factors — the existence of which all sides in the debate acknowledge — rather than the alleged importance of genetic factors.

CONCLUSIONS

In this chapter I have discussed and elaborated upon the observations of several critics of the heritability concept, who have argued against its usefulness in assessing the importance of genetic factors for human trait differences. The concept is widely misunderstood by the public and by some professionals as well. Indeed, in 1985 historian of science Diane Paul looked at genetics textbooks' handling of the "genetics of intelligence" topic and concluded that the discussions of heritability were "confused in the extreme."[558]

The heritability concept should be retained in animal and plant breeding, where it has been widely and successfully utilized. But even in agriculture it is used for the specific purpose of economic gain through increased production.

556. Gottesman & Shields, 1982, p. 225.
557. Uhl & Grow, 2004, p. 225.
558. Paul, 1998, p. 37. For the original study, see Paul, 1985.

Selection does not necessarily produce a "better" cow or chicken, or one that would even survive outside of a controlled environment.[559] Although the vast majority of people utilizing the heritability concept for human traits are opponents of the eugenic program, its only practical application, as Wahlsten pointed out, is to predict the results of a selective breeding program. Ironically, heritability coefficients would have little value even in eugenic selection among humans, because a heritability calculation in the current environment might be very different in another environment.

The heritability concept has little or no value in helping us understand the origins of human trait differences, but to the extent that it is widely misunderstood and misused, it is at best irrelevant[560] and is more often a harmful and misleading concept. Hirsch spoke of the "tidal wave . . . of unjustified human heritability estimates" that "has done so much harm over two decades . . ."[561] As such, its use should be discontinued in the social sciences except when it refers specifically to selective breeding plans. Because even in this limited manner the heritability concept would continue to sow confusion about its true meaning, Stoltenberg recommended using the word "selectability" in its place.

In conclusion, although a researcher might wish to determine whether human trait variation is influenced by genetic factors, the claim that a trait or population has a particular heritability value is meaningless. Therefore, the heritability statistic has no practical purpose in psychology and psychiatry, where its use should be discontinued. Its disappearance would eliminate confusion over the concept, and researchers would utilize other means to show that traits or conditions have a genetic basis. What they would lose is a word (and its corresponding quantification) falsely purporting to gauge the magnitude of genetic influences on a trait, since, contrary to popular belief, "heritability is not a nature-nurture ratio measuring contributions to individual development."[562] "Heritability," therefore, should be returned to plant and animal breeders, for whom it was originally intended.

---

559. McGuire & Hirsch, 1977.
560. Lewontin, 1987.
561. Hirsch, 1990, p. 137.
562. Hirsch, 1997, p. 220.

Chapter 6. The Genetics of Schizophrenia I: Overview

The claim that some people have a disease called schizophrenia (and that some, presumably, do not) was based not on any medical discovery but only on medical authority; that it was, in other words, the result not of empirical or scientific work, but of ethical and political decision making.

— Psychiatry critic Thomas Szasz in 1976.[563]

One can divide the investigators in the [schizophrenia] field into two main types: those who like to look at numbers and those who like to look at patients. The former tend to be hereditarians, the latter environmentalists.

— Schizophrenia researcher David Rosenthal in 1968.[564]

Why don't they take me, too?

— A young Norwegian woman, upon discovering that her identical co-twin sister was being taken away to a mental hospital.[565]

Schizophrenia, the paradigmatic psychiatric disorder, has been studied for more years and in more depth than any other condition in psychiatry, and serves as a model for investigations into the genetics of psychiatric disorders in general. In this chapter and the next we will see that — contrary to what most people in and out of psychiatry believe — there is little scientifically acceptable evidence supporting a genetic basis for schizophrenia. Furthermore, we will see that a "genetic predisposition," even if it existed, is not a meaningful or helpful concept.

In 1990, psychologist Richard Marshall argued that the genetic basis of schizophrenia is a hypothesis, not an axiom.[566] Nevertheless, the "genetics of schizophrenia" debate has been closed for many years; the broad consensus being that genes play an important role. However, the issue has been "resolved" and then reopened many times. Before any twin, adoption, or even family study had been performed, Rosanoff and Orr could write in 1911, after assessing the family histories of a few hospitalized patients, "It would seem, then, that the fact of the hereditary transmission of the neuropathic constitution as a recessive

---

563. Szasz, 1976, p. 3.
564. Rosenthal, 1968, p. 414.
565. Quoted in Kringlen, 1967b, p. 58.
566. Marshall, 1990.

trait, in accordance with the Mendelian theory, may be regarded as definitely established."[567] This view was echoed by Charles Davenport of the Eugenics Record Office in 1920: "The conclusion seems supported that the tendency [for dementia praecox, a term later replaced by "schizophrenia"] is inherited as a simple Mendelian recessive."[568] In 1958, a British psychiatrist would write that "The evidence for a genetic cause of schizophrenia is now generally agreed to be overwhelming, although certain details may not yet be known."[569] David Rosenthal argued in 1970 that the results from schizophrenia adoption research provided evidence "so consistently and so strongly in favor of the genetic hypothesis that the issue must now be considered closed."[570] In 1993, a pair of twin researchers stated that the "substantial hereditary component in schizophrenia is surely one of the two or three best-established facts in psychiatry,"[571] and a group of schizophrenia researchers wrote in 2004, "One of the most well-established findings in schizophrenia research is that vulnerability to the illness can be inherited."[572] The 1987 revised third edition of the American Psychiatric Association's Diagnostic and Statistical Manual (DSM-III-R) concluded that "genetic factors have been proven to be involved in the development" of schizophrenia,[573] although the 1994 edition (DSM-IV) softened this position, stating that "much evidence suggests the importance of genetic factors in the etiology of Schizophrenia."[574]

There are no biological markers or laboratory tests distinguishing people diagnosed with schizophrenia from those not diagnosed. Rather, psychiatrists and others diagnose schizophrenia based on factors such as history, observation, behavior, and self-report. The question of whether schizophrenia is caused by a disease process of the brain is beyond the scope of this book, although the evidence in favor of the disease model, as Breggin and others have demonstrated, is weak.[575] Theoretically, however, schizophrenia could involve a malfunctioning of the brain and yet have no genetic basis whatsoever. Nevertheless, the alleged biological basis of schizophrenia is frequently cited in support of the genetic

---

567. Rosanoff & Orr, 1911, p. 228.
568. Davenport, 1920, p. 305.
569. Smythies, quoted in Jackson, 1960, p. 38.
570. Rosenthal, 1970b, pp. 131-132.
571. Bailey & Pillard, 1993, p. 241.
572. Walker et al., 2004, p. 407.
573. APA, 1987, p. 192.
574. APA, 1994, p. 283.
575. Breggin, 1991. See also Boyle, 1990, 1999, 2002.

position, and the alleged genetic basis of schizophrenia is frequently cited in support of the biological position. As Laing observed, however, "These two theories do not . . . reciprocally validate each other. Rubbing two phantom flints produces only the illusion of fire."[576]

Another matter is whether "schizophrenia" should be viewed as a disease or illness, or whether it is better understood as a set of socially disapproved behaviors labeled as a disease. Szasz described schizophrenia as "a concept wonderfully vague in its content and terrifyingly awesome in its implications,"[577] which served as the model for redefining "the criterion of disease . . . from abnormal bodily structure to abnormal personal behavior."[578] The validity and usefulness of the schizophrenia concept has been questioned by critics such as Boyle, Breggin, Cohen and Cohen, Gosden, Hill, Laing, and Sarbin and Mancuso.[579] The symptoms psychiatrists and others use to diagnose schizophrenia often reflect judgments about the appropriateness of a person's beliefs or behavior. Beliefs, as critics frequently point out, can become "delusions" when they are culturally unacceptable.[580] If someone says that Jesus speaks to him, he is a "good Christian." If another person says that Napoleon Bonaparte speaks to her, she becomes a "schizophrenic." How can a choice of historical figures differentiate between respected members of the community (or even a head of state), and those with "hereditary brain diseases?" Clearly, the schizophrenia diagnosis is sometimes used to label social deviance as a "disease."[581]

A psychiatric diagnosis, wrote psychologist Mary Boyle, gains legitimacy by "adopting the language and assumptions of medical diagnosis."[582] She argued that this position has two essential flaws. First, unlike in physical disease, no relationship has been established between the observed phenomena (in the case of schizophrenia, for example, there is no a priori relationship between withdrawal and hallucination). Second, she argued that it has not been demonstrated that any process, whether psychological or biological, holds together the phenomena under investigation. Psychiatric diagnoses thus fail the scientific test of

---

576. Laing, 1981, p. 97.
577. Szasz, 1976, p. xiv.
578. *Ibid.*, p. 12.
579. Boyle, 1990, 2002; Breggin, 1991; Cohen & Cohen, 1986; Gosden, 2001; Hill, 1983; Laing, 1967; Sarbin, 1990; Sarbin & Mancuso, 1980; Szasz, 1961, 1970, 1976.
580. See Gosden, 2001, for a discussion of this point.
581. Sarbin & Mancuso, 1980.
582. Boyle, 1999, p. 79.

validity. If psychiatric diagnoses are not valid, then they cannot be reliable, and research based on them cannot be regarded as sound.

Another issue is that people diagnosed with schizophrenia have low reproductive rates. According to the DSM-IV-TR (the 2000 "Text Revision" of the DSM's Fourth Edition), approximately 60%-70% of people diagnosed with schizophrenia never marry, and most have limited social contacts.[583] The persistence of a "genetic disorder" in which the gene carriers reproduce at low rates does not seem possible. Supporters of the genetic theory of schizophrenia sometimes acknowledge this, yet offer few plausible explanations. In 2001, biological psychiatrist Nancy Andreasen could only observe that it is "fascinating" that "schizophrenia persists in the human population despite the fact that the majority of people who develop it do not marry or procreate."[584] Psychiatrists Torrey and Yolken were more to the point, writing in 2000:

> Since the beginning of the [American] deinstitutionalization movement in the 1960s, the vast majority of individuals with severe schizophrenia have been released from hospitals or have been hospitalized for brief periods only. They have been allowed to live in the community and are reproducing at a much higher rate than they did prior to deinstitutionalization . . . At the same time, there are suggestions from several European studies that the incidence of schizophrenia may be decreasing . . . If schizophrenia is truly a genetic disease, therefore, it is the only known genetic disease that becomes more common with less reproduction by those affected and less common with more reproduction by those affected.[585]

Today, many researchers have moved "beyond" twin and adoption studies and are engaged in molecular genetic research. However, the search for "schizophrenia genes" is based on the assumption that family, twin, and adoption studies have settled the question of the genetic nature of schizophrenia, with the next step being the identification of the predisposing gene or genes. In Chapter 10 I discuss the continuing failure to find genes for schizophrenia, as well as the likelihood that this search will turn up nothing.

---

583. APA, 2000, p. 302. Also see Slater et al., 1971.
584. Andreasen, 2000, p. 109.
585. Torrey & Yolken, 2000, p. 114.

Schizophrenia carries a lifetime population expectancy rate of 0.8%-1.0%.[586] Of the people currently living in the United States, it has been estimated that close to 4 million will be diagnosed with schizophrenia at some point in their lives.[587] Although it is often reported that prevalence rates are similar in most countries, some have argued that schizophrenia is found much more frequently in the industrialized parts of the world.[588] Men and women are diagnosed in roughly equal numbers.[589]

The term *dementia praecox* (dementia of early life) was popularized by the Swiss-German psychiatrist Emil Kraepelin in the late 19th century. Kraepelin developed his concept by combining three previously reported conditions into one syndrome. An author sympathetic towards his work described Kraepelin's method:

> The major contribution of Kraepelin was the inclusion, in the same syndrome, of catatonia, already described by Kahlbaum, hebephrenia, and "vesania typica," also described by Kahlbaum, characterized by auditory hallucinations and persecutory trends. . . . Kraepelin was able to discern the common characteristics in these apparently dissimilar cases.[590]

In 1911, Eugen Bleuler of Switzerland published his famous *Dementia Praecox or the Group of Schizophrenias*, which was translated into English in 1950. Bleuler proposed that the condition's name be changed to *schizophrenia*, meaning "split mind." Although Bleuler based his conception of schizophrenia on Kraepelin's work, he saw the condition as being characterized by "primary symptoms" such as autism (turning inward), blocking of thoughts, and loose associations (vague connections between thoughts). Bleuler regarded hallucinations and delusions as "secondary symptoms," the result of a primary disturbance in thought.

---

586. Rosenthal, 1970b; Slater & Cowie, 1971. Gottesman (1991, p. 75).
587. Torrey, 1995.
588. Torrey, 1980.
589. Gottesman, 1991; Rosenthal, 1970b; Slater & Cowie, 1971.
590. Arieti, 1974, p. 10. Commenting on Arieti's description, Thomas Szasz (1976, p. 10) wrote,

> The point I want to emphasize here is that each of these terms refers to a behavior, not disease; to disapproved conduct, not histopathological change; hence, they may loosely be called "conditions," but they are not, strictly speaking, medical conditions. If none of these three items is a disease, putting them together still does not add up to a disease.

Both Kraepelin and Bleuler believed that dementia praecox/schizophrenia had a strong inherited component, in spite of the lack of empirical studies supporting their beliefs. In 1899 Kraepelin wrote, "An inherited predisposition to mental disturbances was apparent in approximately seventy per cent of those cases in which data could be evaluated."[591] And Bleuler wrote in his monograph that "heredity does play its role in the etiology of schizophrenia, but the extent and kind of its influence cannot as yet be stated."[592] These views prompted Boyle to write,

> Thus, before any attempt at systematic data collection was ever made, the two most prominent users of the concepts of dementia praecox and schizophrenia were disseminating the view that whatever phenomena they included under these terms were largely inherited.[593]

Following the publication of Rüdin's 1916 schizophrenia family study, Bleuler wrote, "No one denies that an inherited predisposition plays an important role."[594] Bleuler also advocated the eugenic sterilization of people diagnosed with (and viewed as predisposed to) schizophrenia. In a rarely-cited passage from his book on schizophrenia, Bleuler wrote,

> Lommer and von Rohe have again recommended castration which, of course, is no benefit to the patients themselves. However, it is to be hoped that sterilization will soon be employed on a larger scale in these cases as in other patients with a pathological *Anlage* [predisposition] for eugenic reasons.[595]

Interestingly, as Boyle has persuasively argued, many of the people Kraepelin and Bleuler studied may have suffered from the viral infection *encephalitis lethargica*. The features of this disease, whose symptoms are remarkably similar to "schizophrenia" and "dementia praecox," were not described until 1917, well after Kraepelin and Bleuler had published their major works.[596]

591. E. Bleuler, quoted in Boyle, 1990, p. 118.
592. E. Bleuler, 1950, p. 337.
593. Boyle, 1990, p. 118.
594. Quoted in Rosenthal, 1978, p. 477.
595. E. Bleuler, 1950, p. 473.
596. See Boyle, 1990, pp. 65-71.

The Genetic Theory of Schizophrenia

Unknown to the general public and to many in the mental health field is the following empirical finding: Approximately 63% of people diagnosed with schizophrenia have a "negative family history," meaning no schizophrenia diagnoses among their first- and second-degree biological relatives, including nieces and nephews.[597] The fact that almost two-thirds of those diagnosed with schizophrenia lack similarly diagnosed biological relatives is not consistent with genetic theories of schizophrenia.

Family, twin, and adoption studies record the expression of a postulated schizophrenia *phenotype*, which refers to a person's observable behavior. Molecular genetic studies attempt to identify a schizophrenia *genotype*, which refers to a person's genetic makeup with reference to a particular trait. Several genetic theorists have posited the existence of a single gene, while others have argued in favor of a polygenic basis for schizophrenia. The polygenic theory holds that the interaction of many genes is necessary to produce a predisposition for a trait or disease, for which individuals are said to have a genetic vulnerability. The theory postulates the existence of a liability threshold, which has been defined as "the point on the scale of liability above which all individuals are affected and below which none are affected."[598]

The genetic predisposition concept. The most accepted causal framework for schizophrenia is the "diathesis-stress theory," as it was called by Rosenthal in 1963. According to Rosenthal, "what is inherited is a constitutional predisposition to schizophrenia."[599] A person is viewed as inheriting a predisposition which will develop into schizophrenia in the presence of the necessary (though possibly unknown) environmental triggers, which might include psychological factors, abuse, viruses, toxins and so on. The diathesis-stress theory is acceptable to everyone who believes in a role for both genetic and environmental influences on schizophrenia, excluding only those who take a purely environmental or purely genetic position, as well as those who reject the schizophrenia concept entirely. Unfortunately, members of the public as well as many professionals frequently misinterpret "genetic predisposition" as being synonymous with "it's genetic."

597. Gottesman, 1991, p. 103.
598. Falconer, 1965, p. 53.
599. Rosenthal, 1963, p. 507.

An early criticism of common interpretations of the genetic predisposition concept is found in a 1932 discussion by American psychiatrist Abraham Myerson:

> Let us suppose that a brick falls from a building and fractures a man's skull. There must be some genetic background in his make-up that gives him a skull of such thinness and such disposition of bony forces as to permit it to be fractured by a brick. Yet it is a poor analysis in every way that says it is the genetic condition of the skull that gave him the fracture and disregards the brick.[600]

Myerson argued that it is "perfectly possible" that if environmental conditions are changed, mental disorders "would not show up, even though there were a genetic background for them."[601]

As Myerson's example shows, the emphasis on genes or environment is related to how one approaches the problem, and, more importantly, whether the environmental trigger or triggers are easily identifiable. Richard DeGrandpre drew an analogy between ideas about genetic predisposition and a fire that destroys houses on a city block.[602] In the fire, all of the wooden houses are destroyed, while the concrete houses remain intact. Genetic researchers would likely blame the destruction on the builders of the (predisposed) wooden houses and pay little attention to the person who started the fire. On the other hand, advocates of a psychosocial intervention would seek to prevent fires from starting in the neighborhood, thereby protecting all of the homes. Interestingly, the importance some would place on the material used to build houses (a genetic predisposition for schizophrenia) is proportionate to our knowledge of who started the fire (the environmental trigger). If we are able to identify the person

---

600. Myerson, 1932, p. 491. Although Myerson was known as one of his era's foremost critics of genetic determination, he, like those he criticized, supported compulsory eugenic sterilization. In the final chapter of *The Inheritance of Mental Diseases*, Myerson claimed society's "right to ask for sterilization of those types of feeblemindedness which we know to run in families," and called for the sterilization of second-generation institutionalized patients who came from families with a history of "mental disease" (Myerson, 1925, p. 320). In 1942 Myerson claimed that people with serious mental disorders "have more intelligence than they can handle," and that "the reduction of intelligence is an important factor in the curative process" (quoted in Whitaker, 2002, p. 73).
601. Myerson, 1932, p. 491.
602. DeGrandpre, 1999.

or persons responsible for setting fires, the material used to build the houses becomes much less important.

In Tienari's Finnish adoption study (see Chapter 7), the investigators found that virtually all adoptees they considered genetically predisposed to schizophrenia, and who ultimately received this diagnosis, were raised in "severely disturbed adoptive families."[603] One could conclude that schizophrenia would be eliminated by preventing people from being raised in severely disturbed families, yet the Finnish study is more often cited in support of genetic factors. Paradoxically, the genetic predisposition concept speaks more to what we *don't* know (or are unable or unwilling to change) about the environment than it does to what we *do* know about genetics. Had the brick in Myerson's example been invisible, the person suffering a fractured skull may have been viewed as having a mysterious genetic disorder.

The reception given to Tienari's Finnish study reveals the frequently *pessimistic* nature of genetic theorizing. It is far easier to emphasize genetic predisposition than it is to create a society which ensures that children do not grow up in seriously disturbed families. In this sense, the *decision* to emphasize genetic endowment over environment is often based on vested interests and beliefs. It is not predestined that disturbed families will always exist or that children will be forced to endure them. I am talking about preventing psychological damage before it happens and not, as typically occurs in psychotherapy, simply trying to undo damage that has already been done. Genetic and biological theories in psychiatry tend to blame the victim and let society off the hook, and in doing so help support the social and political status quo. As Richard Lewontin once observed, the question comes down to whether we see an individual as a problem for society, or society as a problem for the individual.[604]

The knowledge that some people are genetically predisposed to a condition could help them avoid known environmental triggers. As Myerson pointed out, however, when the trigger is known, the condition may well disappear. Whether or not some people are genetically predisposed to develop pellagra, a healthy diet protects *everyone* from pellagra. Thinking of people diagnosed with schizophrenia as being "genetically predisposed" promotes their unwarranted stigmatization and, even if true, delays the discovery of schizophrenia's environmental triggers.

603. Tienari, Sorri, et al., 1987, p. 483.
604. Lewontin, 1991.

Turning point. The June/July, 1967, Second Research Conference of the Foundations' Fund for Research in Psychiatry, held in Dorado Beach, Puerto Rico, saw the presentation of the Danish-American adoption series preliminary findings. The results of these studies carried considerable influence (see Chapter 7). In particular, the psychiatric geneticists viewed the Kety et al. Adoptees' Family report as a "steamroller," which had "obviously abolished a number of more or less fragile observations and assumptions."[605] According to Gottesman, a major consequence of the Dorado Beach conference was that "the entire field of schizophreniology was converted, at least in public pronouncements, to some kind of [gene-environment] interactionist stance."[606] The papers and speeches from this conference were published the following year in a book entitled *The Transmission of Schizophrenia*.[607]

Family studies. The genetic theory of schizophrenia found early support in the discovery of disturbed or psychotic people in the family histories (pedigrees) of people diagnosed with schizophrenia, which led to the hasty conclusions of Kraepelin, Bleuler, Rosanoff and others. Given the non-systematic nature of such evidence, researchers studied the families of a large group of people diagnosed with schizophrenia in order to determine whether the condition was found in greater numbers among their biological relatives than would be expected in the general population.

The first systematic family study was published by Rüdin in 1916,[608] and more than two dozen have been published since. Most of these studies were carried out by strong proponents of the genetic position, many did not diagnose blindly, and some used hearsay or sketchy information in making diagnoses. Several modern family studies have used control groups and blind diagnoses based on structured interviews. Most found lower first-degree relative rates than the older studies, and at least three found no significant difference between the first-degree relatives of people diagnosed with schizophrenia versus the expected population rate, or versus the rate among the first-degree relatives of controls.[609]

---

605. Strömgren, 1993, p. 11.
606. Gottesman, 1991, p. 83.
607. Rosenthal & Kety, 1968.
608. Rüdin, 1916.
609. For family studies finding no significant clustering of schizophrenia among first-degree biological relatives, see Abrams & Taylor, 1983; Coryell & Zimmerman, 1988; Pope et al., 1982.

As we saw in Chapter 1, most genetic researchers now agree that familial clustering is consistent with both genetic and environmental explanations. According to Rosenthal,

> Although [family] studies are well worth undertaking for their own sake, the inference of a genetic basis must be held in abeyance until it can be shown that the association between incidence and consanguinity cannot be explained on some other basis. We might conceivably find a similar association with respect to some infectious diseases or with respect with a trait like poverty, where environmental factors may be of overriding importance. As a matter of fact, just such an association would be predicted by many clinical psychologists and psychiatrists who hold that the occurrence of functional behavior disorders results from peculiar or unusual behavior that takes place in certain families.[610]

The inconclusive nature of schizophrenia family studies led to the study of twins and adoptees in an attempt to control for environmental confounds.

SCHIZOPHRENIA TWIN RESEARCH

In Chapter 3 we saw that the twin method, like a family study, is unable to disentangle possible genetic and environmental influences. For this reason, I will not review the schizophrenia twin studies individually. Instead, I discuss them collectively while highlighting important trends. I also discuss other types of twin research cited in support of the genetic position. Chapter 7 is devoted to an analysis of schizophrenia adoption research, the results of which are usually presented (albeit incorrectly, as we will see) as conclusive proof that genetic factors operate.

The results of the 15 published schizophrenia twin studies are summarized in Table 6.1. All concordance rates are based on the "pairwise" method, which is equal to the percentage of twin pairs concordant for schizophrenia. For example, if both twins are diagnosed with schizophrenia in 3 out of 10 pairs, pairwise concordance for schizophrenia would be 3/10 = 30%. Since the 1960s, the "proband method" has gained acceptance among several schizophrenia twin researchers. (In the psychiatric genetic literature, a person through whom other relatives are identified is often referred to as a "proband.") This method, which yields the "probandwise" concordance rate, has been described as follows: "In the proband

---

610. Rosenthal, 1970b, p. 37.

method a pair of twins concordant for schizophrenia is counted twice if each of them was ascertained independently in the course of identifying subjects for the study."[611] In this case, if both twins are diagnosed in 3 out of 10 pairs, probandwise concordance for schizophrenia would be 6/13 = 46%. The pairwise method asks, "In what proportion of pairs are both called schizophrenic"; the probandwise method asks, "In what proportion of probands is there a schizophrenic co-twin?"[612] Because the number of concordant pairs is doubled, the proband method always produces higher concordance rates. Generally speaking, leading genetic theorists such as Kendler and Gottesman use the proband method, while those more interested in environmental factors use the pairwise method. It is difficult to understand how the proband method has gained the level of acceptance that it has, however, considering that twin subjects and their co-twins usually are not identified independently.[613] Typically, the co-twin of a person diagnosed with schizophrenia is identified because his or her co-twin had already been located, which does not constitute independent ascertainment.

As seen in Table 6.1, the pooled pairwise concordance rates for schizophrenia are 40.4% identical, and 7.4% fraternal. The older "classical" studies, using non-blind diagnoses and resident hospital samples (e.g., Rosanoff, Kallmann, Slater), reported higher rates than the more recent studies. Not only did researchers make diagnoses non-blinded in these early studies, in most cases they did not even define schizophrenia. In a table listing schizophrenia twin studies used in their 2003 meta-analysis, Sullivan and Kendler, in a column listing the "diagnostic criteria" used in each study, wrote "unstated" for the studies of Essen-Möller, Inouye, Kallmann, Rosanoff et al., Slater, and Tienari. They justified the inclusion of four of these studies failing to define schizophrenia on the grounds that they were performed by "highly respected researchers" using "accepted research practices of their era," and because of the "tradition" established by previous reviewers.[614]

---

611. Torrey et al., 1994, p. 10.
612. Boyle, 1990, p. 131.
613. *Ibid.*

\multicolumn{10}{c}{*Table 6.1. Pairwise Concordance Rates in Published Schizophrenia Twin Studies*}
AUTHOR(S)	YEAR	COUNTRY	IDENTICAL PAIRS	NUMBER CONCORDANT	%	SAME-SEX FRATERNAL PAIRS	NUMBER CONCORDANT	%	
\multicolumn{10}{c}{"Classical Studies"}
Luxenburger [a]	1928	Germany	17	10	59%	13	0	0%	
Rosanoff et al.	1934	USA	41	25	61%	53	7	13%	
Essen-Möller [b]	1941/1970	Sweden	7	2	29%	24	2	8%	
Kallmann	1946	USA	174	120	69%	296	34	11%	
Slater	1953	UK	41	28	68%	61	11	18%	
Inouye	1961	Japan	55	20	36%	17	1	6%	
\multicolumn{10}{c}{"Contemporary Studies"}
Tienari	1963/1975	Finland	20	3	15%	42	3	7%	
Gottesman & Shields [c]	1966b	UK	24	10	42%	33	3	9%	
Kringlen [d]	1967	Norway	45	12	27%	69	3	4%	
NAS-NRC [e]	1970/1983	USA	164	30	18%	268	9	3%	
Fischer [f]	1973	Denmark	25	9	36%	45	8	18%	
Koskenvuo et al.	1984	Finland	73	8	11%	225	4	2%	
Onstad et al.	1991	Norway	24	8	33%	28	1	4%	
Franzek & Beckmann	1998	Germany	9	6	67%	12	2	17%	
Cannon et al. [g]	1998	Finland	—	—	—	—	—	—	
\multicolumn{10}{c}{POOLED RATES}
"Classical"			335	205	61%	464	55	12%	
"Contemporary"			384	86	22%	722	33	5%	
TOTAL			719	291	40%	1,186	88	7%	

Concordance rates based on the authors' narrow ("strict") definition of schizophrenia; age correction factors not included. When two dates are stated, the first indicates the year results were first published, the second indicates the final report, whose figures are reported in the table.

[a] Based on figures from Gottesman & Shields (1966a). Hospitalized co-twins only.

[b] Identical twin figures from Essen-Möller (1970). Essen-Möller did not report fraternal twin figures in this 1970 paper. Fraternal twin concordance rate based on 1941 definite cases among co-twins, as reported in Gottesman & Shields (1966a, p. 28).

[c] Cardno et al. (1999) reported probandwise rates of identical = 41%, fraternal = 5% for an enlarged and re-diagnosed sample based on Gottesman & Shields's study.

[d] Based on a strict diagnosis of schizophrenia; hospitalized and registered cases.

[e] National Academy of Sciences-National Research Council. Original report by Pollin et al., (1969); final report by Kendler & Robinette (1983).

[f] Final report of an expanded sample originally collected by Harvald & Haugue (1965).

[g] Cannon et al. reported probandwise concordance rates of 46% identical and 9% same-sex fraternal. The pairwise equivalents of these figures are not listed in Table 1 because the number of pairs in each group was not given by Cannon et al.

Kallmann's sample constitutes about 25% of the pooled rates in Table 6.1. If we limit the analysis to the more modern register and consecutive admissions studies beginning with Tienari, the pooled pairwise concordance rates fall to identical 22.4% (86/384), and fraternal 4.6% (33/722). Furthermore, identical twin concordance is 18% or less in three of the last nine published studies, as well as in two of the three studies with the largest identical twin sample. Most

---

614. Sullivan & Kendler, 2003, p. 1188.

investigators concluded that their results were consistent with a genetic influences, although some pointed to the importance of (usually unstated) environmental factors as well.

Three Low Concordance Rate Studies

As seen in Table 6.1, the Tienari, NAS-NRC (National Academy of Sciences-National Research Council), and Koskenvuo et al. studies reported low identical twin schizophrenia concordance rates (15%, 18%, and 11% respectively). Koskenvuo and colleagues' results were based on the computer files of Finnish hospital discharge records from the years 1972-1979. In spite of having produced the third largest identical twin sample, this study is almost never mentioned by contemporary reviewers or by the authors of psychiatry and psychology textbooks. Neither Kendler nor Gottesman, for example, ever mentioned it in their numerous reviews of schizophrenia twin research published through 2003.[615] In fact, I am aware of only one review that has mentioned the Koskenvuo et al. study, and of no review that has included its results in a table.[616] Apart from Tienari's early findings, Koskenvuo's study produced the lowest identical twin schizophrenia concordance rates ever reported, yet strangely is missing (without explanation) from textbooks and other literature reviews.[617]

The NAS-NRC and Tienari results, on the other hand, have been *misreported* by influential proponents of the genetic position. Mary Boyle wrote about the "selective criticism of low-rate studies" in *Schizophrenia: A Scientific Delusion?*,[618] and I will elaborate upon her discussion here.

Tienari. In 1963 Tienari published the results of his Finnish schizophrenia twin study, which at that time included 16 identical and 21 fraternal pairs. Tienari, who located twins through the use of population registers, reported *zero* identical pairs and one fraternal pair concordant for schizophrenia. This report took the schizophrenia twin studying world by storm, considering that Kallmann had reported an 85.8% age-corrected identical twin concordance rate just 17 years earlier. Tienari published follow-ups in 1968 and 1971, and a final

---

615. For example, see Sullivan & Kendler, 2003.
616. The review mentioning the Koskenvuo et al. study was by Murray & Reveley, 1986.
617. Because I only became aware of the Koskenvuo et al. 1984 study in 2001, it is not mentioned in any of my writings published through that year.
618. Boyle, 1990, p. 134

paper in 1975.[619] In this 1975 final report, Tienari increased his identical twin total to 20 pairs, and his fraternal total to 42 pairs. His final results were as follows — identical twins: 3/20 concordant, pairwise concordance rate = 15%; fraternal twins: 3/42 concordant, pairwise rate = 7.5%.[620]

Commenting on these findings in 1982, Gottesman and Shields wrote, "one might have expected a low concordance rate from a small sample of male twins."[621] Four other schizophrenia twin studies had similar identical twin pair sample sizes (including their own, N = 24), but escaped Gottesman and Shields's criticism. It is also puzzling that Gottesman and Shields would expect a sample of *male* identical twins to produce lower concordance, when twelve pages later they argued that non-resident hospital based studies found no significant concordance rate differences between male and female identical twins.[622] They presented a table listing the figures from Tienari's 1971 update, rejecting his 1975 final results on the grounds that he added three new identical pairs, and that it "lacked detail."[623] They claimed that Tienari's identical twin probandwise concordance rate stood at 36%, and suggested that this figure should be age-adjusted upwards. According to Boyle, "it is not clear why Tienari's research should be singled out in this way, given the haphazard and often unreported information on which other researchers based their diagnoses."[624] She questioned Gottesman and Shields's suggestion of adjusting for age, given that Tienari's twins were all over 40-years-old and were therefore past the risk period for schizophrenia. (They ranged in age from 46-55 in 1975.)

Tienari's results show how the proband method can be used to artificially inflate relatively low concordance rates. Let's take Tienari's 1975 identical twin pairwise figure of 3/20 = 15% as an example. Using the proband method, this figure becomes 6/23 = 26%, which represents a 73% concordance rate increase from one method to the other. Now suppose the pairwise concordance rate had been 15/20 = 75%. Utilizing the proband method, this figure becomes 30/35 = 86%. In this case, concordance has risen only 14.6%.

In 1991, Gottesman wrote that "Tienari on follow-up (1975), reported a pairwise concordance rate of 36 percent for MZ twins and 14 percent for DZ

619. Tienari, 1968, 1971, 1975.
620. Tienari, 1975, p. 35.
621. Gottesman & Shields, 1982, p. 103.
622. See Gottesman & Shields, 1982, pp. 114-115. Also see Joseph, 2001b.
623. Gottesman & Shields, 1982, p. 104.
624. Boyle, 1990, p. 134.

twins and is no longer the odd man out."[625] In fact, Tienari reported no such rates. According to Tienari's 1975 paper, "The concordance rate for the MZ group is 3:20 (15 per cent) and that for the DZ group is 3:42 (7.5 per cent)."[626] In 1987, Kendler calculated a probandwise 7/21 = 33% concordance rate for Tienari's identical twins. He failed to explain how he came up with this figure because, even using the proband method, the rate is 6/23 = 26%.[627] Unfortunately, Kendler's readers were not informed that Tienari reported 3 out of 20 identical pairs concordant for schizophrenia, not 7 out of 21.

The NAS-NRC study. Pollin and colleagues' 1969 study of the NAS-NRC veterans sample was based on 80 identical and 146 same-sex fraternal pairs identified through a US Armed Forces veteran twin register. Because diagnoses were made by military doctors, this study had the advantage of not having to rely on diagnoses matched against (potentially biased) hospital records. Pollin and colleagues reported pairwise concordance rates of identical = 13.8%, fraternal = 4.1%, which are among the lowest rates ever reported. However, these figures would not be allowed to stand.

In 1972, Allen and colleagues reassessed the original sample, plus additional twins from the register, for a total of 220 pairs.[628] They calculated new pairwise concordance rates of identical = 27%, fraternal = 4.8%. However, as Boyle noted, they arrived at these figures by broadening their diagnostic criteria and rediagnosing each twin on the basis of the original medical charts. And while the armed forces doctors/psychiatrists may have been aware of the twins' zygosity status, they had no research hypothesis to test when making their diagnoses. Allen et al., who *were* testing a hypothesis, were aware of each pair's zygosity status. Acknowledging their genetic bias, the authors wrote that "the reviewer may have been influenced toward seeing more similarities in MZ pairs than in DZ pairs. This factor would tend to increase the MZ concordance rate."[629] In 1982 Gottesman and Shields converted Allen and colleagues' 27% pairwise figure to a probandwise rate of 43%. Thus, in the space of 12 years the

---

625. Gottesman, 1991, p. 111.
626. Tienari, 1975, p. 33.
627. The 7/21 identical twin concordance rate Kendler reported for Tienari's study was not a typographical error, since he used the same figure in other reviews published during this period.
628. Allen et al., 1972.
629. *Ibid.*, p. 944.

NAS-NRC identical twin concordance rate rose — on paper — from 13.8% identical (pairwise, uncorrected) to 43%.[630]

Kendler and Robinette updated and reanalyzed Pollin and colleagues' original results in 1983. They reported probandwise schizophrenia concordance rates of identical = 30.9%, fraternal = 6.5%. They also created a "broad concordance" category to include any psychiatric disorder. Employing this enlarged definition, they found that the "broad probandwise concordance for schizophrenia for monozygotic twins is 103 of 194 (53.1% ± 1.8%); for dizygotic twins it is 66 of 227 (23.8% ± 2.6%)."[631] What Kendler and Robinette mentioned only in passing was that, even though all of the pairs had passed through the schizophrenia risk period (> 45-years-old), only 30 out of 164 pairs were concordant for schizophrenia (pairwise concordance = 18.3%).[632]

The Tienari, NAS-NRC, and Koskenvuo et al. studies comprise three of the nine "modern" schizophrenia twin studies, yet influential partisans of the genetic position have artificially inflated the concordance rates of two of these studies, and have ignored or were unaware of the other. But the fact remains that in the three studies combined, 84% of identical twin pairs were judged *discordant* for schizophrenia.

Schizophrenia Twin Researchers' Interpretations of Concordance Rate Differences

Interestingly, most schizophrenia twin researchers have recognized that — contrary to the equal environment assumption — environmental factors *do* influence concordance rate differences. This is an important point because the logic of the twin method holds that identical-fraternal concordance rate differences are attributable only to genetic factors. Once twin researchers allow that environmental factors influence these differences, it is difficult for them to deny the possibility that environmental factors *completely* explain identical-fraternal concordance differences. Unfortunately, reviewers and textbook authors discussing these studies usually fail to mention that most of the original investigators did not believe that genes are the only factor influencing concordance. The following quotations are listed in chronological order, and I have added all emphases.

---

630. Gottesman & Shields, 1982.
631. Kendler & Robinette, 1983, p. 1554.
632. *Ibid.*, p. 1553.

Kallmann

[Different rates among same- and opposite-sex fraternal twins] are by no means extensive enough to permit a non-genetic explanation for the entire difference, or a major part of the difference, between the concordance rates of monozygotic and dizygotic twin pairs.[633]

Rosenthal

The above findings do not rule out the possibility of other psychological hypotheses which might account for higher familial incidence, higher concordance in MZ than DZ twins, and no higher frequency of the illness in twins than in nontwins. *An hypothesis like "identification" might account for the higher concordance rate of illness in MZ twins.* ... One could imagine a co-twin being drawn toward schizophrenic behavior if his twin had become schizophrenic and if he felt himself to be so much like his twin that he was completely convinced that the fate which had befallen his twin would befall him as well. He might be unable to resist this conviction, and presumably could behave in accordance with it.[634]

Even though the bodies of data presented have suggestive value in an accounting of the sex-concordance ratios found in studies of schizophrenia, the role of genetic factors cannot be excluded. However, if the found sex-concordance ratios are valid, *it seems reasonable to conclude that some psychological factors are influencing these ratios in good part.*[635]

Essen-Möller

Quite obviously, then, the logical evidence furnished by the classical twin method is not unambiguous, as originally believed. *A greater concordance in monozygotics must not invariably depend on their genetic identity*, since also their environment may have been more similar.[636]

Tienari

It is doubtful, moreover, whether the difference in concordance rates between identical and fraternal groups of twins can, as such, be ascribed to hereditary factors. In all likelihood, the environment, too, is more similar in the case of identical than in the case of fraternal twins. ... Furthermore, it is

---

633. Kallmann, 1946, p. 316.
634. Rosenthal, 1960, p. 303.
635. Rosenthal, 1962a, p. 419.
636. Essen-Möller, 1963, p. 69.

obvious that the intensity of the mutual relationship of identical twins is considerably greater than that of siblings in general and, also, of fraternal twins. . . . *It is apparent that differences in concordance rates between groups of identical and fraternal twins, as well as between female and male pairs, are partly attributable to environmental (psychological) effects.*[637]

Inouye

The question is what will be the possible cause of concordance of neurosis in these twins. Theoretically there are several possible causes. First is an environmental influence identically shared by two members of a twin pair. This mechanism was seen in one female monozygotic pair in dissociative reaction type. In this pair the direct etiologies of the neurosis of both twins were diverse psychogenic factors and the common primitive cultural background, which coincidentally influenced the emotion of the twin subjects. The second possible cause of concordance is a susceptibility to neurosis in both twin members resulting from a particular situation to twins. This possibility was pointed out by not a few psychoanalysts. . . . We pointed out the significance of mutual relationship between two members of each pair, and concluded that the particular situation is a probable cause of ego immaturity, which in turn is the probable cause of concordance of the neurosis or neurotic personality.[638]

Gottesman & Shields

Higher female concordances in two particular studies (Rosanoff and Slater) could be an artifact of sampling; or it could be associated with more environmental variability for males or with some aspect of the process of identification.[639]

Rosenthal considers that a psychological hypothesis such as identification might be used to explain differential concordance rates in MZ and DZ twins without implying a higher incidence of illness in MZ twins. *This would appear to be so* provided that the same proportion of potential schizophrenics are held back from overt illness by identifying with a normal twin as *those who became ill by identifying with an abnormal one.*[640]

---

637. Tienari, 1963, pp. 119-21.
638. Inouye, 1965, p. 1172.
639. Gottesman & Shields, 1966b, p. 815.
640. Gottesman & Shields, 1966a, p. 55.

Kringlen

There is clearly brought out by the data a trend toward higher concordance rates for females, both in monozygotic and dizygotic same-sexed twins. . . . It is difficult to explain this tendency. . . . With an increasing female emancipation this sex difference in upbringing and attitudes toward boys and girls has diminished. This could offer an explanation of the disappearance of the higher female concordance rates in more recent studies. This phenomenon might also be related to national culture. *One would then expect higher female concordance to be particularly pronounced in cultures where girls and women are most restricted in their activities, whereas the phenomenon would vanish in cultures where females enjoy equal rights with males.*[641]

Shields

It is established that the MZ co-twins of schizophrenics are at least twice as often and, in many types of samples, 4 or 5 times as often schizophrenic as DZ co-twins of the same sex. This difference will be accounted for by influences from two sources: by the effects of the greater genetic similarity of MZ twins, and by *greater similarity in environmental factors relevant to schizophrenia shared by MZ twins and not by DZ twins.*[642]

The total difference in concordance rate between MZ and DZ twins cannot be ascribed to genetic factors only. A series of studies of both normal and abnormal twins show that the environment of the MZ twin pair is more similar than the environment of the DZ twin pair.[643]

Slater

Concordance rates are higher in pairs of female twins than in pairs of male twins. If this is confirmed, one might well conclude that the environmental contribution to causation could be responsible. . . . Anything which tends to diminish environmental variance will tend to magnify the apparent contribution made by heredity, and this will show up in twin concordance rates just as much as in other measures.[644]

---

641. Kringlen, 1967a, pp. 91-92.
642. Shields, 1968, p. 100.
643. Kringlen, 1976, p. 431.
644. Slater, 1968, pp. 23-24.

Pollin et al.

> An hypothesis like "identification" might account for the higher concordance rate in monozygotic twins without implying a higher incidence of schizophrenia in monozygotic twins. . . . Because there are not only genetic but also environmental differences between the monozygotic group and the dizygotic group, *differences in concordance rates may be explained by environmental as well as by genetic hypotheses.*[645]

Fischer

> The assumption that a number of environmental factors are similar in MZ and DZ pairs may be correct only to some degree.[646]

> The result that there is a significant difference between concordance for schizophrenia in same sexed DZ pairs and same sexed sibling pairs might be an expression of a real difference, but it might be kept in mind that a statistically significant difference only expresses the likelihood of an event, in this case the low likelihood that no difference exists.[647]

Koskenvuo et al.

> Greater environmental similarity between MZ and DZ twin partners can also bias conclusions. . . . The intrapair relationship seems to differ significantly in MZ and DZ twins. How much of this is due to different intrapair influences or parental treatment, and how much to different genetic similarity, is unknown.[648]

These quotations show that in addition to Rosenthal, who "provided a blueprint for improving the state of the art of schizophrenia twin studies,"[649] the authors or co-authors of at least 12 of the 15 schizophrenia twin studies have written that environmental factors influence concordance rate differences. Their opinions stand counterposed to Kendler's position (later modified, see Chapter 3) that the behavioral similarity of identical versus fraternal twins is caused only by genetic factors.

645. Hoffer & Pollin, 1970, p. 476.
646. Fischer, 1973, p. 10.
647. *Ibid.*, p. 69.
648. Koskenvuo et al., 1984, p. 322.
649. Gottesman, 1991, p. 109.

Same Sex- and Opposite-Sex Fraternal Twin Concordance Rates

As we have seen, the twin method makes comparisons between identical and *same-sex* fraternal twins. However, several schizophrenia twin researchers collected concordance rate data for opposite-sex fraternal pairs as well. Although this was not the original investigators' intention, opposite-sex fraternal pairs can help assess the validity of the equal environment assumption. Jackson posed the question quite well in 1960:

> Obviously same-sexed and different-sexed fraternal twins have the geno-typical relationship of ordinary siblings. Therefore, because it is not claimed that schizophrenia is a sex-linked disorder, one would not expect a differ-ence in concordance for schizophrenia on a hereditary basis. On the other hand, if the hypothesis is correct that identical twins are more concordant for schizophrenia because of their "twinness," one would expect a higher incidence of concordance for schizophrenia in same-sexed fraternal twins because they are more alike from the identity standpoint than different-sexed fraternal twins.[650]

The pooled figures from schizophrenia twin studies reporting concor-dance rates for both types of fraternal twins are: same-sex fraternal = 59/523 (11.3%), and opposite-sex fraternal = 20/422 (4.7%).[651] Thus, the pooled same-sex rate is 2-3 times greater than the opposite-sex rate.[652] Typically, supporters of schizophrenia twin research either acknowledge, deny, or ignore this dif-ference; rarely if ever do they attempt to explain it in terms of genetics. As an example of denial, psychiatrists Richard Keefe and Philip Harvey, the authors of 1994's *Understanding Schizophrenia*, wrote that "same-sex and opposite sex fra-ternal twins have exactly the same concordance rate for schizophrenia."[653] Behavior geneticists Plomin and colleagues, on the other hand, recognized that "opposite-sex twins have a consistently lower concordance."[654]

---

650. Jackson, 1960, pp. 64-65.
651. Joseph, 1999b.
652. This finding is consistent with the theory that twin identification is largely respon-sible for the differing rates. In a study of 90 pairs of six-year-old twins from the Chicago area (including 36 same-sex fraternal and 19 opposite-sex fraternal pairs), Koch (1966, p. 234) asked individual twins to answer yes or no to the question of whether they desired to be like their co-twin. Among the same-sex twins, 57% answered "yes," whereas only 16% of the opposite-sex twins answered this way. Although these twins were young, Koch's study went straight to the question of how much each twin identified with his or her co-twin.
653. Keefe & Harvey, 1994, pp. 82-83.

Like Keefe and Harvey, in 1982 Gottesman and Shields denied the existence of differences between same-sex and opposite-sex fraternals, claiming "that opposite-sex twins, when studied, are no *less* concordant than same-sex fraternal twin-pairs."[655] They later qualified this statement by writing that "opposite-sex DZ pairs are as concordant as same-sex DZ pairs in recent studies."[656] By referring to "recent studies" they created the impression of a trend, while failing to mention that significant differences were found in the older investigations. In fact, the "recent studies" they cited was actually one study (Kringlen's 1967 investigation).

In 1966, however, Gottesman and Shields had a much different analysis of fraternal same-sex and opposite-sex concordance rate differences, writing, "One significant and consistent difference that emerged from our analyses was a lower concordance rate for opposite-sex fraternal pairs than same-sex fraternal pairs for studies giving information on this point."[657] Only one study reported opposite-sex fraternal concordance rates in the intervening years, and it therefore is puzzling how Gottesman and Shields, on the basis of very similar numbers, could have reached such different conclusions in 1966 and 1982.

At the pivotal 1967 Dorado Beach schizophrenia conference, Shields observed,

> The pooled crude concordance rates in six studies of 430 DZ pairs of oppo-
> site sex is only 5-6%, which is about half the concordance rate usually
> reported in DZ pairs of the same sex, and the difference is consistent in all
> studies except Kringlen's.[658]

In 1967, Shields saw Kringlen's finding as the exception that proved the rule; it would only be in later years that this exception would become, according to Gottesman and Shields, a "trend." Their 1982 claim is also striking when one considers that these investigators were leading defenders of Kallmann's and Slater's studies, each of whom found a statistically significant concordance rate difference between their same- and opposite-sex fraternal twin pairs.[659]

In another publication, Gottesman and Shields compared Kringlen's findings to those of Kallmann:

654. Plomin, DeFries et al., 1990, p. 339.
655. Gottesman & Shields, 1982, p. 114. Emphasis in original.
656. *Ibid.*, p. 243.
657. Gottesman & Shields, 1966a, p. 76.
658. Shields, 1968, p. 98.

Kringlen's overall pattern of results is impressively like that of Kallmann's: there were no significant sibling-DZ, male-female, or same-sex DZ/opposite-sex DZ differences in concordance rates, all three "nulls" leading to an emphasis on autosomal genetic factors and a deemphasis on sex role and identification factors.[660]

In fact, Kallmann's 1946 same-sex fraternal twins *were* significantly more concordant than his opposite-sex fraternal pairs. Gottesman and Shields also overlooked the statistically significant differences between Slater's fraternal twin versus sibling, and same-sex fraternal versus opposite-sex fraternal comparisons, as well as the significant differences in Fischer's 1973 fraternal twin versus sibling comparison.

The 2–3 times higher pooled schizophrenia concordance rate among same-sex versus opposite sex fraternals is difficult to explain on genetic grounds, but is consistent with Jackson's view (see Chapter 2) that, from the environmental perspective, we should find those pairs more alike from the identity (and environmental) standpoint to be more concordant for schizophrenia.

Selected Case History Excerpts of Identical Twins Concordant for Schizophrenia

The case histories of identical twins judged concordant for schizophrenia support Jackson's thesis that the close association and social isolation of identical twins play a major role in concordance. Here, I will quote from case histories published both before and after Jackson's 1960 critique. Most of these histories were written by people with little interest in psychodynamic processes, and who probably failed to capture the essence of the bond between the twins. Whether the pairs I will highlight should be viewed as concordant for schizophrenia is not our concern at the moment. More important is the twins' relationship to each other.

---

659. In his 1962a paper, Rosenthal displayed the statistically significant difference between Kallmann's 1946 fraternal same-sex and opposite-sex pairs in a table, to emphasize the point. Gottesman & Shields (1966a) also presented Kallmann's figures in a table, without calculating a probability figure for the difference. The figures from Kallmann's 1946 schizophrenia twin study are: same-sex fraternals = 34/296 (11%), vs. opposite-sex fraternals = 13/221 (6%), p = .019, Fisher's Exact Test, one-tailed. The figures from Slater's 1953 study are: same-sex fraternals = 11/61 (18%), vs. opposite-sex fraternals = 2/54 (4%), p = .014, Fisher's Exact Test, one-tailed.

660. Gottesman & Shields, 1976a, p. 373.

In 1934, Rosanoff et al. provided a sketch of identical twins George and Foster, whom they counted as concordant for schizophrenia "but in a manner quantitatively dissimilar."[661] The twins were born in Ohio in 1906. George was committed to a mental hospital in June of 1930, receiving a diagnosis of dementia praecox (schizophrenia). According to the case history, "The twins had always been together and had always been devoted to each other. When George was taken to the hospital Foster began to worry greatly. . . . He lost all ambition to work, sat around at home 'just thinking,' [and] paid no attention to any one who spoke to him." Foster was finally convinced to seek voluntarily admission to a mental hospital "for a couple of weeks' rest and change," but he would not remain long:

> While he was there word came that George was better and could come home on a trial visit. *"The minute that Foster heard that, he snapped right out of it, brightened up, and the two boys came home the same day."* He resumed his work and seemed as well as ever. At the time of our observation of the case he appeared quite normal mentally [emphasis added].[662]

This story describes a young man so attached to his identical co-twin that he was willing to follow him into a mental hospital. Upon hearing that his twin brother had been released, Foster's "subacute schizophrenia" symptoms vanished and he happily rejoined his twin brother in the family home. Perhaps Galton would have interpreted this story as showing that "the clocks of their two lives move regularly at the same rate, governed by their internal mechanism" (see Chapter 2). But would the clocks have been set to run *backwards* at the same rate? Foster "snapped out of it" because he found out that his twin brother was leaving the hospital. The "mechanism" was the identical twin relationship, not heredity.

Slater's 1953 study was the last to be published before the appearance of Jackson's critique. Case history excerpts from some of the most interesting pairs of identical twins concordant for schizophrenia are noted below:

661. Rosanoff, Handy, Plesset, & Brush, 1934, pp. 257-8.
662. *Ibid.*, 1934, p. 258.

Pair #20, females, born 1894. (Lily and Mary). "Following disagreements at home Lily and Mary left in 1914 to live together, and have done so since. The first signs of their deafness came on about then. They rather shut themselves up together, with little outside contact. . . . According to a neighbor, they were well known in the neighborhood for their likeness to one another. . . . Both have paranoid illnesses coming on at the time of the menopause and within 2 years of one another."[663]

Pair #110, females, born 1872. (Marjorie and Cynthia). "Until they were 40 they lived at home. . . . Marjorie was the first to become peculiar. . . . Cynthia became ill very suddenly, about two years after Marjorie had begun to be peculiar. . . . Eventually they both barricaded the flat, which led to their certification and admission to [the mental hospital] on the same day in June, 1918, aged 46."[664]

Pair #198/199, females, born 1912. (Irene and Maureen). "They were always much alike, short and thin with dark eyes. And dark straight hair. They were frequently mistaken for one another, and at one time one twin deputised for the other at the factory where she worked without anyone noticing. They always remained intimate and never troubled to make separate friends. Many 'telepathic' experiences are reported; if one fell down at one spot the other would too, and hurt the same knee. Maureen is said to have had pains when Irene had her baby. . . . At the present time both twins are attending a psychiatric clinic where observation has shown that in both symptoms are much worse just before the periods. . . . The first remark [in an interview], on being introduced by their mother, was: 'Yes, we're the twins, and we're still ill. It's a mysterious illness, ours.' They seemed to share one illness between them."[665]

Pair #287, males, born 1915. (Leslie and Reginald). "In April 1940, [Leslie] was admitted to a general hospital. . . . Reggie attended the psychiatric clinic in 1940 with his twin, and he said that he himself had bad nerves, got into arguments, had few interests. . . . When Leslie was certified in 1940, Reggie was very upset and would say: 'I've murdered my brother, I've poisoned him'. . . . He wandered about at night picking at biscuits and would call his parents names. . . . In April 1941 he became violent, attacked his parents and was taken to the observation ward and to the mental hospital."[666]

---

663. Slater, 1953, pp. 127-129.
664. *Ibid.*, p. 150.
665. *Ibid.*, pp. 166-168.
666. *Ibid.*, pp. 175-176.

Kringlen provided detailed case histories in his 1967 study. The following excerpts describe the social isolation and mutual association of several concordant identical pairs (referred to as "Twin A" and "Twin B"):

Case 1, males, born 1913. "As children, they were mistaken for the other, even by their parents. . . . They were brought up alike, dressed in the same way, and were much together as children. . . . [Twin A] was transferred to a mental hospital [in 1934]. . . . [Twin] B had a short period of 'nervousness' at age 19 when his brother became ill. . . .Then he seemed to have been all right until the age of 21, when his brother was admitted to the hospital. He visited his brother often, and said to his family after some time that he felt he himself had to go to the hospital. . . . Both twins were seen in 1965 in the same hospital where they have stayed. Both are living in the same room."[667]

Case 2, males, born 1921. "Mistaken for each other as children by the teacher. Extremely strong attachment to each other. . . . Before the age of 25 they had been separated for only very short periods of time. . . . [Twin A] admitted to a mental hospital in 1947. . . . [Twin] B became psychotic in 1949."[668]

Case 3, males, born 1916. "They were always together as children, and made the same few friends. They were, however, living in an isolated area and consequently had slight contact with other children. If one was home from school because of illness, the other one also stayed home, even if he was not sick. . . . They were not separated until they were 20 years old. . . . [Twin A at age 30] was living with his twin brother on the isolated family farm up in the hills and they had practically no contact with other people. . . [Twin] A was placed in a mental hospital 1957-1958. . . . [Also in 1957, twin B] began being more overly [sic] disturbed, knocking at the walls, scolding into the air, apparently hallucinated. [Admitted to a mental hospital, 1959]."[669]

Case 4, males, born 1930. "As children the twins were mistaken for one another by their teachers, and considered as alike as two drops of water. . . . They were inseparable as children, and even more so as adults. . . . They were never separated from one another. . . . [Twin A was admitted to a mental hospital in 1949]. At admittance he was accompanied by his parents and

---

667. Kringlen, 1967b, pp. 4-7.
668. Ibid., pp. 10-11.
669. Ibid., pp. 15-19.

181

his twin brother. . . . [Twin] B was attending a technical school when his brother got sick. Some time afterwards he lost interest in the school work, started sleeping poorly and stayed up at night . . . [Twin B hospitalized in 1951]."[670]

Case 6, males, born 1908. "As children, their siblings could not tell them apart — they were like two drops of water. They had only slight contact with the other children and clung together most of the time. Their attachment was extremely strong in childhood and even after puberty. They were never separated before the age of 17."[671]

Case 9, females, born 1907. "They always clung together, and were extremely dependent on one another also as adults. They seemed to have 'normal' contact with other children, but they never made separate companions among their friends. . . . [Twin A] started getting ill when she was 39 years old. . . . When [twin] A was hospitalized in 1952, [twin] B stayed home with marginal social functioning. She was longing intensely for her sister's company. . . . [In 1959, after being placed in a sanitarium, twin B] was soon transferred to the same mental hospital in which her twin sister was (1959) and has since stayed there together with [twin] A. . . . Both twins were seen in the hospital in 1965. They came to my office dressed in the same way, walking and behaving in much the same way — a remarkable couple. The nurses told me that the twins were together all day long."[672]

Case 10, females, born 1930. "As children, they were greatly dependent on each other and virtually inseparable. . . . [Twin A's] first hospitalization period was from December 1953, to April 1954. . . . When [twin] A was hospitalized, [twin] B said: 'Why don't they take me, too'. . . . [Twin] B became psychotic when she was 25 years old."[673]

Case 12, females, born 1923. "The twins grew up on a small, lonely farm. . . . Neither twin went about with other girls or boys much. . . . At the age of 22-23, twin A developed a catatonic-like remittent schizophrenia. . . . [Twin] B began to be 'nervous' when [twin] A was admitted to the hospital in 1944. . . . Twin B fell ill shortly after [twin] A, with approximately the same symptoms."[674]

670. *Ibid.*, pp. 22-25.
671. *Ibid.*, p. 34.
672. *Ibid.*, pp. 48-53.
673. *Ibid.*, pp. 56-58.
674. *Ibid.*, pp. 65-69.

Like Rosanoff's "George and Foster," several of Kringlen's twins were apparently so distraught over the hospitalization of their co-twins that they were willing to follow them into a mental hospital in order to continue the relationship. In Case 10, when Twin A was being taken to the hospital her identical twin sister cried out, "why don't they take me, too." A more compelling (though tragic) human drama could hardly be imagined.

I have selected the following case histories of concordant identical twins from Gottesman and Shields's study:

Pair 1, males, born 1893. (Twins A and B both diagnosed with schizophrenia at age 30).

> "[Mother] idolized twins, gave them most similar names she could think of; overprotected them in that she did not allow them to mix with other children. . . . The twins were inseparable until 18 years of age. . . . Taken in isolation this pair should *not* give more comfort to genetic enthusiasts than to environmental ones with respect to weighting the etiological importance of heredity and intrafamilial factors. . . . The influence of one twin upon the other could account for the similar *content* of their symptoms."[675]

Pair 7, females, born 1940. (Twin A diagnosed with schizophrenia at age 18, twin B diagnosed with schizophrenia at age 19).

> "Twins very close — 'couldn't make a move without the other.' They had few friends . . . [The twins] were extremely dependent on their [mother] who encouraged their dependence. . . . *Psychiatric history of B.* Behavior gradually became unpredictable and irrational after transfer to another library and [twin] A became ill (at 17); accused [mother] of causing twin's illness . . . . later increasingly withdrawn. . . . This East End London family parallels those described in the psychodynamic family study literature of Lidz and others."[676]

Pair 17, males, born 1934. (Both twins were diagnosed with schizophrenia at age 22. The twins were separated at birth but lived together for one year at age five).

---

675. Gottesman & Shields, 1972, pp. 72-75.
676. *Ibid.*, pp. 88-91.

"The arrival, for the first time since his adoption, of the biological mother at A's home early in October preceded his manifestly odd behavior by about 2 months. B was taken to visit his sick twin on December 22 in the belief that this might improve A's condition. Instead it appears to have precipitated B's illness, for the same evening he was disturbed and on January 5 was admitted to hospital, just before A."[677]

Pair 19, females, born 1926. (Twin A diagnosed with schizophrenia at age 31, twin B diagnosed with schizophrenia at age 35).

"Both stubborn children; later, as adolescents, backward 'but nice'; regarded as moody, suspicious, 'keeping together'. . . . Both remained single. . . . [Twin B had no symptoms until she] joined 'wild scene' when A became disturbed (*onset*): felt persecuted, thought food poisoned, tried to jump through window; became equally disturbed as A and talked about attempts to murder A. Together admitted to observation ward (at 31) where wept and wailed continuously. . . . Socialized only with twin, upset by any suggestion of separation, once responded immediately by breaking vase. . . . It is reasonable to raise questions of *folie à deux* and culture shock in this pair. Clearly their twinship was responsible for their shared delusions and simultaneous onset of psychosis."[678]

The descriptions by Rosanoff, Slater, Kringlen, and Gottesman and Shields are consistent with Jackson's observation that social isolation, mutual association, and identity confusion are common themes among concordant identical twin pairs. Throughout these case histories we encounter observations such as, "they rather shut themselves up together," "never troubled to make separate friends," "no contact with other people," "they seemed to share one illness between them," "were never separated from one another," "longing intensely for her sister's company," "did not like to mix too much with others," "always clung together," "inseparable," "couldn't make a move without the other," and so on. Although some closely associated and isolated pairs have been judged discordant for schizophrenia, these conditions likely contribute to higher concordance rates among identical versus fraternal twins.

Twins Reared Apart

Due to the rarity of the occurrence, no researcher has undertaken a systematic study of schizophrenia among reared-apart identical twins. However,

677. *Ibid.*, pp. 114-118.
678. *Ibid.*, pp. 122-124.

several individual case histories of ostensibly separated pairs have been reported. In Farber's 1981 review of these cases she concluded that, according to her "lenient criteria," nine identical pairs warranted consideration as legitimately separated twins.[679] However, in all of these cases (6 pairs were considered concordant by Farber) the twins were aware of each other's existence and had periodic contact.

Regardless of how many individual pairs are reported concordant for schizophrenia, however, they do not constitute scientifically acceptable evidence in favor of genetics. As we saw in Chapter 4, pairs often come to the attention of researchers because of their similarities. In the case of schizophrenia, researchers or hospital administrators might become aware of a pair of identical twins hospitalized for schizophrenia, whereas a discordant pair, where only one twin is hospitalized, would not come to their attention as often. Moreover, most cases were reported by genetically-oriented investigators, whose bias influenced which pairs they chose to report, how they reported them, and how they diagnosed the twins. In any case, a basic principle of science is that a collection of anecdotes does not equal data.

Craike and Slater's well-known 1945 pair, "Edith and Florence," illustrates the bias and arguable claim of "separation" common to most of these cases. Farber considered this pair to be "the best-separated set in the literature."[680] Edith and Florence were British identical twins separated 9 months after birth. Florence was adopted by a maternal aunt, while Edith stayed with her father until the age of 8, when she was placed in a children's home. Although they did not meet again until age 24, each was aware of the other's existence. Edith reported that while living with her father, "Florence was making trouble for her by writing to her father and telling him Edith had told her that he was a drunkard."[681] Edith also believed that Florence was watching her house and had been plotting against her. Their supposed delusional systems, which contributed to their schizophrenia diagnosis, centered on mutual distrust:

679. Farber, 1981, p. 165.
680. *Ibid.*, p. 156.
681. Craike & Slater, 1945, pp. 214-215.

Each twin occupies for the other an over-valued position: *to each the other is supremely important*, although the circumstances of their lives touch at few points. Edith at first sight places Florence at the center of her persecutors; Florence, with her own inborn tendency to paranoia, reacts to this by coming in turn to regard Edith as her chief enemy [emphasis added].[682]

Although Craike and Slater claimed that these twins "were brought up along entirely different lines,"[683] it is clear that this "best separated set" had, in the words of Craike and Slater themselves, centered their "delusions around the other,"[684] meaning that the twin relationship played a major role in the delusional systems of these "reared-apart" twins. Moreover, some have argued that neither twin's symptoms warranted a diagnosis of schizophrenia.[685]

The investigators' genetic bias, as we have seen, frequently influences the way they write about and diagnose twins. R. D. Laing wrote of Craike and Slater's "linguistic tricks used to establish concordances . . . in respect of [Edith and Florence's] 'childhood neurosis.'"[686] Although Craike and Slater reported that "Both were nervous as children; Edith walked in her sleep and had nightmares, Florence had fainting attacks and was nervous of the dark,"[687] Laing pointed out that they could just as easily have written, "Edith walked in her sleep, Florence did not. Edith had nightmares, Florence did not. Florence had fainting attacks, Edith did not."[688]

In closing, I agree with the observations of many schizophrenia twin researchers that the case histories of "separated" identical twins concordant for schizophrenia represent little more than "fascinating curiosities."[689] In 1982 Gottesman concluded, "After a quarter century of experience with twins reared together and twins reared apart, it is my conviction that twins reared apart are a wonderful source of hypothesis generation, but not a useful source for hypothesis testing."[690]

---

682. *Ibid.*, p. 220.
683. *Ibid.*, p. 219.
684. *Ibid.*, p. 221.
685. See Pam, 1995; Ratner, 1982.
686. Laing, 1981, p. 124.
687. Craike & Slater, 1945, p. 219.
688. Laing, 1981, p. 124.
689. Gottesman, 1991, p. 121.
690. Gottesman, 1982, p. 351.

Fischer's study. Another method of studying twins compares schizo-phrenia rates among the offspring of *discordant* identical twin pairs (that is, one member of the pair is diagnosed with schizophrenia, and the other is not). Review articles and textbooks published since the early 1990s frequently cite these studies, and Gottesman and Bertelsen's 1989 study in particular, as strongly supporting the genetic position. Because identical twins share the same genes, genetic theory predicts that, among discordant pairs, the offspring of twins diagnosed with schizophrenia would have the same schizophrenia risk as the offspring of their *non*-schizophrenia diagnosed identical co-twins, and that both risks would be significantly higher than population expectations. Environ-mental theories, according to proponents of this method, would expect an ele-vated schizophrenia risk among the offspring of the co-twins diagnosed with schizophrenia, but would expect the risk among the offspring of the non-diag-nosed co-twins to be the same as the population risk. Margit Fischer introduced this method into twin research in 1971. Her Danish sample included three groups:

> Group 1 — The offspring of the schizophrenia-diagnosed members of discordant identical twin pairs. Age corrected rounded-off N = 32.

> Group 2 — The offspring of the non-diagnosed members of discordant identical twin pairs. Age corrected rounded-off N = 24.

> Group 3 — The offspring of the non-diagnosed members of discordant *fraternal* twin pairs. Age corrected rounded-off N = 9.

Fischer used the Group 3 offspring of non-diagnosed fraternal co-twins to control for the supposed environmental expectation that there would be no dif-ference between their rate of schizophrenia and that of the Group 2 offspring.

Fischer reported three cases (9.4%) of schizophrenia and "schizophrenia-like psychosis" among Group 1 children, three cases (12.3%) among Group 2 children, and zero cases (0%) among Group 3 children.[691] She concluded that while these results "could probably best be explained" by genetic factors,[692] "the present material is of such limited size that no conclusion can be drawn."[693] In

---

691. *Ibid.*, p. 45.
692. *Ibid.*, p. 46.
693. *Ibid.*, p. 51.

1973 Fischer reported similar numbers, but failed to add the caveat that no conclusion could be drawn from the data.[694]

Before continuing, I should emphasize two points. (1) In neither Fischer's
study nor Gottesman and Bertelsen's 1989 follow-up did the researchers state
that offspring diagnoses were made blindly. This introduced a major bias into
the diagnostic process because, among other reasons, these psychiatric genetic
researchers were partisans of the genetic cause. One could invalidate their conclusions on the sole basis of non-blind diagnoses. (2) It is doubtful that Fischer's
strategy adequately separates genetic and environmental influences. Her model
assumes that a purely environmental etiology would show (a) that Group 1
children would develop schizophrenia at a higher rate than Group 2 children,
and (b) that Group 2 children would develop schizophrenia near the population
rate of 1%. However, this study was carried out in Denmark at a time when
schizophrenia was widely viewed as a hereditary disease (see Chapter 7). The
children *or nieces/nephews* of people diagnosed with schizophrenia were therefore
stigmatized because of their perceived genetic heritage, and were considered at
risk for schizophrenia on this basis alone. A pregnant Danish woman diagnosed
with schizophrenia was encouraged to abort her pregnancy for eugenic reasons,
and if she gave birth, the social service agencies often kept her child out of the
adoption process.[695] Such restrictions were based on the belief that the relatives
of people diagnosed with schizophrenia were genetic carriers of the disorder,
and is therefore a good indication of the prevailing sentiments in Denmark. Thus,
the biological offspring of the normal identical co-twins of people diagnosed
with schizophrenia would have been considered schizophrenia "taint carriers,"
which could have influenced their rearing environments.

In addition, the discordant co-twins grew up in the same family and likely
experienced similar environments and treatments. Thus, many of the *environmental* factors leading to a schizophrenia diagnoses in one of the twins were also
experienced by their "well" co-twin, who could have manifested deviant rearing
patterns toward his or her children. Fischer recognized this possibility, but dismissed it as being "a little far-fetched."[696] Moreover, Group 2 offspring may have
had considerable contact with their schizophrenia-diagnosed aunts and uncles.
We will never know to what extent such influences existed, in part because

694. Fischer, 1973.
695. Kety et al., 1994.
696. Fischer, 1973, p. 65.

practically none of these offspring was investigated personally by Fischer.[697] According to Gottesman and Bertelsen, "For both twin types the offspring of the normal co-twins share little to none of the within-family exposure to a schizophrenic parent."[698] However, it has been shown that deviant rearing patterns of "normal" parents can predict schizophrenia in their children.[699] Therefore, "exposure to a schizophrenic parent" is not a necessary component of a schizophrenia-producing environment. In general, about 90% of people diagnosed with schizophrenia were reared by two non-diagnosed parents.[700]

Fischer reported only three cases of definite schizophrenia among the offspring (Group 1 = 2 cases; Group 2 = 1 case).[701] Other diagnoses included "acute psychotic episode; schizoid personality disorder," "? paranoid schizophrenia/reactive psychosis," and "psychosis of an uncertain type."[702] Moreover, Fischer gave one of her two Group 1 definite schizophrenia diagnoses to the daughter of a *concordant* identical twin (Pair 18).[703] Fischer included several children of concordant pairs in Group 1 in order to increase her sample size. The other Group 1 offspring diagnosis (paranoid schizophrenia) was given to the daughter of Twin 9A. However, Twin 9B (Birgit) had died when she was 18-years-old, indicating that they were not a true discordant pair. As Fischer wrote, "On account of the early death of Birgit it was not possible to decide whether this pair was concordant or discordant."[704] Based on these cases and questions about the certainty of Group 2 offspring diagnoses, Kringlen and Cramer wrote in 1989 that "there were no definite schizophrenic cases in [Fischer's] total sample."[705]

Fischer's small sample and handful of offspring schizophrenia cases meant that, although Group 1 offspring were not diagnosed at a significantly higher rate than Group 2 offspring, *neither group had schizophrenia rates significantly higher than general population expectations.* For example, her Group 1 schizophrenia rate of 3/32, when compared to a general population expectation of 0/32 or 1/132, is not statistically significant.[706] Therefore, one cannot conclude that a non-significant dif-

697. Kringlen & Cramer, 1989.
698. Gottesman & Bertelsen, 1989, p. 868.
699. See Wynne et al., 1976.
700. Gottesman, 1991.
701. Fischer, 1971, Table 6
702. *Ibid.*, pp. 47-49.
703. Case #18; see Fischer, 1973, p. 148.
704. Fischer, 1973, p. 140.
705. Kringlen & Cramer, 1989, p. 877.
706. p = .12, Fisher's Exact Test, one-tailed.

ference between these two groups supports the genetic hypothesis, or that either group had a schizophrenia risk higher than the risk in the general population. In 1971 Fischer correctly cautioned against drawing conclusions from her data, while adding that "the question raised might be evaluated in other similar studies."[707]

Gottesman and Bertelsen's 1989 follow-up. In 1989, Gottesman and Bertelsen published an 18-year follow-up of Fischer's study.[708] According to them,

> If the observed risk to the offspring of schizophrenic identical twins were like that shown for offspring of probands generally and if the risk to the offspring of their genetically identical but clinically normal nonschizophrenic co-twins were close to the population base rate of 1%, it would indicate that nongenetic within-family factors have the power to produce schizophrenia. . . . If, on the other hand, the risks to the two classes of MZ offspring were both high and equal to each other, it would undermine any significant etiologic role for the unshared factor of a schizophrenic parent. . . . The corresponding risks for the offspring of same-sex fraternal twins concordant and discordant for schizophrenia will serve as control values for first- and second-degree relatives of schizophrenics.[709]

Gottesman and Bertelsen gave the rates of "schizophrenia and schizophrenialike psychosis" and their accompanying age-corrected "morbid risks" (MR) as follows (for reasons of clarity, these groups reflect Fischer's designations): *Group 1* — 6/47, MR = 16.8 ± 6.6; *Group 2* — 4/24, MR = 17.4 ± 7.7; *Group 3* — 1/52, MR = 2.1 ± 2.1. Arguing that Group 1 and Group 2 offspring risks were similar and elevated above the population risk, and that the Group 2 risk was significantly higher than the Group 3 risk, Gottesman and Bertelsen concluded in favor of "a strong role for genetic factors, still unspecified, in the etiology of schizophrenia."[710] However, as in Fischer's original study, Group 2 and Group 3 (and possibly Group 1) diagnoses did not significantly exceed the expected population risk. Moreover, all six of their Group 1 offspring diagnoses are questionable. Three were the offspring of Twin 9A (whose co-twin had died at age 18), two were the offspring of *concordant* Pair 18, and the other was diagnosed with schizophreniform disorder.[711]

---

707. Fischer, 1971, p. 50.
708. Fischer died in 1983.
709. Gottesman & Bertelsen, 1989, p. 868.
710. *Ibid.*, p. 871.
711. *Ibid.*, p. 870, Table 7.

Gottesman and Bertelsen made a Group 2 schizophrenia diagnosis on the daughter of twin 405b ("Karl"), who was the non-schizophrenic identical co-twin of schizophrenia-diagnosed twin 405a ("Morten"). According to Fischer's original description,[712] Morten had been an "unremarkable" and "quite ordinary" 23-year-old until the day that he fell from a height of 13-16 feet onto a stone cornice and lay unconscious for five minutes, suffering "severe cranial trauma."[713] Morten grew up in an early 20th century "very poverty-stricken" home in rural Denmark, and probably did not receive adequate medical attention for his cranial trauma. Although he recovered from the injury, "the family dated his psychic change to this time."[714] Diagnosed with schizophrenia, he spent much of the rest of his life in mental hospitals. Like Fischer, Gottesman and Bertelsen decided to retain Morten and Karl even though organic brain damage caused by the fall was a plausible explanation for Morten's subsequent "delusional" behavior. However, I would argue that they (and Fischer) should have removed pair 405 from the study because of this possibility. This is another example supporting Rosenthal's 1962 observation that in order to avoid small sample sizes, twin researchers tend to retain pairs for whom they have inadequate information.

Gottesman and Bertelsen commented that Morten's "cranial injury appears to have been a sufficient but not a necessary releaser for the predisposition to schizophrenia in proband MZ 405A; the nonschizophrenic co-twin never encountered a releaser for his diathesis but transmitted it to his daughter." They continued, "Without this type of pedigree, the schizophrenia in the proband might have passed as a nongenetic phenocopy of schizophrenia caused by cranial insult."[715] A criticism of twin study diagnostic procedure has been that a person has a greater chance of being diagnosed with schizophrenia by a doctor or a researcher on the basis of having a family history of schizophrenia, or a co-twin diagnosed with schizophrenia, and this problem reappears in Gottesman and Bertelsen's argument. It is circular reasoning to offer a diagnosis made on the *assumption* that schizophrenia is genetic as *evidence* that schizophrenia is genetic.[716] If we remove this pair and their offspring, Gottesman and Bertelsen's significantly higher Group 2 versus Group 3 rate, to the extent that it means anything, becomes statistically non-significant.[717]

712. Fischer, 1973, pp. 124-125.
713. *Ibid.*, p. 46.
714. *Ibid.*, p. 124.
715. Gottesman & Bertelsen, 1989, p. 871.

Biological psychiatrist E. Fuller Torrey considered the case material handed down by Fischer to be "markedly unsatisfactory,"[718] and found that none of the identical twins on whom Gottesman and Bertelsen based their findings "met the basic criteria for twins with clearly diagnosed schizophrenia in the index twin and verifiable (i.e., based on interview) normality in the co-twin."[719] Torrey also observed that the "schizophrenic" twins from Fischer's sample reproduced at nearly the same rate as their "well" co-twins, which does not square with the generally low reproductive rate among people diagnosed with schizophrenia. Gottesman and Bertelsen disputed Torrey's claim that reproductive rates were similar,[720] but their 1989 data showed clearly that both "schizophrenic" and "normal" co-twins averaged roughly four offspring each, which calls into question the validity of the original hospital schizophrenia diagnoses. Torrey argued that Gottesman and Bertelsen's decisions about which twins to retain or remove were "entirely arbitrary," and that their conclusion in favor of genetic factors "is premature at best."[721]

The results of Gottesman and Bertelsen's 18-year follow-up, therefore, are as inconclusive as Fischer's original study.

Kringlen and Cramer. In another study using the Fischer design, but making blind diagnoses on the basis of structured interviews, Kringlen and Cramer found a higher but statistically non-significant Group 1 versus Group 2 schizophrenia rate difference.[722] They properly drew no important conclusions from these results, and their data further illustrated the problems that small samples create for this design. Even if none of Kringlen and Cramer's 45 Group 2 offspring had received a schizophrenia diagnosis, the difference between the groups would have remained statistically non-significant.[723] Clearly, this result

---

716. I might add that on the basis of the case history material, the schizophrenia diagnosis of 405b's daughter is by no means clear-cut, which is reflected by the "provisional diagnosis" given by Gottesman & Bertelsen.

717. According to Fischer (1973, p. 124), Group 2 twin 405b (Karl) had three offspring. If we remove these offspring from the figures shown in Gottesman & Bertelsen's Table 5 (1989, p. 869), the comparison becomes, Group 2: 3/21 vs. Group 3: 1/52, $p = .069$, Fisher's Exact Test, one-tailed.

718. Torrey, 1990, p. 976.

719. *Ibid.*, p. 977.

720. Gottesman & Bertelsen, 1990.

721. Torrey, 1990, p. 977.

722. Kringlen & Cramer, 1989. Group 1: 3/28 = 10.7% vs. Group 2: 1/45 = 2.2%, $p = > .05$.

723. 3/28 vs. 0/45, $p = .053$, Fisher's Exact Test, one-tailed.

would not support the genetic hypothesis on the basis of a non-significant difference between Groups 1 and 2.

***

The studies of Fischer, Gottesman and Bertelsen, and Kringlen and Cramer provide no scientifically acceptable evidence in support of genetic influences on schizophrenia. As Torrey concluded, "There may be genetic factors in schizophrenia but they cannot be elicited from a study of the offspring of MZ twins in the Fischer study."[724] The design itself appears unable to adequately separate genetic and environmental influences, and small samples make it unlikely that differences will be found between the offspring of discordant identical pairs. In order to guard against bias, future researchers using this design must ensure that diagnoses are made by blinded raters.

CONCLUSIONS

Genetic influences on schizophrenia cannot be established by the results of family studies, twin studies, published studies of the offspring of discordant identical pairs, or individual cases of "reared apart" identical twins. Taken together, this body of research points merely to the *possibility* that genes influence schizophrenia, and nothing more. We saw in Chapter 2 that adoption studies have replaced twin studies as the most frequently cited evidence in favor of the genetic basis of schizophrenia. Let us now turn our attention to a detailed review of these studies.

---

724. Torrey, 1990, p. 977.

Chapter 7. The Genetics of Schizophrenia II: Adoption Studies

Someday in the twenty-first century, after the human genome and the human brain have been mapped, someone may need to organize a reverse Marshall plan so that the Europeans can save American science by helping us figure out who really has schizophrenia or what schizophrenia really is.

— Biological psychiatrist Nancy Andreasen, in 1998.[725]

Once the presence of a mental disease has been established, the specific diagnosis of schizophrenia offers further difficulties. Only a few isolated psychotic symptoms can be utilized in recognizing the disease, and these too, have a very high diagnostic threshold value. Manic and depressive moods may occur in all psychoses; flight of ideas, inhibition and — as far as they have not assumed specific characteristics — hallucinations and delusions, are partial phenomena of the most varied diseases. Their presence is often helpful in making a diagnosis of a psychosis, but not in diagnosing the presence of schizophrenia.

— Eugen Bleuler, originator of the schizophrenia concept, in 1911.[726]

It should be apparent now that if we had included in our comparisons of index and control relatives only those who clearly had process [chronic] schizophrenia, we would have found no difference between the two groups of relatives.

— Schizophrenia adoption researcher David Rosenthal, in 1972.[727]

If schizophrenia is a myth, it is a myth with a strong genetic component!
— Schizophrenia adoption researcher Seymour Kety in 1974.[728]

The decision to perform adoption studies, which played a crucial role in establishing schizophrenia as a genetic disorder, was based on their authors' belief that family and twin results were potentially confounded by environmental factors. According to adoption researchers Seymour Kety and colleagues, the evidence from schizophrenia family and twin studies is "inconclusive . . . in that it fails to remove the influence of certain environmental factors. . . . In the case of monozygotic twins it has been pointed out that such individuals usually

---

725. Andreasen, 1998, p. 1659.
726. E. Bleuler, 1950, p. 294.
727. Rosenthal, 1972, p. 68.
728. Kety, 1974, p. 961.

share a disproportionate segment of environmental and interpersonal factors in addition to their genetic identity."[729] Adoption research, they argued, would finally provide a "means of disentangling hereditary and genetic influences on schizophrenia."[730]

In addition to schizophrenia, the early studies performed in Oregon and Denmark helped pave the way for the acceptance of an important role for genetic factors in shaping human psychological and behavioral trait differences in general. Behavior geneticists Rowe and Jacobson summarized the impact of the Oregon and Danish-American studies, writing, "When a single theory is monolithic in a field, contrary findings can break paradigms . . . It is just this role, we believe, that the first adoption studies of schizophrenia played in the 1960s." They added that these studies "stirred the Zeitgeist and gave impetus to the formation of the Behavior Genetics Association at the Institute for Behavior Genetics in Boulder, Colorado."[731] Thus, the results from the Oregon and Danish-American investigations not only strengthened the genetic theory of schizophrenia, but helped establish the behavior genetics field, whose theories have gained widespread acceptance. They also helped revive psychiatric genetics, which was still reeling from its founders' collaboration with German National Socialism.

I will argue in this chapter, however, that this paradigm shift rested, and continues to rest, on very shaky foundations. I will attempt to show that the adoption research that helped bring it about is flawed on several critical dimensions rarely discussed in mainstream accounts. This task has been left to critics, who are generally ignored or dismissed by mainstream sources intent on demonstrating the definitive nature of this research. In this sense, the largely uncritical acceptance of these studies and of the genetics of schizophrenia in general is, to paraphrase Wahlsten's comments about heritability, more the result of a power struggle than of a debate among scientists.[732]

---

729. Kety et al., 1968, p. 345.
730. *Ibid.*, p. 346.
731. Rowe & Jacobson, 1999, pp. 14-15.
732. Wahlsten, 1994.

ADOPTION STUDY RESEARCH METHODS

Until the 1960s, adoption studies had been used primarily to assess genetic influences on IQ. In Chapter 2 we found Galton discussing studying adoptees as early as 1865, and the issue was taken up again in 1913 by L. F. Richardson in *Eugenics Review*. Richardson proposed an investigation into whether the "off-spring of poor parents, adopted when newly born into well-to-do and well educated families, turn out markedly different from the birthright members of those families . . ." He continued,

> a thorough study of a hundred such cases of adopted children would do more to reveal the nature of the poorer classes than statistics of 100,000 poor persons brought up in poverty. The author hopes that a co-operative study of this kind can be arranged among the adherents of the Eugenics Education Society.[733]

I will now describe the three main designs used in schizophrenia adoption research.

(1) The Adoptees' Family Method

The Adoptees' Family method begins with children given up for adoption who are later diagnosed with a "schizophrenia spectrum disorder" (abbreviated here as "SSD"). These disorders include chronic schizophrenia, as well as other psychiatric disorders considered by the researchers to be related to chronic schizophrenia. I will discuss the spectrum concept in more detail later. In the 1968 and 1975 Danish-American adoption studies, Kety and colleagues defined the schizophrenia spectrum as diagnoses B1, B2, B3, D1, D2 , D3, and C, and they made several "outside the spectrum" (non-SSD) psychiatric diagnoses. The 1968 and 1975 schizophrenia spectrum, as defined by Kety et al., is seen in Table 7.1.

Table 7.1. Kety and Colleagues' 1968 & 1975 Schizophrenia Spectrum Disorders (SSDs)

B1. Chronic Schizophrenia
B2. Acute Schizophrenia Reaction
B3. Borderline Schizophrenia
D1. Uncertain Chronic Schizophrenia
D2. Uncertain Acute Schizophrenia Reaction
D3. Uncertain Borderline Schizophrenia
C. Inadequate Personality (Includes Schizoid Personality)

Sources: Kety et al.,1968, p. 352; Kety et al.,1975, p.154.

733. L. F. Richardson, 1913, p. 394.

A control group of non-SSD diagnosed adoptees is also established. The investigators then attempt to identify and diagnose the biological and adoptive relatives in each group. A statistically significant difference between the index and control biological relatives is seen as evidence in favor of the genetic hypothesis, although we will see that there are questions about whether this was the original design of the Danish-American Adoptees' Family study. The Adoptees' Family design is shown in Figure 7.1.

Figure 7.1

The Adoptees' Family Method

Kety et al. 1968, 1975, 1994

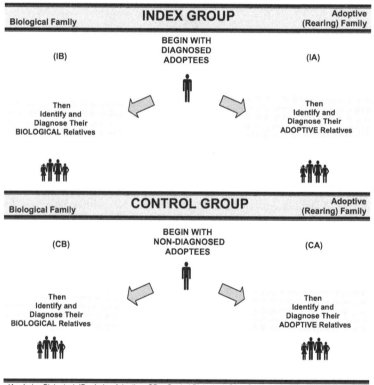

IA = Index Biological, IB = Index Adoptive, CB = Control Biological, CA = Control Adoptive. See Figure 7.3.

The results of the early Danish-American Adoptees' Family studies, which were limited to the population of the greater Copenhagen area, were published by Seymour Kety, David Rosenthal, Paul Wender and colleagues in 1968 and 1975. The study was extended to the rest of Denmark in 1975. These "Provincial study" final results were published in 1994. The 1968 Copenhagen study assessed the diagnostic status of the biological and adoptive relatives of 34 adopted children diagnosed with an SSD, versus the diagnostic status of the relatives of 33 matched control adoptees not diagnosed with an SSD.

(2) The Adoptees Method

In contrast to the Adoptees' Family method, which begins with adoptees, the Adoptees method begins with parents (usually mothers) diagnosed with an SSD. Researchers then determine the SSD rate among their adopted-away biological offspring. This rate is then compared to a control group consisting of the adopted-away biological offspring of parents not diagnosed with an SSD. Figure 7.2 shows the basic Adoptees model.

The first published schizophrenia adoption study (by Heston) utilized the Adoptees method. The first publication from the Danish sample, by Rosenthal and colleagues, appeared in 1968. In 1971, the Rosenthal group reported a statistically significant SSD difference between their index and control adoptee groups.

A third region of investigation has been Finland. The research team is headed by Pekka Tienari, who holds the unique distinction of having performed both a schizophrenia twin and adoption study. Tienari and colleagues are distinguished from Heston and the Danish-American group by their interest in family interaction effects as well as genetics. While presenting evidence which they believe supports the genetic hypothesis, they have found that a high level of adoptive family disturbance can predict which of their adoptees will be diagnosed with an SSD.

(3) The Crossfostering Method

The Crossfostering method looks at the adopted-away children of non-SSD biological parents, who are raised by an adoptive parent eventually diagnosed with an SSD. The Danish-American Crossfostering study was published by Wender and colleagues in 1974, whose purpose was to "explore the hypothesis that rearing by or with schizophrenic parents will produce schizophrenic psychopathology among persons who carry a normal genetic load."[734]

Figure 7.2

The Adoptees Method

Heston, 1966; Rosenthal et al., 1968, 1971;
Tienari et al., 1987, 2003

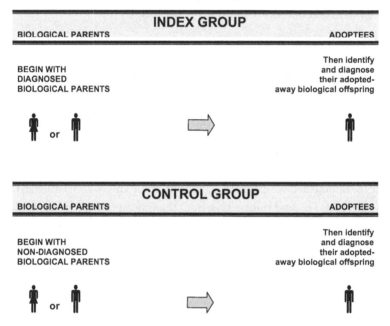

The diagnostic status of this "crossfostered" group was compared with the status of a group of adopted-away children of normal biological parents (who were reared by normal adoptive parents), and with a group of adopted-away offspring of SSD biological parents, who were reared by normal adoptive parents.

I will not review Wender's Crossfostering study in this chapter, since its methodological problems are numerous and are similar to those I will discuss in detail as they relate to the Kety et al. and Rosenthal et al. studies. Wender and colleagues recognized that their results were "modest,"[735] and critics such as Lidz, Lewontin et al., and Boyle have documented the glaring problems with the study.[736] These include (1) the use of global mental health ratings in place of

734. Wender et al., 1974, p. 122.
735. Kety et al., 1976, p. 416.

diagnosing schizophrenia, (2) the use of post data-collection comparisons to support the genetic position, (3) the failure to find statistically significant differences between important comparison groups, (4) the investigators' failure to consider non-genetic explanations of their results, and (5) that the mean age of the crossfostered adoptees at the time of their parent's diagnosis was 12-years-old.[737]

By the 1980s, Wender would recognize that his "study was particularly difficult" because

> adoption agencies do their best to place their infant charges with normal parents; only an unusual kind of schizophrenic — whom the adoption agency cannot detect — would receive a child for adoption. Such schizophrenic parents might become ill only later in life, long after having adopted a child, or they might be able to conceal their illness from prying adoption-agency eyes. So the group of adopting schizophrenic parents is different from the group of schizophrenic parents who have their own children. Thus, the question of what would happen if children born of normal parents were placed in the homes of typical schizophrenics *cannot be answered* [emphasis added].[738]

Wender's belated recognition that his investigation was unable to test his main hypothesis is reason enough to refrain from performing a detailed analysis of his study.

* * *

This chapter focuses on two broad themes relating to schizophrenia adoption research: (1) That genetic inferences from these studies' results are confounded by environmental factors; and (2) that to varying degrees, most are methodolgically unsound and were subject to the investigators' genetic biases.

In a 1980 review article, French critics Cassou, Schiff, and Stewart repeatedly referred to adoption research as consisting of "studies of abandoned children" ("les études d'enfants abandonnés").[739] As Alvin Pam noted, this phrase indicates "the irony of using throwaway kids as proof that schizophrenia is genetically transmitted."[740] Cassou and associates reminded us that the

---

736. Boyle, 1990; Lewontin et al., 1984; Lidz, 1976.
737. Van Dyke et al., 1975, p. 227.
738. Wender & Klein, 1981, p. 175.
739. Cassou et al., 1980.
740. Pam, 1995, p. 31.

decisive "evidence" in favor of the genetic theory of schizophrenia is based the families of 150 or so abandoned Danish children.

Like the twin method, adoption studies are based on a critical theoretical assumption: the "No selective placement assumption." Researchers must assume that factors relating to the adoption process (including the policies of adoption agencies) did not lead to the placement of index (experimental) adoptees into environments contributing to a higher rate of the condition or trait in question. They must assume that children were not placed into homes correlated with the status of their biological family. In the various studies, I will argue in support of the likelihood that, from a psychological standpoint, index adoptees experienced more harmful rearing environments than those experienced by control adoptees. I will show that most adoptees in these studies were born into societies where eugenic ideas were strong and the compulsory eugenic sterilization of people diagnosed with schizophrenia was permitted by law. This suggests that children whose bio- logical family had a history of mental disorders were seen as inferior potential adoptees and were placed with adoptive families more likely to produce serious mental disturbance. Thus, adoption studies' theoretical ability to disentangle genetic and environmental influences may not be true in many cases.

Other problems relating to methodology and bias in schizophrenia adoption research include:

(1) The questionable use of a "schizophrenia spectrum of disorders."

(2) Inconsistent and biased methods of counting relative diagnoses.

(3) The frequent failure to adequately describe the basis upon which a schizophrenia diagnosis was arrived at.

(4) The counting of first- and second-degree relatives with the same weighting.

(5) The lack of case history material, which would allow reviewers to assess the environmental conditions experienced by adoptees and relatives.

(6) The bias introduced by counting relatives individually in the Adoptees' Family studies, which violates an assumption of the statistical measures used.

(7) The use of late-separated or late-placed adoptees.

(8) The evidence that the investigators' bias in favor of genetic explana- tions had an important influence on their methods and conclusions.

Table 7.2 lists the presence or absence of important methodological problems in schizophrenia adoption research. A detailed exploration of these and other issues encompasses the remaining part of this chapter.

TABLE 7.2 Important Methodological Problems in Schizophrenia Adoption Research

Author (Year)	Country	Significant Chronic Schizophrenia First-Degree Bio Relative Rate vs. Controls? [a]	Significant Chronic Schizophrenia First-Degree Bio Family Rate vs. Controls? [b]	Used Late Separated Adoptees?	Probable Selective Placement Bias?	Blind Diagnoses?	Adequate Definition of Schizophrenia?	Used a Schizophrenia Spectrum?	Studied Adoptive Family Rearing Environments?	Adequate Case History Information Provided?	Used Fabricated Interviews to Make Diagnoses?	Counted First- and Second-Degree Relatives Equally?
Heston (1966)	USA (Oregon)	Yes	N/A	No	Yes	No [c]	No	No	No	No	No	N/A
Kety et al. (1968)	Denmark	No	No	N/A	Yes	Yes	No	Yes	No	No	N/A	Yes
Kety et al. (1975)	Denmark	No	No	N/A	Yes	Yes	No	Yes	No	No	Yes [f]	Yes
Kety et al. (1994)	Denmark	Yes	No	N/A	Yes	Yes	No	Yes	No	No	?	Yes
Rosenthal et al. (1968, 1971)	Denmark	No	N/A	Yes	Yes	Yes	No	Yes	No	No	No	N/A
Wender et al. (1974)	Denmark	No	N/A	Yes	Yes	Yes	No	Yes	No	No	No	N/A
Tienari et al. (1987, 2003)	Finland	No [d]	N/A	Yes	Yes	Yes	Yes [e]	Yes	Yes	No	No	N/A

[a]Using the traditional one-tailed .05 level of statistical significance utilized by all schizophrenia adoption researchers. Based on a comparison between either (1) the first-degree biological relatives of chronic

schizophrenia index adoptees vs. the first-degree biological relatives of control adoptees (Kety); (2) the chronic schizophrenia rate among the adopted-away biological offspring of people diagnosed with chronic schizophrenia vs. the biological offspring of controls (Heston, Rosenthal, Tienari); or (3) the chronic schizophrenia rate of crossfostered adoptees vs. controls (Wender).

[b]Based on a comparison between the biological families of index adoptees diagnosed with chronic schizophrenia, with at least one first degree- biological relative diagnosed with chronic schizophrenia, vs. the

biological families of controls with at least one first-degree biological relative diagnosed with chronic schizophrenia.

[c]Although two of the three raters made diagnoses while unaware of the adoptees' group status, the third rater, Heston, was not blind to their status.

[d]See the discussion of Tienari's results in this chapter.

[e]Although the validity of DSM-III-R criteria (used by Tienari) is debatable, his is the only study to use clear and accepted diagnostic guidelines.

[f]For documentation of Kety et al's use of fabricated "pseudo-interviews," see Lewontin et al., 1984, p. 224; Kendler & Gruenberg, 1984, p. 556. It is not clear whether Kety et al. used pseudo-interviews in the 1994 study.

N/A - Not Applicable.

The Danish-American investigators began their work in the early 1960s, but Heston was the first to publish a schizophrenia adoption study. Thus, we begin with Heston's Oregon study.

HESTON'S 1966 OREGON SCHIZOPHRENIA ADOPTION STUDY

Behavior geneticists have called Leonard Heston's 1966 Oregon adoption study a "remarkable report,"[741] and a "classic" study that "turned the tide towards acceptance of genetic influence on schizophrenia."[742] However, although Heston's study helped shift the prevailing views of the causes of psychiatric disorders in the genetic direction, there are many reasons to disagree with Heston's and most subsequent commentators' conclusions in favor of genetics.

Heston (who, like Gottesman and Shields, was mentored by Munich-trained Eliot Slater at London's Psychiatric Genetics Research Unit, Maudsley Hospital) identified 47 adopted-away biological offspring of institutionalized women diagnosed with schizophrenia. These children had been separated from their mothers in the first days of life and had been placed with family members or in an orphanage (foundling home). These index group adoptees (Heston used the term "experimental group"), who were born between 1915 and 1945, were matched against a control group of 50 adopted-away biological offspring of mothers having no record of being admitted to an Oregon state psychiatric hospital. Heston matched his index and control adoptees on the basis of sex, type of eventual placement (adoptive, foster family, or institutional), and the length of time adoptees spent in child care institutions. He compiled dossiers for each adoptee, which contained psychiatric hospital records, personal interviews, and psychological test results. These dossiers were evaluated "blindly and independently" by two psychiatrists, in addition to Heston.[743] The results showed that five index and zero control adoptees received a diagnosis of schizophrenia, a statistically significant finding. Heston concluded, "The results of this study support a genetic etiology of Schizophrenia."[744]

---

741. Gottesman & Shields, 1976a, p. 364.
742. Plomin, DeFries et al., 1990, p. 354.
743. Heston, 1966, p. 821.
744. *Ibid.*, p. 823.

Bias in the Diagnostic Process

According to Heston, "No attempt was made to assess the psychiatric status of the father."[745] This meant that one-half of the index group gene pool was unaccounted for, which, according to Rosenthal,

> it is equivalent to Gregor Mendel beginning his investigation with one selected type of sweet pea plant in the F1 generation, crossing it with other plants whose characteristics he knew nothing about, and then trying to relate all the characteristics he finds in the F2 generation to those of the F1 parent whose special characteristics were known to him.[746]

Rosenthal noted the "confusion and folly of this procedure."[747] In addition, Heston checked the control biological parents' mental health status only by searching the records of Oregon state mental hospitals. Thus it is fair to say that he knew little of the people who contributed over 75% of the genes in his study.

Heston personally interviewed 72 of the 97 adoptees, describing this process as follows:

> The interview was standardized, although all promising leads were followed, and was structured as a general medical and environmental questionnaire which explored all important psychosocial dimensions in considerable depth. Nearly all of the interviews were conducted in the homes of subjects, which added to the range of possible observations.[748]

Heston was familiar with the histories and psychiatric records of his 97 index and control adoptees, and was aware of the group status (index or control) of the 72 people he interviewed. Although the interviews were standardized, Heston's decision to follow "all possible leads" (and in what direction he was looking for "leads") may have been influenced by his knowledge of the adoptee's group status. The dossiers of these 97 adoptees were then rated "blindly and independently by two psychiatrists. A third evaluation was made by the author."[749] Thus, because Heston was one of the three people making diagnoses, the diagnostic procedure was contaminated because Heston *did not* make diagnoses blindly.[750]

---

745. *Ibid.*, p. 819.
746. Rosenthal, 1974a, p. 168.
747. *Ibid.*, p. 168.
748. Heston, 1966, p. 821.
749. *Ibid.*, p. 821.
750. Cassou et al., 1980.

Although the other raters initially evaluated each case independently, according to Heston "several differences arose in the assignment of specific diagnoses. In disputed cases, a fourth psychiatrist was asked for an opinion and differences were discussed in conference."[751] A major bias is evident in this description, since Heston, as the lead investigator, knew whether each subject was an index or control adoptee. He was also aware that five index adoptees had received a hospital diagnosis of schizophrenia. It is possible that the other raters were swayed by Heston's superior knowledge of these adoptees, and had they been prepared to diagnose a control adoptee with schizophrenia, Heston might have been tempted to convince them not to. Clearly, Heston should not have been involved in the diagnostic process. As Rosenthal once commented, "it is essential that the diagnostician not know whether the individual examined is related to an index or control proband."[752]

As previously noted, Heston was unable to interview 25 of the 97 adoptees. Of these 25, 6 refused to be interviewed, 8 were dead, 7 were inaccessible, and 4 were not approached because Heston felt that their status as adoptees might have been revealed in the process. Rather than remove them from his study, Heston decided to retain these 25 adoptees on the grounds that "considerable information was available for most of them."[753] This information often consisted of second-hand reports, prison or military records, etc. In the Danish-American studies, the name of a non-interviewed adoptee or relative could be checked against a national psychiatric register. Given the lack of such a register in the United States, the best Heston could do was to check the records of public psychiatric hospitals in three West Coast American states, plus those in areas in which adoptees were thought to be living. It is possible that several control adoptees were hospitalized for schizophrenia without anyone in the study

751. Heston & Denney, 1968, p. 368.
752. Rosenthal, 1975b, p. 20. In a review article published in *Science* four years after his original investigation, Heston obscured the fact that he had participated in the diagnostic process while not blind to the status of his adoptees. In a discussion of the adoption studies performed in the 1960s, Heston (1970, p. 250) wrote, "in these newer studies diagnoses either were made by raters who did not know the genetic background of the subjects or were taken unchanged from medical records." Heston then cited his 1966 investigation as one of these "newer studies." However, Heston *was* aware of the genetic background of his subjects, and his diagnoses were not simply "taken unchanged from medical records." Had this been the case, there would have been no reason to conduct interviews or to have reached diagnoses on the basis of a conference of psychiatrists.
753. Heston, 1966, p. 821.

knowing about it, which could have changed the study's outcome. Clearly, Heston should have removed these 25 adoptees, but "In a conference the raters agreed that it would be misleading to discard any cases, and that all subjects should be rated by forced choice."[754] One might argue that it was "misleading" to retain adoptees for whom insufficient information was available.

How Heston Defined Schizophrenia

Although Heston was the only adoption researcher not utilizing a schizophrenia spectrum, he provided little information on how he defined schizophrenia, writing only that diagnoses were based on "generally accepted standards,"[755] and that the diagnosis of schizophrenia "was used conservatively."[756] This vague definition made it impossible for independent reviewers to determine whether the five adoptees diagnosed with schizophrenia meet accepted diagnostic standards (DSM or otherwise). A rare case history presented by Heston described a man who would not be diagnosed with chronic schizophrenia using the criteria of any edition of the DSM (DSM-I through IV-TR).[757] This individual received a hospital diagnosis of DSM-I "schizophrenic reaction," but it is not clear which type of "reaction" it was. Of DSM-I's nine "schizophrenia reactions," "schizophrenia reaction, acute undifferentiated type" seems most appropriate for this adoptee.[758] Using DSM-IV-TR classifications, he probably would be diagnosed with schizophreniform disorder or brief psychotic disorder. Changing this one adoptee's diagnosis is enough to render Heston's index-control comparison statistically non-significant. One can only wonder, as did Cassou and colleagues in 1980, what else could have been learned had Heston chosen to publish more case history information.

Evidence of Selective Placement

The social and political conditions in *all* regions where schizophrenia adoption studies have been performed make it probable that their results were confounded by selective placement factors. Almost all of Heston's adoptees were born and placed at a time when eugenic ideas were dominant and the com-

---

754. Heston & Denney, 1968, p. 368.
755. Heston, 1966, p. 822.
756. Heston & Denney, 1968, p. 369.
757. See Heston, 1966, p. 823 for a description of this case.
758. APA, 1965, p. 27.

pulsory sterilization of people diagnosed with schizophrenia was sanctioned by law.

In 1909, the Oregon legislature passed a eugenic sterilization law by a wide margin, but it was vetoed by the Governor. However, a new law was passed by the legislature and signed by the Governor in 1917, establishing the Oregon State Board of Eugenics, the nation's first official eugenical organization. The duty of the Board, which was comprised mainly of superintendents of mental and penal institutions,[759] was to authorize, in the words of the law, the compulsory sterilization of

> all feeble-minded, insane, epileptic, habitual criminals, moral degenerates and sexual perverts, who are persons potential to producing [sic] offspring who, because of inheritance of inferior or antisocial traits, would probably become a social menace, or a ward of the state.[760]

Clause 2889 of the 1917 law gave the Board of Eugenics power to examine the "family traits and histories" of such persons,

> and if in the judgment of the majority of the said board procreation by any such person would produce children with an inherited tendency to feeble-mindedness, insanity, epilepsy, criminality or degeneracy . . . then it shall be the duty of said board to make an order directing the superintendent of the institution in which the inmate is confined to perform or cause to be performed upon such inmate such a type of sterilization as may be deemed best by said board.[761]

An additional law passed in 1919 stipulated that the mere fact that a person had been admitted to a mental hospital constituted "prima facie evidence that procreation by any such person would produce children with an inherited tendency to feeble-mindedness, insanity, epilepsy, criminality or degeneracy."[762] These eugenic sterilization laws were on the books until 1983, almost 40 years after the last of Heston's adoptees was born. The author of a 1925 article in *Eugenical News* wrote that, while sterilization laws in many states were limited to "inmates of institutions," in Oregon "there is a Eugenics Commissioner, who has authority to comb the state for degenerates and enforce sterilization."[763] And far from being a little known or discussed statute, leading newspapers championed

759. Largent, 2002.
760. Olson, 1920, p. 1487.
761. *Ibid.*, p. 1487.
762. *Ibid.*, p. 3176.
763. Anonymous, 1925, p. 71.

its cause and published summaries of eugenic lectures and statements by pro-sterilization authorities.[764] According to a 1935 article in the *Oregon Journal*, "Taking a tip from Nazi Germany, Oregon today considered embarking on a far-reaching program of sterilization of its unfit citizens."[765]

On December 3, 2002, Oregon Governor John Kitzhaber issued an official apology for the "great wrong done to more than 2,600 Oregonians over a period of about 60 years," who endured "forced sterilization in accordance with a doctrine called eugenics." He continued that "most of these Oregonians were patients in state-run institutions. The majority of them suffered from mental disorders and disabilities." Writing on behalf of the people of Oregon, he ended by saying that "The time has come to apologize for public policies that labeled people as 'defective' simply because they were ill, and declared them unworthy to have children of their own."[766]

The prevailing social conditions in Oregon, which Heston never mentioned, suggest that the offspring of institutionalized women diagnosed with schizophrenia were viewed as the carriers of an inherited predisposition for "insanity" and "degeneracy." As Kringlen commented, "Because the adoptive parents evidently received information about the child's biological parents, one might wonder who would adopt such a child."[767] In Oregon circa 1915-1945 it was unlikely that such children would have been placed into, or would have been accepted by, qualified adoptive homes. And little help could be expected from Oregon state hospital physicians, since they were strident supporters of the sterilization laws.[768] A 1929 article by eugenicist Paul Popenoe captured the sentiment of the times. He believed, in general, that American adoptees' "ancestry" was not "up to par," although "the best are sorted out early by the child-placing agencies. The remainder, collected in orphanages, represent predominantly the inferior levels and usually show up badly in tests." Popenoe advised prospective adoptive parents to "pick out a child with as good ancestry as possible," which in

764. Largent, 2002.
765. *Ibid.*
766. John Kitzhaber (2002). *Proclamation of Human Rights Day, and apology for Oregon's forced sterilization of institutionalized patients*, December, 3, 2002. Retrieved on 3/5/2004 from http://www.people1.org/eugenics/eugenics_article_3.htm
767. Kringlen, 1987, pp. 132-133.
768. Reilly, 1991.

"the mother is always known."[769] Clearly, the "inferior" biological offspring of "insane" mothers were not seen as desirable potential adoptees.

It is therefore unlikely that Heston's index and control group adoptees were randomly placed into available adoptive homes, which violates adoption studies' critical "no selective placement" assumption. Leaving aside other methodological problems, Heston's results are plausibly explained by the likelihood that index adoptees were reared in more psychologically harmful environments than control adoptees.

According to Heston's figures, significantly more index than control adoptees had spent over a year in a penal institution or were diagnosed with "sociopathic personality."[770] In addition, alcoholism ("problem drinking") was concentrated "almost exclusively" in the index group.[771] These findings suggest that the index adoptees experienced inferior rearing environments compared with control adoptees. Heston, however, believed that genetic factors explained the higher rate of "psychosocial disability" in his index group.

Abandoned Children

In his discussion of the early environmental deprivation experienced by adoptees who had spent years in orphanages, Heston recognized that poor conditions, and the lack of adult nurture they experienced, had a negative impact on these children's psychological development:

> The actual extent to which these children were deprived of maternal or emotional nurture beyond that implicit in group care must be inferred largely from indirect evidence. . . . the later history of these children strongly supports the hypothesis that significant deprivation did occur. Random observations recorded in school or nursery records and recollections of foster parents and subjects describe several children as shy, withdrawn, demanding excessive attention, or sad; few as happy, spontaneous, or normal. Some were requiring sedation at night. . . . It is certain that most of these children were unhappy and probable that there was significant deprivation of emotional nurture [emphasis added].[772]

From a psychosocial perspective we would expect that many such children, who spent an average of two years in an orphanage, would grow up to

769. Popenoe, 1929.
770. Heston, 1966, p. 822.
771. Eight cases were reported; Heston & Denny, 1968, p. 370.
772. Heston & Denney, 1968, p. 366.

be "neurotics," felons, "sociopaths" . . . or the recipients of schizophrenia diagnosis. Most psychodynamic and developmental theories stress the importance of the first few years of a child's life in his or her subsequent psychological development, and Heston recognized that some of his adoptees had been emotionally damaged during this period. In fact, three of the five schizophrenia diagnoses were given to index adoptees who had spent months or years in an orphanage.[773]

By observing the "later history" of children raised in orphanages, Heston inferred a lack of emotional nurture in their earliest years. He astutely saw their sadness and withdrawal as symptomatic of their deprivation, and not as the result of a genetic predisposition for childhood depression. Unfortunately, Heston failed to infer a lack of early emotional nurture from the significantly more troubled lives of his index group *adults* — seeing instead a genetic relationship between psychosocial disability and schizophrenia. In Heston's view, the emotional damage caused by early deprivation associated with institutional care "can be spontaneously reversed by the time adulthood is achieved."[774] What a convenient theory for reconciling two radically different approaches to the understanding of human suffering.

Heston's Oregon study contrasts sharply with later behavioral and IQ adoption studies that were able to minimize deprivation and attachment disturbance. In the Colorado Adoption Project, for example, all adoptees were separated from their biological mothers at or shortly after birth and were placed into qualified adoptive homes within a few weeks.[775] In contrast, almost half of Heston's adoptees spent a good deal of time in an orphanage, which, as Heston acknowledged, could not provide a nurturing environment and which contained many sad and withdrawn children. Heston also admitted, and I will add the emphasis, that *none of the subjects were reared in typical or 'normal' circumstances.*"[776] This was especially true for the index adoptees, who may have been treated differently by their adoptive parents on account of their biological background. Although Heston claimed that there were no important differences in adult psychological disturbance between adoptees placed in orphanages versus those

773. *Ibid.*
774. Heston et al., 1966, p. 1110.
775. Plomin & DeFries, 1985.
776. Heston & Denney, 1968, p. 374. In striking contrast to Heston & Denney's original description, biological psychiatrist Nancy Andreasen (2001, p. 199) wrote that in Heston's study, "adoptees were reared in families that were considered to be 'normal' or 'healthy'"

placed with families,[777] it remains likely that his adoptees experienced more psychological damage than non-adopted children. One therefore must question whether the results of this study (already biased by methodological problems and selective placement) can be generalized to the non-adoptee population. Unfortunately, the diagnostic status of these 97 "abandoned children" has been the basis (sometimes combined with the results of other studies) for sweeping yet unwarranted claims about the importance of genetic factors in schizophrenia.

Conclusion

There are five major problems with Heston's study. Each one by itself is reason enough to seriously question his and others' conclusions in favor of genetics. (1) The evidence that selective placement occurred in the sample. (2) A diagnostic process that was contaminated because one of the raters (Heston) was aware of the group status and personal history of the adoptees. (3) Heston's failure to provide case history material, which would allow independent analysis of adoptees' history and mental status. (4) About 26% of the adoptees were not interviewed, yet were retained in the study. (5) Schizophrenia was not defined.

The evidence suggests that Heston's 1966 Oregon schizophrenia adoption study was methodologically unsound and was subject to the confounding influence of placement bias.

THE DANISH-AMERICAN ADOPTEES' FAMILY STUDY: 1968

Although Kety and associates' Danish-American schizophrenia adoption studies are perhaps the most frequently cited evidence in support of a genetic basis for schizophrenia, they have been the subject of several critical reviews. My objective is to take a closer look at the key issues raised in these reviews in order to determine whether Kety and colleagues' conclusions are supported by the evidence.[778]

The first phase of the study was conducted in the greater Copenhagen area. The most important papers coming out of this work were published in 1968

---

777. Heston et al., 1966.
778. This discussion of the Kety et al. schizophrenia adoption studies is based on a revised version of a previously published article (Joseph, 2001a).

and 1975.[779] The 1968 study was based entirely on institutional records. The 1975 study, which used the 1968 adoptees and relatives, was based on interviews conducted with some of these relatives. Following the publication of their 1975 Copenhagen study, the investigators extended their research to encompass the rest of Denmark. Preliminary reports on this "Provincial" sample were based on institutional records only, with the final results, published in 1994, based on interview-derived diagnoses. In 1994 the investigators also reported results for the Danish National sample, meaning the combined diagnoses of the Copenhagen and Provincial studies.

Kety, Rosenthal, Wender and their Danish associates presented their first paper on the Danish-American adoption work at the June/July, 1967 Dorado Beach conference. This paper was published the following year.[780] The investigators established an index group of 34 adoptees diagnosed with "chronic schizophrenia" (designated "B1," N = 16), "acute schizophrenia" (designated "B2," N = 7), or "borderline schizophrenia" (designated "B3," N = 10).[781] They selected index adoptees from the records of all adoptions granted in the City and County of Copenhagen between 1924 and 1947. In all, 507 adoptees from this group were recorded as having been admitted to a psychiatric facility. The investigators were granted access to the Danish Adoption Register of the State Department of Justice, the *Folkeregister* (Population Register), the Psychiatric Register of the Institute of Human Genetics, records of the Mother's Aid Organization, and police, court, and military records. These 507 adoptees' records were screened, and when a consensus B1, B2, or B3 diagnosis was made by Kety, Rosenthal, and Wender, that person became an index case. A control group of 33 adoptees having no record of admission to a psychiatric facility was established as a comparison group.

The researchers matched index and control adoptees on the basis of age, sex, age at transfer to the adoptive parents, and the socioeconomic status of the adoptive family. After establishing these groups, they identified 463 of these adoptees' biological and adoptive relatives. Kety and colleagues then searched for any existing psychiatric records for these relatives, and when such records were found, a blind consensus diagnosis was made by Kety, Rosenthal, Wender, and Schulsinger. The code concealing the identity and group status of the rela-

779. Kety et al., 1968, 1975.
780. Kety et al., 1968.
781. The index group consisted of 34 subjects but only 33 cases because two of the index adoptees were B3 identical twins.

tives was then broken, and each case was assigned to the biological or adoptive relative category of the appropriate group (index or control). All diagnoses were based on information obtained from institutional records. There was no personal contact between the investigators and any of the adoptees or relatives in the 1968 study.

In comparisons between index and control relatives, Kety et al. counted all SSDs (schizophrenia spectrum disorders) as "schizophrenia" (see Table 7.1). They made diagnoses according to a global (or consensus) diagnostic system, describing this method, which they used in the entire series spanning more than 25 years, as follows:

> Four copies of the edited summary were prepared and distributed to the four authors who served as raters and who independently characterized each subject according to the classification described below. The individual ratings were then tabulated and those cases in which there was disagreement among the raters were discussed at a conference of all four authors where an effort was made to review additional edited information which it was possible to obtain and to arrive at a consensus diagnosis acceptable to all. In 4 cases there remained an evenly split opinion regarding the presence of schizophrenia or doubtful schizophrenia, and these were not included in those categories.[782]

Although Kety claimed high diagnostic inter-rater reliability, each investigator approached the diagnostic process somewhat differently.[783] According to Kety, each rater's "individual definitions of schizophrenia varied by virtue of [their] training and experience, from a substantial reliance on Kraepelin and Bleuler to the broader psychodynamic concepts which were taught in the 50s."[784]

Kety and colleagues first reported the distribution of SSDs among index and control adoptive and biological relatives in their 1968 paper, finding significantly more SSDs among index biological relatives versus control biological relatives:

> Of 150 biological relatives of index cases 13, or 8.7%, had a diagnosis of schizophrenia, uncertain schizophrenia or inadequate personality compared to 3 of 156, or 1.9%, with such diagnoses among the biological relatives of the controls. The difference is highly significant.[785]

782. Kety et al., 1968, pp. 351-352.
783. Kety, 1974.
784. Kety, 1987, p. 424.

They concluded that "genetic factors are important in the transmission of schizophrenia."[786]

In the following sections, I discuss six important topics relating to the 1968 and all subsequent Danish-American Adoptees' Family studies: (1) the schizophrenia spectrum concept, (2) questions relating to the design of the study, (3) the failure to study environmental variables, (4) counting half-siblings in statistical calculations, (5) the researchers' emphasis on counting individual relatives as opposed to counting families, and (6) the evidence of selective placement in the Danish adoption process.

The Schizophrenia Spectrum Concept

Although traditionally only chronic B1 "process" cases were counted as schizophrenia by Danish psychiatrists,[787] Kety and colleagues broadened the definition of schizophrenia considerably to include what they considered related diagnoses. I have argued against the validity of the Kety et al. schizophrenia spectrum in detail elsewhere, and here I review the main points. [788]

In 1963, Kety, Rosenthal, and Wender began a collaboration with Fini Schulsinger, a Danish psychiatric geneticist whose expertise and language abilities would enable them to obtain the records of a large number of adoptees through the extensive population and psychiatric registers then existing in Denmark. Schulsinger has been credited with convincing the Danish Ministry of Justice to open up national adoption records for the purpose of scientific research. His persuasiveness convinced the authorities to allow access to these records, even though earlier Danish researchers had made similar, unsuccessful requests.[789]

The researchers obtained information on 5,483 adoptees from the greater Copenhagen area. Based on rough population expectations, they probably expected this sample to produce a sufficient number of adoptees diagnosed with chronic schizophrenia (B1) to compare to a control group of adoptees with no record of a schizophrenia diagnosis. The results of prevalence studies conducted

---

785. Kety et al., 1968, p. 353.
786. *Ibid.*, p. 361.
787. Kety, 1978. Also see Munk-Jørgensen, 1985.
788. Joseph, 2000a.
789. Strömgren, 1993.

215

in Western countries have placed the lifetime expectancy rate for schizophrenia at between 0.8 and 1%.[790] Thus, Kety and associates were probably expecting to find 50-55 B1 adoptees in their sample (5,483 x .01= 54.8). Epidemiological studies conducted in Denmark produce a lower age-corrected Danish population prevalence of 0.69%.[791] Based on this relatively low rate, the expected number of B1 adoptees from the greater Copenhagen sample would have been 38 (5,483 x 0.0069 = 37.8). However, contrary to genetic expectations, Kety and colleagues diagnosed only 16 adoptees as B1. This rate of about 3/1,000 is less than one-half of the expected 7-10/1,000 rate in the general population.

Thus, at the very beginning, the investigators were confronted with powerful evidence in favor of the environmentalist position — for the numbers suggested that merely being raised by parents screened for mental health by adoption agencies had reduced the chance of a person being diagnosed with schizophrenia by over 50%. This suggests that Danish adoptees constituted a distinct group when compared to the general population. According to critics Theodore Sarbin and James Mancuso, the results "show, quite simply, that the social services agencies have suitably done their job."[792]

Finding only 16 adoptee chronic schizophrenia diagnoses, as Rosenthal recognized, meant that the investigators did not have enough cases to be able to conduct their study:

> The fact that the number of 16 hard-core, process [B1] schizophrenic, index cases *is too small to give the heritability of such disorders the proper opportunity to express itself*, and the possibility that by relying only on psychiatric records rather than on personal examinations we were missing a number of cases [emphasis added].[793]

And a year earlier Rosenthal had made a similar point:

> The second [important] feature [of the research] has to do with the fact that we have included a broad spectrum of disorders in the ones I am calling schizophrenic. These include not only the classical chronic, process types of cases, but patients called doubtful schizophrenic, reactive, schizo-affective, borderline or pseudoneurotic schizophrenic, or schizoid or paranoid. *If we dealt only with hardcore schizophrenia, our ns [number of subjects] would be too small to make any of these studies meaningful* [emphasis added].[794]

790. Rosenthal, 1970b; Slater & Cowie, 1971. Few people over 45 receive their first schizophrenia diagnosis.
791. Quoted in Slater & Cowie, 1971, p. 13.
792. Sarbin & Mancuso, 1980, p. 138.
793. Rosenthal, 1972, p. 68.

He continued,

> However, a more positive reason for including the spectrum of disorders is that in the process, we hope to be able to determine whether any or all of these disorders, which phenotypically have strong resemblances to hard-core schizophrenia, are genetically related to it as well.[795]

This passage suggests that, whereas the researchers believed that assessing the relationship between B1 and the other SSDs was a "positive reason" for employing a spectrum, a "negative reason" was their need to broaden the definition of schizophrenia in order to obtain enough adoptees to be able to conduct their study.[796]

Clearly, had Kety and colleagues restricted their definition of schizophrenia to B1 only, they would have had no possibility of achieving statistically significant results in the genetic direction. Rosenthal was clear on this point also: "It should be apparent now that if we had included in our comparisons of index and control relatives only those who clearly had process schizophrenia, we would have found no difference between the two groups of relatives."[797]

The evidence suggests that Kety, Rosenthal, Wender and colleagues created the schizophrenia spectrum because — contrary to genetic expectations — Danish adoptees were diagnosed with schizophrenia 50% less often than non-adoptees (who were reared by their biological parents). Thus, they might have concluded that their results were not generalizable to the non-adoptee population. In any case, the investigators were confronted with the choice of giving up on several years of hard work, or broadening the definition of schizophrenia in order to obtain more cases. This does not mean that Kety and associates, who were diagnosing blindly, knew into which group (index or control) these cases fell. But wherever they might have fallen, there were not enough to allow the possibility of finding significant results.

It appears that the main reason that Kety, Rosenthal, and Wender created the schizophrenia spectrum was to have enough cases to conduct their study, and not, as they maintained in their major publications, in order to assess the relationship between B1 and the other SSDs. This was suggested in a revealing 1975 description of the spectrum's origins by Rosenthal:

---

794. Rosenthal, 1971b, p. 194.
795. *Ibid.*, p. 194.
796. This scenario was suggested by Lewontin & colleagues in 1984, and Pam in 1995.
797. Rosenthal, 1972, p. 68.

It seems somewhat ironic that while representing the US Field Center in the WHO International Pilot Study of Schizophrenia, John Strauss and Will Carpenter were working upstairs at the National Institute of Health Clinical Center, trying to hone a definition of schizophrenia as sharply as they could possibly make it, while Paul Wender and I were working downstairs, in concert with Seymour Kety, in effect *broadening the concept of schizophrenic disorder as widely as it may have ever been reasonably conceived before* [emphasis added].

While Carpenter and Strauss emphasized the limits or boundaries of the process [B1] schizophrenia concept, *our group strained* to encompass all disorders that shared salient clinical and behavioral manifestations with process schizophrenia and to group these as a spectrum of schizophrenic disorder [emphasis added].

Of course, we did not know which disorders, if any, should be included in such a spectrum to meet the criterion of genetic or familial commonality. Nevertheless, we selected the ones that we thought had the highest probability of meeting this criterion, and introduced them into our research studies as a hypothesis to be tested. It was easy to read through the *APA Diagnostic and Statistical Manual*, second edition, to make such selections.[798]

Because the spectrum was based on DSM-II definitions (published in 1968), Rosenthal seemed to date its creation to no earlier than 1967, since it was in February of that year that a DSM-II draft version was distributed to psychiatrists.[799] Assuming that the investigators did not have access to the draft before it was distributed to psychiatrists for commentary, as late as 1967 they were "straining" to broaden the concept of schizophrenia "as widely as it may have ever been reasonably conceived before" — well after they had identified their adoptees. One can therefore ask: What circumstances compelled Kety et al. to broaden the definition of schizophrenia several years after they began their study?

I will now briefly mention several other problems with Kety and colleagues' schizophrenia spectrum. (1) The investigators' rationale for the inclusion of SSDs was based on finding a significantly greater number of these disorders among index versus control biological relatives. However, with the exception of the vaguely defined "uncertain borderline schizophrenia" (D3), they found no individual SSD in significantly greater numbers among index biological relatives versus controls in either the 1968 or 1975 studies.[800] (2) The

---

798. Rosenthal, 1975b, p. 19.
799. APA, 1968, p. ix.

researchers' conclusion that B1 was related to the other SSDs was based on invalid methods of counting diagnoses.[801] (3) In spite of Kety's reliance on E. Bleuler's description of latent schizophrenia,[802] Bleuler did not believe that it was possible to distinguish "milder cases of schizophrenia" from people who were merely "whimsical." He therefore called on clinicians to use a "very high diagnostic threshold value" in making a schizophrenia diagnosis.[803] (4) As the researchers admitted, they often had difficulty distinguishing SSDs from comparable non-SSDs.[804] (5) The spectrum concept was based on the questionable assumption that diagnoses are genetically related to each other simply because they are (allegedly) found together. As Stephen J. Gould noted, "the invalid assumption that correlation implies cause is probably among the two or three most serious and common errors of human reasoning."[805] (6) Although the BI, B2, and B3 diagnostic formulations included references to homosexuality, a "disorder" subsequently dropped by the American psychiatric establishment in 1974,[806] there is no mention of homosexuality in the schizophrenia descriptions or diagnostic criteria of any edition of the DSM. (7) It is questionable whether the investigators should have counted diagnoses labeled "uncertain," and there is no evidence that they made the decision to count these diagnoses before they assigned them to their respective categories. According to Kety, "in the case of the relatives, questionable or uncertain schizophrenia had to be added if relatives with less certain diagnoses were not to be lost."[807] However, Kety could have prevented these relatives from becoming "lost" without counting them in his study's statistical calculations.

Because of the doubtful validity of Kety and colleagues' schizophrenia spectrum, chronic schizophrenia is the only diagnosis they should have counted as "schizophrenia" in their studies.

---

800. See Joseph, 2000a.

801. *Ibid.*

802. Kety, 1985.

803. E. Bleuler, 1950, p. 294.

804. See, for example, Kety, Rosenthal, & Wender, 1978.

805. Gould, 1981, p. 242.

806. Kutchins & Kirk, 1997.

807. Kety, 1987, p. 424.

Questions Concerning the Design of the Study

As seen in their 1968 publication, Kety and colleagues concluded in favor of genetics on the basis of a diagnostic comparison between index and control bio-logical relatives. However, questions remain about whether this comparison reflected their original research design. The four groups of relatives they studied are represented in Figure 7.3 (which is based on the groups represented in Figure 7.1).

FIGURE 7.3 *Groups of Relatives Receiving Diagnoses in the Danish-American Adoptees' Family Studies (Kety et al. 1968, 1975, 1994)*

	Biological Relatives	Adoptive Relatives
Index Adoptees	IB (Index Biological)	IA (Index Adoptive)
Control Adoptees	CB (Control Biological)	CA (Control Adoptive)

The investigators based their conclusions in support of genetics on finding a statistically significant individual relative SSD difference between groups IB and CB, and finding no significant SSD difference between groups IA and CA. According to Rosenthal's 1970 description,

> A higher incidence among the biological relatives of index cases than of controls indicates that heredity is contributing significantly to the disorder. A higher incidence among the adoptive relatives of index cases than of controls supports the view that rearing by, of, or with schizophrenics contributes to the development of the disorder.[808]

Critics Theodore Lidz and Sidney Blatt addressed the origins of Kety and colleagues' 1968 design in their important 1983 critique, making the unsupported claim that the researchers' original intent had been to compare the schizo-phrenia rates of index biological versus *index adoptive* relatives (IB vs. IA). According to Lidz and Blatt, "The investigators sought to differentiate genetic from intrafamilial environmental factors by comparing the occurrence of such

---

808. Rosenthal, 1970b, p. 57.

disorders in the biological and adoptive relatives of schizophrenic patients who had been adopted at a very early age."[809] They also argued that, because Danish adoptive parents were screened for psychiatric disorders, Kety was compelled to create a control group of non-SSD adoptees and their relatives. Lidz and Blatt concluded that "the major interest of the study became the comparison of the biological relatives of the schizophrenic and control adoptees rather than the original purpose of the project."[810]

Kety replied that Lidz and Blatt misunderstood the aim and logic of his study. Furthermore,

> We anticipated that there would be differences between adoptive and biological relatives in age, socioeconomic status, life style, and other variables. For that reason we planned not to make comparisons between these two groups of relatives but, instead, as described fully in the original publications and outlined above, to compare each group with their respective controls in evaluating separately the significance of genetic or family-related environmental factors.[811]

In their 1968 study, which contained the first published description of their methods, Kety and colleagues did indeed discuss the importance of making direct comparisons between index and control relatives from each group. However, three months prior to the investigators' first public presentation of their study in 1967, Rosenthal wrote a conference paper which supports Lidz and Blatt's contention. As Rosenthal described it,

> In Denmark, with the collaboration of Dr. Fini Schulsinger and others, we began with adoptees who are now schizophrenic. *We compare the incidence of schizophrenic disorders in their biological and adoptive families.* The same procedure is carried out for a matched group of normal adoptees, who serve as controls [emphasis added].[812]

Rosenthal is therefore on record as stating, in March of 1967, that he and his colleagues intended to compare groups IB versus IA relatives, and then groups CB versus CA relatives. However, they did not make their diagnoses until

---

809. Lidz & Blatt, 1983, p. 426.

810. *Ibid.*, p. 427.

811. Kety, 1983a, p. 721.

812. Rosenthal, 1967, p. 25. I am unaware of any other published description of the design and method of the 1968 study prior to its publication. Apart from Kety's 1959 article discussing the need to study adopted children, none is cited in any Danish-American publication.

April of 1967.[813] If the investigators had made their diagnostic comparisons using Rosenthal's March, 1967 description, the SSD difference would not have been statistically significant in either relative group. Kety et al. found 13 SSDs out of 150 index biological (IB) relatives (8.7%), and two such diagnoses out of 74 index adoptive (IA) relatives (2.7%). The difference, however, is not statistically significant.[814] Thus, had the investigators compared their groups as described by Rosenthal in March, 1967, they would have concluded that there were no significant differences in either comparison, and that their study had found no evidence supporting a genetic transmission of schizophrenia. Interestingly, neither Rosenthal nor Kety cited Rosenthal's 1967 paper in any of their subsequent publications on the genetics of schizophrenia.[815]

The Failure to Study Environmental Variables

In 1966, Kety wrote, "although genetic factors undoubtedly operate in schizophrenia, they do not constitute a sufficient explanation for the genesis of this disorder."[816] If Kety had no doubt that genetic factors were operating, he might have concentrated on identifying environmental causes of schizophrenia, for why spend so much time and money in order to investigate something that is "undoubtedly" true? Kety followed these lines by writing, "we must continue to look for environmental factors that operate to produce what we call schizophrenia in the genetically vulnerable individual. Fortunately, this seems to be the attitude of most investigators in the field."[817] Unfortunately, Kety assessed only

---

813. Kety et al., 1968, p. 346.
814. *Ibid.*, 1968, p. 355 (p = .076, Fisher's Exact Test, one-tailed).
815. Psychiatric geneticists Faraone & Tsuang (1995, p. 92) described the Adoptees' Family method, which they call the "adoptee-as-proband design," as follows:

> As its name suggests, the adoptee-as-proband design starts with ill and well adoptees and examines rates of illness in both biologic and adoptive relatives. If the biologic relatives of ill adoptees [IB in Figure 7.3] have higher rates of illness than the adoptive relatives of ill adoptees [IA], then a genetic hypothesis is supported. In contrast, if the adoptive relatives show higher rates of illness, then an environmental hypothesis gains support.

> The likelihood that Kety & colleagues (1968) abandoned this comparison in favor of another (IB vs. CB) which produced significant results did not hinder Faraone & Tsuang, two paragraphs later, from citing this study as evidence in favor of the genetic basis of schizophrenia.

816. Kety, 1966, pp. 230-231.
817. *Ibid.*, p. 231.

one environmental variable, for the likely purpose of answering anticipated objections to his research design. In 1970 he wrote,

> We are really examining only one environmental factor and that is the presence of a person in the adoptive family with a mental illness. There are thousands of environmental factors which we have not examined: the personality of the family, their child-rearing practices, the diet which the individual has had, the lead in the drinking water and many perhaps undreamed of. Therefore, these data by no means rule out the operation of environmental factors. They simply indicate that at least one of these factors, namely, having a mentally ill person with a schizophrenic form of illness in the immediate environment, is not an important or significantly operating variable.[818]

Kety wrote of the family rearing environment as just another possible environmental variable, on par with lead in the drinking water, diet, etc. But it is also a major psychosocial explanation of why people are eventually diagnosed with schizophrenia, and deserved to be looked at more closely. A subsequent schizophrenia adoption study by Tienari and associates in Finland (reviewed later in this chapter) found that "all adoptees who had been diagnosed either as schizophrenic or paranoid had been reared in seriously disturbed adoptive families."[819] Kety and colleagues' studies would have been substantially more interesting and important had they, like Tienari, taken a closer look at the psychosocial environments of the people they studied.

The Validity of Counting Half-siblings

In their statistical calculations the investigators decided to count SSDs among first- and second-degree relatives with equal weighting. Most critics of the 1968 study observed that Kety et al. made only one B1 diagnosis (a half-sib) among their 150 index biological relatives, and that the statistically significant index SSD rate they reported was dependent on diagnosing several non-B1 SSDs among their index biological half-siblings.[820] The researchers found nine SSDs in the index biological half-sibling group, whereas none was found among the control half-siblings. In the 1968 study, half-siblings constituted 57% of all biological relatives (173/306) and 69% of index biological relative SSD diagnoses (9/13). As we will see, in 1975 the investigators rested their case for "compelling evi-

818. Kety, 1970, p. 240.
819. Tienari, Sorri, et al., 1987, p. 482.

dence" in favor of genetics on the distribution of SSDs among index and control biological paternal half-siblings.

The mid-1970s saw a discussion among several prominent schizophrenia researchers over the importance of the study's half-sibling SSD diagnoses. Gottesman and Shields noted that half-siblings had a higher SSD rate than full-siblings, even though "genetic theory predicts a much higher risk for full siblings."[821] Kringlen argued that the high rate among half-siblings "is, from a

___

820. The mythological status of the Danish-American adoption studies continues unabated in psychiatry, as seen in the following examples. (1) The authors of a 1995 American Psychiatric Association *Annual Review of Psychiatry* article claimed that the 1968 Kety et al. study "found that the prevalence of schizophrenia was significantly higher in the biological parents," and that the "Danish adoption studies also found that the biological relatives of schizophrenic persons had elevated rates of 'borderline schizophrenia' ..." (Byerley & Coon, 1995, p. 366). In fact, in 1968 Kety diagnosed *zero* biological index parents with chronic schizophrenia, and only one with borderline schizophrenia (see Kety et al. 1968, p. 354, Table 4a). (2) In a 2001 tribute to Kety (who died in 2000), his colleague Philip S. Holzman wrote that one of the two "incontrovertible" conclusions of the Danish-American adoption studies was that "children born to a schizophrenic mother and reared in an adoptive family become schizophrenic at the same rate as those reared by the biological mother, who is schizophrenic" (Holzman, 2001). However "incontrovertible" this conclusion may have appeared to Holzman, no such comparisons were made by the Danish-American investigators. The Kety et al. Adoptees' Family studies *began* with adoptees diagnosed with an SSD, and *then* recorded diagnoses for their biological and adoptive relatives. As it turned out, only 2 of these adoptees' 66 biological mothers (3%; 1975 N = 1, 1994 N = 1) were diagnosed with chronic schizophrenia. (3) According to Neale & Oltmanns (1980, p. 197), "Kety (1974) suggested that a comparison of the prevalence of schizophrenia in the paternal half-siblings" would eliminate the possibility that *in utero* experiences led to greater index biological relative SSD rates. However, because Kety made this suggestion *after* he had collected and reviewed the data (see Kety, 1974, p. 961; Neale & Oltmanns, 1980, p. 197) there is no evidence that he determined the importance of the paternal half-sibling comparison *before* the collection of data. (4) According to Lyons & colleagues (1991, p. 131), "Kety et al. ... predicted that 'if schizophrenia were to some extent genetically transmitted, there should be a higher prevalence of disorders in the schizophrenia spectrum among the biological relatives.' ... Kety did, indeed, find a concentration of schizophrenia spectrum diagnoses in the biological relatives of schizophrenic adoptees." Kety & associates were thus credited with "predicting" and "finding" such a concentration in the same 1968 publication. (5) The Tenth Edition of *Modern Clinical Psychiatry* listed the authors of the 1968 Adoptees' Family study — one of the most famous investigations in the history of psychiatry and performed by Kety, Rosenthal, Wender, & Schulsinger — as "Kety, Rosenfeld, Winther, and Scholfinger" (Kolb & Brodie, 1982, p. 351). One wonders how many psychiatry textbook authors actually read the original adoption study source material they cite (see Joseph, 2000b; Leo & Joseph, 2002; Paul, 1998).

genetic point of view, meaningless."[822] Kety et al. admitted that they had "toyed with the idea of giving them half-weight as soon as we realized how many were being identified but rejected it as being too pretentious."[823] Boyle would later comment, "it is difficult not to wonder whether the idea was also rejected because it leads to a quite different set of conclusions."[824] In 1976 critic Lorna Benjamin wrote, "this finding is peculiar and contradictory. It shows, in effect, that the *less* consanguinity, the *greater* the 'genetic' effect. Differences should be weakest, not strongest, in the half-sibling category."[825] Kety replied that the difference was not statistically significant, and even if it had been, it could be explained by "socioenvironmental" and other factors.[826]

Lidz and Blatt argued that Kety and colleagues should have counted half-siblings weighted on the basis of the common parent's status, whether the other parent was known, and the type of genetic model used. They complained that the investigators provided no information about a half-sibling's other parent, or about his or her rearing environment. Kety replied that his 1975 results would have reached statistical significance even if half-siblings had been counted one-half, and that in his opinion there was just as much reason to give them a weighting of two. He accused Lidz and Blatt of giving half-siblings no weight at all, because they focused on diagnoses among first-degree relatives.[827]

Although the investigators had information for only one parent of each half-sibling in question, Rosenthal wrote in another context that a "proper genetic study must be based on who mates with whom. In fact, although genetics has been defined in various ways, the simplest and perhaps best definition of genetics is: *the science of matings*."[828] Regarding the Danish-American half-siblings, we know little about "who," and *nothing* about "whom," yet Kety and Rosenthal placed great importance on the products of such "matings."[829]

From an environmental standpoint it is unacceptable to count people for whom little or nothing of their family and social environment is known. Kety and colleagues provided no information about the half-siblings' rearing environ-

---

821. Gottesman & Shields, 1976a, p. 370.
822. Kringlen, 1976, p. 430.
823. Kety et al., 1976, p. 416.
824. Boyle, 1990, p. 144.
825. Benjamin, 1976, p. 1131.
826. Kety, 1976, p. 1135.
827. Kety, 1983a.
828. Rosenthal, 1974a, p. 168.

ments, or how many actually lived with their biological parents. An index father, for example, could have sired the half-sibling in question without having been the child's rearing father. An SSD biological half-sibling could have grown up eating out of garbage cans on the streets of Copenhagen, whereas non-diagnosed half-siblings could have been raised in exceptionally nurturing and loving environments.

Even Gottesman and Shields recognized the problem of assessing the environments of the Danish half-siblings:

> The high degree of psychopathology seen in the half sibs could have a major environmental component and could reflect cultural transmission. That is, the half sibs might have stayed with a disturbed biological parent or been subjected to various kinds of institutional care. Although such happenings may not be sufficient to produce schizophrenia, they may have produced "inadequate personalities" of the kind that might have been diagnosed as definite or uncertain latent or borderline schizophrenia (B3 or D3).[830]

---

829. Kety was once asked the following question after giving a talk on the genetics of schizophrenia and the results of his studies: "In your data on the paternal half siblings, how can we be sure that those fathers might not have sought out a schizophrenogenic mate repeatedly?" Kety replied to this question as follows:

> That is a very good question. Frankly, we cannot rule it out; but this hypothesis requires the assumption that the fathers have some uncanny ability, greater than that of any psychiatrist I know, to pick out schizophrenogenic mothers. Furthermore, if the father had the propensity for picking schizophrenogenic mothers, this would still not be as effective as *being* the schizophrenogenic mother. On the basis of the hypothesis you propose, we would expect the biological maternal half siblings to have more schizophrenia than the paternal half siblings. That is not what we find. ... so this hypothesis finds no support. (Kety, 1978, p. 67)

Kety's response contained one clear error: The rate of schizophrenia among the *biological* maternal half-siblings is irrelevant to the questioner's hypothesis because they were not raised by the same mother who raised the adoptee. Kety should have responded that children in the *adoptive homes* were the group of interest, because this is where the theorized schizophrenogenic environment was located. There were only four full- and two half-siblings raised in the adoptive homes of B1 index adoptees, so this question could not be explored in the 1975 study. However, the evidence from schizophrenia family studies indicates that the siblings of people diagnosed with schizophrenia are more likely to be diagnosed with schizophrenia than randomly selected members of the general population. Theoretically, this increased risk could be due to genes, environment, a combination of both, or methodological error. From a psychosocial perspective, family studies demonstrate that schizophrenigenic environments do exist, and nothing in Kety's data suggests that such environments did not also exist in the homes (or institutions?) of the paternal half-sibs.

Gottesman and Shields brought up an important point: How many of these half-siblings were raised in institutions? To my knowledge, Kety and colleagues never wrote about this.

According to Rosenthal, in order to "demonstrate that genes have anything to do with schizophrenia," an investigator must show that "the frequency of schizophrenia in relatives of schizophrenics [is] positively correlated with the degree of blood relationship to the schizophrenic index cases."[831] The Adoptees' Family series failed this test, suggesting that environmental factors played a major role in the reported elevated SSD rate among the biological half-siblings.

Who Gets Counted, Families or Individuals?

In 1976, Benjamin observed that counting biological relatives separately and combining them into index and control totals is a violation of the assumption of independent observations:

> The procedure of counting up all the possible relatives of each index case and pooling them as if they were independent samples . . . would allow some families to disproportionately affect the results. . . . This group of people, who were likely to have been reared together and to have had the same health care system supplying records used in the study, could falsely inflate the "significance" of the difference.[832]

According to Benjamin, the 1975 study should have reported an N of 67 (the number of index + control biological relative *families*), rather than Kety and colleagues' N of 347 (the number of biological *relatives*). Benjamin's observation applies to the 1968 study as well. Kety replied that the 1975 study did indeed recognize the importance of counting families in addition to individuals.[833] By his calculations, SSDs were found in 17 of 33 index biological families, but in only 5 of 34 control biological families.[834] However, he had to redefine the spectrum in order to find statistical significance in this comparison.[835]

---

830. Gottesman & Shields, 1982, pp. 144-145.
831. Rosenthal, 1974b, p. 589.
832. Benjamin, 1976, p. 1130.
833. Kety, 1976.
834. Kety et al., 1975, p. 163.

According to Loren Mosher, then editor of *Schizophrenia Bulletin* and the Director of Schizophrenia Research at the National Institute of Mental Health:

> The actual sample size of all the adoption studies is, at least for genetic purposes, the number of index probands [adoptees in this case], *not* the number of relatives identified. The power of the adoption methodology is its separation of heredity and environment for genetic analysis; therefore, when N's are reported as the number of biological relatives seen, it leaves the misleading impression that the genetic/environmental separation is applicable in this group, whereas, in point of fact, it is not. Basically, studying either biological or adoptive relatives is just a special family study.[836]

Benjamin and Mosher suggested that a familial clustering of SSDs could be explained by exposure to common environmental influences, or by siblings receiving similar hospital diagnoses because hospital psychiatrists viewed them as sharing a common genetic heritage. For Mosher in particular, counting all biological relatives individually in statistical comparisons transformed the Adoptees' Family method into little more than a "special family study." Thus, Kety and colleagues' practice of emphasizing SSD diagnoses among individual relatives, as opposed to the proper procedure of counting *families* with at least one SSD-diagnosed member, violated the independence assumption of the statistical procedures they used.

The Selective Placement of Adoptees

In this section I present evidence that the crucial "no selective placement assumption" of adoption studies was violated, suggesting the unlikelihood that Danish agencies randomly placed prospective adoptees into the range of available families. According to criminality adoption researchers Hutchings and Mednick, who carried out research in Denmark,

---

835. As noted by Lidz & Blatt (1983), SSD category C, which in 1975 was still in the spectrum, was omitted from this comparison. If category C is included, the results were: 23/33 index families affected vs. 16/34 affected control families. Using a Fisher's Exact Test the probability for this difference is not significant (p = .05108). In another paper, Kety et al. (1976, p. 418) acknowledged that the comparison was not significant, writing "NS" below the totals (see my discussion of the 1975 study).
836. Mosher, 1975, p. 3.

> The most important limit of the adoption method is the possibility that the adoption procedure results in selective placement, promoting correspondence between the adoptive home and the characteristics of the biological parents. . . . The Danish organization which arranged many of the adoptions examined in this study states clearly that they do aim at matching in certain respects.[837]

Hutchings and Mednick found a statistically significant correlation between the social class of the biological and adoptive fathers. Genetic investigator T. W. Teasdale studied all 14,427 National sample adoptees identified by Kety and associates, and found a significant correlation between the socioeconomic status of the adoptees' biological mothers and fathers and the adoptees' adoptive fathers. Teasdale concluded that "some selective placement has occurred in the sample; i.e., the adoption agencies tended to place children into adoptive homes as a function of their biological background."[838]

However, Kety and colleagues believed that placement policies in Denmark did not create a problem for their study:

> Since the etiological role of environmental variables remains obscure at present, it is not likely that a social agency, even if it set about doing so deliberately, could find sufficient of the unknown variables in the prospective adoptive parents to materially affect the risk of [schizophrenic] illness in the adoptee.[839]

In their view, it was unlikely that agency policies could have led to the placement of index adoptees into more schizophrenia-producing environments.

The potentially confounding influence of selective placement can be seen in the example of pellagra. As we saw in Chapter 1, because of its tendency to cluster in certain families, pellagra was once believed to carry an important genetic component. It was later shown to be caused by a niacin deficiency. Pellagra was found mainly in poor families, whose diet did not provide enough niacin.[840] According to Kety's logic, an early 20th century adoption agency, unaware of the true causes of pellagra, could not have systematically placed certain classes of adoptees into more "pellagragenic" environments. However, if

837. Hutchings & Mednick, 1975, p. 115.
838. Teasdale, 1979, p. 108.
839. Kety et al., 1994, p. 452.
840. Joseph, 2000d.

an agency had placed adoptees into homes corresponding to the socioeconomic status of their biological families, then adoptees born into poor families would have been placed into poorer adoptive homes, where they would have been more likely to develop pellagra. All of this could have occurred even though the adoption agency was unaware of the "unknown variables" of pellagra. And Kety has acknowledged that schizophrenia, like pellagra, is correlated with lower socioeconomic status.[841]

It is critically important to understand that Denmark had a long history of governmental and social support for eugenic practices. As early as 1925, Denmark passed a law forbidding "insane and highly feeble-minded persons" from marrying.[842] In 1929 Denmark became the first country in the world to pass national legislation for the purpose of promoting eugenic sterilization, predating the Nazi sterilization law of 1933 by four years. The 1929 legislation legalized sterilization in cases of mental retardation or mental disorder. Although the word "eugenic" did not appear in the law, it allowed sterilization "where suppression of reproduction must be regarded of being of great importance to society."[843] The law passed easily in the Danish Parliament and, according to historian Bent Hansen, "The Danish version of eugenics seemed to command agreement among all political parties."[844]

Denmark passed another law in 1935 allowing compulsory sterilization in certain cases, typically for people labeled "mentally abnormal."[845] Compulsory eugenic sterilization was now widely supported in Denmark: "While everybody up to and during the passing of the 1929 law had recommended caution, they now spoke of eugenics legislation as something that was urgently needed."[846] According to Hansen, Denmark performed nearly 6,000 eugenic sterilizations between 1929 and 1950, and did not abolish compulsory eugenic sterilization until the 1960s.

Manfred Bleuler, who was well acquainted with European attitudes toward the "hereditary taint" of schizophrenia, gave the following description of the effect of these attitudes on his patients' families:

---

841. Kety et al., 1994, p. 452.
842. S. Hansen, 1925, p. 81.
843. Quoted in B. Hansen, 1996, p. 38.
844. *Ibid.*, p. 39.
845. *Ibid.*, p. 41.
846. *Ibid.*, p. 45.

> If one knows schizophrenics and their families well, it is sometimes a matter for despair to see how much they suffer under the terrible concept of "familial tainting." Like a sinister shadow it darkens the lives of many people and of entire families. The stifling, uncertain fear of coming from an "inferior breed," of carrying within one's self the seeds of something pathological, morbid, and evil (I am speaking in the jargon the afflicted apply to themselves), like a curse that you must pass on to someone else, causes oppressive feeling of inferiority.[847]

And there is good reason to believe that these attitudes prevailed in early- to mid-20th-century Denmark.

There were many more available Danish children than families wanting to adopt them during much of the period when adoptees were placed. According to Mednick and colleagues,

> Many of these adoptions took place during the Great Depression and World War II. It was more difficult to find willing adoptive homes in these periods owing partly to the relative unavailability of adoptive parents and to the additional number of adoptees available.[848]

Apparently, the adoption process was a "buyers' market" during this period of Danish history, and in the words of Mednick and Hutchings, "serious deviance in the biological parents was routinely reported to the prospective adoptive parents unless they refused the information."[849] Mednick discussed case reports of children who were put up for adoption but were never placed: "Every weekend (at least in the 1930s), Danish people who wished to adopt would visit the orphanages and pick children. . . . Children whose selection by an adoptive parent is delayed may be less attractive physically and behaviorally."[850] It is likely that many children were also less attractive *genetically*. Each weekend, Danish prospective adoptive parents arrived at the orphanage and selected children as they might have selected apples at the market. The most qualified parents picked healthy, attractive, and non- "hereditarily tainted" children, as they would have chosen the most attractive apples. The "best" families would have gotten the most desirable children, leaving the less desirable either unadopted or placed with families providing less favorable environments.

847. M. Bleuler, 1978, p. 473.
848. Mednick et al., 1987, p. 78.
849. Mednick & Hutchings, 1977, p. 161.
850. Mednick, 1996, p. 134.

Continuing with the analogy, we might expect that all buyers would have an equal chance to select the best apples. It is unlikely that this was the case in the Danish adoption process, however, because placement officials had the ultimate authority to determine where children would be placed. Most likely, they placed non genetically-stigmatized children into more stable and nurturing family environments, leaving the rest to less stable families.

Kety and colleagues claimed that adoption agencies' knowledge of the parents' psychiatric status did not influence their results because "in practically every case," index adoptees (all of whom were placed between 1924 and 1947) were born to *parents* with no record of mental disorder at the time of adoption.[851] But as we are about to see, a prospective adoptee's biological parents were not the only relatives checked by the agencies for a history of mental disorder. As seen in the 1946-47 annual report of the Mother's Aid Organization of Copenhagen, which was the largest adoption service in the country, the genetic background of a potential adoptee was of great concern to the Danish authorities:

> Before a child is cleared for adoption, it is investigated with respect to health, and an attempt is made to obtain detailed information on the child's family background and to form an impression of its developmental potential. Not only for the adoptive parents, but also for the child itself, these investigations are of great importance for its correct placement. Information is obtained on the child's mother and father; *on whether or not there are serious physical or mental illness in the family background*; criminal records are obtained for the biological parents; and in many cases school reports are obtained. By means of personal interview with the mother an impression of her is formed. Where information is uncovered on convicted criminality or on mental retardation, *mental illness, etc. in the family background*, the case is referred to the Institute of Human Genetics of Copenhagen University, with whom there exists a valuable cooperation for advice on the advisability of adoption [emphasis added].[852]

Thus, potential adoptees were carefully screened for a family history of mental disorders, which went well beyond the biological parents. Where the agencies suspected a family history of mental disorder, they turned the case over to the Institute of Human Genetics, which was the keeper of the National Psychiatric Register. The Psychiatric Register, established in 1937, registered "all

---

851. *Ibid.*, p. 453. No mention of the adoptees' biological parents' diagnostic status at the time of adoption was made in the Kety et al. 1968 or 1975 papers.
852. Mother's Aid Organization for Copenhagen, Copenhagen County and Frederiksborg County, Annual Report for 1946-47. Quoted in Mednick & Hutchings, 1977, p. 163.

mental disorders; psychiatric inpatient services and partial hospitalization."[853] (The world's first psychiatric register was established in Norway in 1936.) Note that these registers were established at a time when eugenic ideas were at their high point in these countries, which suggests that they were intended for use in eugenic programs.[854] It is therefore ironic that psychiatric registers, which enabled the Danish-Adoption studies to be performed, may have been used in determining adoption placements.

It is likely that the Institute of Human Genetics performed an assessment of recorded psychiatric diagnoses among a potential adoptee's biological relatives. Therefore, the fact that an adoptee's biological *parents* had no record of mental disorder at the time of adoption does not diminish the likelihood that placements were influenced by the prevalence of mental disorders *among other family members*. Because adoption agencies viewed mental disorders as having an important genetic component, thus branding a potential adoptee as "tainted," they considered a potential adoptee's family background "of great importance for its correct placement." The agencies believed that children with a family history of mental disorders had poor "developmental potential" — that is, poor *genetic* potential — and probably placed them into less qualified adoptive families.

An indication of selective placement's impact was reported by Lewontin et al. in 1984, who were granted access to the Danish-American raw data. It turned out that a notation had been made in each relative's file if he or she had ever been placed in a mental institution. Lewontin et al. found that in 8 of the 33 index adoptive (rearing) families, but in none of the 34 control adoptive families, a parent had been admitted to a mental hospital at some point. Thus, Kety and colleagues' finding of higher SSD rates among index versus control *biological* relatives might reflect little more than the agencies' placement of children with "tainted" biological relatives into more psychologically harmful adoptive homes. Lewontin and colleagues concluded,

> A credible interpretation of the Kety et al. results [is] that the schizophrenic adoptees, who indeed had been born into shattered and disreputable families, acquired their schizophrenia as a result of the poor adoptive environments into which they were placed. The fact that one's adoptive parent goes into a mental hospital clearly does not bode well for the psychological health of the environment in which one is raised.[855]

853. Häfner & der Heiden, 1986, p. 29.
854. Broberg & Roll-Hansen, 1996.
855. Lewontin et al., 1984, p. 223.

It appears that the single environmental variable assessed by Kety and associates — the prevalence of mental disorders among adoptive family members — showed a significantly higher psychiatric hospitalization rate among index versus control adoptive parents. Mednick, who had intimate knowledge of the Danish adoption process, recognized that in "Denmark, the national adoption agency had a policy of selective placement," although he believed that "the agency was not successful."[856] However, the evidence suggests that the agency was "successful" in many respects.

It therefore is likely that the socioeconomic and psychiatric background of a potential adoptee's biological family influenced Danish adoption agencies' decisions about where to place available children, and potential adoptive parents' decisions about whether to adopt them, leading to the placement of genetically-stigmatized children into more psychologically harmful rearing environments. Thus, Kety and colleagues' 1968, 1975, and 1994 results might be explained on this basis alone.

THE DANISH-AMERICAN ADOPTEES' FAMILY STUDY: 1975

In 1975, Kety and colleagues published the final results of their Copenhagen study, which was based on interviews with the 1968 adoptees and relatives (they added one additional control adoptee and his or her relatives). Although the task of interviewing adoptive and biological relatives proved difficult, the investigators sought to determine if there were SSDs among relatives who had no record of being admitted to a Danish psychiatric facility.

The Danish researchers conducted the interviews, and English language transcripts were sent to Kety, Wender, and Rosenthal in the United States, who made blind global diagnoses. According to the investigators, the interviews

> were extremely exhaustive, 35 pages in length, including many check lists and much narrative material, and covered the major aspects of the life experience: sociological, educational, marital, occupational, and peer relationship history from birth, medical background, and a careful mental status examination. These interviews were transcribed in English . . . and the transcripts were edited to remove any clues which a sophisticated reader might use to guess that this was a biological or adoptive relative of an index case or of a control.[857]

856. Mednick, 1996, p. 134.

There were now 347 identified biological relatives (173 index, 174 control). Kety et al. re-diagnosed a 1968 B3 index adoptee (S3) as B1, meaning that index adoptee diagnoses now consisted of 17 B1, 9 B3, and 7 B2. They identified a "screened control" group of 23 control adoptees (out of 34), which consisted of interviewed control adoptees judged free "from the suggestion of schizophrenic disorder."[858]

The investigators found a statistically significant clustering of SSDs in the index biological relative group versus the control biological relative group (37/173 vs. 19/174, p = .006). Index/control differences for the various SSDs were given as follows: B1 index versus control = 5 to 0; B3 index versus control = 6 to 3; D1+ D2 + D3 index versus control = 13 to 3; and C index versus control = 13 to 13.[859] Of these SSD diagnoses, nearly one-half were "uncertain," and 65% were given to half-siblings. Of the five B1 index biological relatives, only one was a first-degree relative. The remaining 4 B1 diagnoses were half-siblings. The investigators found no significant SSD difference between index and control adoptive (rearing family) relatives.

Although they found significantly more SSDs among their index biological relatives, this finding was, according to Kety et al., "compatible with a genetic transmission for schizophrenia, but it is not entirely conclusive, since there are possible environmental factors such as *in utero* influences, birth trauma, and early mothering experiences which have not been ruled out." They continued that "one cannot, therefore, conclude that the high prevalence of schizophrenia illness found in these biological relatives of schizophrenics is genetic in origin."[860] Unfortunately, this statement did not prevent a generation of textbook authors and commentators from concluding that the higher prevalence of "schizophrenia illness" among these index biological relatives did indeed show that schizophrenia is "genetic in origin."[861]

However, Kety and colleagues pointed to another comparison which, they argued, provided "compelling" evidence in favor of genetic factors:

---

857. Kety et al., 1975, p. 150.
858. *Ibid.*, p. 155.
859. *Ibid.*, p. 154.
860. *Ibid.*, p. 156.
861. Joseph, 2000b.

The largest group of relatives which we have is, understandably, the group of biological paternal half-siblings. Now, a biological paternal half-sibling of an index case has some interesting characteristics. He did not share the same uterus or the neonatal mothering experience, or an increased risk in birth trauma with the index case. The only thing they share is the same father and a certain amount of genetic overlap. Therefore, the distribution of schizophrenic illness in the biological paternal half-siblings is of great interest.[862]

They diagnosed 16 biological paternal half-siblings with a record- or interview-based SSD, finding a "highly unbalanced" distribution (14 index vs. 2 control). Kety and colleagues concluded, "We regard this as compelling evidence that genetic factors operate significantly in the transmission of schizophrenia."[863]

As we have seen, the investigators realized that counting individual relatives might not be "entirely appropriate" due to the clustering of cases "into a limited number of biological and adoptive families."[864] They presented a table which showed, based on their criteria, that index families were more affected in statistically significant numbers.[865] They concluded that the weight of the evidence provided strong evidence in favor of the genetic basis of schizophrenia.

Category C

Kety et al. defined their schizophrenia spectrum in different ways in various statistical comparisons. For example, Lidz and Blatt observed that, although Kety and colleagues *did not* remove category C (schizoid and inadequate personality) from their 1975 schizophrenia spectrum, they decided not to count C diagnoses in what they viewed as their most important statistical comparisons.

---

862. Kety et al., 1975, p. 156.
863. *Ibid.*, p. 156.
864. *Ibid.*, p. 156.
865. When the comparison is limited to the B or "definite" cases, Kety & colleagues reported that the rate of affected index families was significantly higher than the control rate (index 14/33 vs. control 3/34, p = .002; Kety et al., 1975, p. 163). However, this comparison reflected an error on the part of Kety et al. because there were only eight index biological families with a member receiving a 1968 or 1975 B diagnosis, rendering the comparison statistically non-significant (8/33 vs. 3/34, p = .08, Fisher's Exact Test, one-tailed). See Boyle (1990) for more details.

After breaking the code and assigning interview diagnoses to their respective categories, Kety et al. discovered that C diagnoses were equally distributed among index and control biological relatives (index 13, control 13). That they still considered this diagnosis part of the schizophrenia spectrum can be seen in a table showing category C under the "Schizophrenia spectrum" heading, with its totals included in a column entitled "Total in schizophrenia spectrum."[866] Kety and colleagues wrote that although this diagnosis did not differentiate index from control biological relatives, thereby casting doubt on its relationship to B1, they "were not prepared to dismiss the possibility that there is a schizoid or inadequate personality which is genetically related to schizophrenia."[867]

As we have seen, Kety and colleagues found "compelling evidence" for the genetic transmission of schizophrenia in their statistically significant biological paternal half-sibling comparison. However, they had to *exclude C* diagnoses in order to achieve this result. Table 7.3, which I have adapted from data Kety et al. published in 1976, shows that the biological paternal half-sibling SSD comparison — including C — was index 29% (18/63) versus control 17% (11/64), yielding a statistically non-significant difference.[868] The shaded totals reflect spectrum diagnoses as Kety and colleagues defined them in their 1975 study. The "Total Spectrum (Excluding C)" column reflects the spectrum diagnosis distribution *after Kety et al. had removed spectrum category C from the paternal half-sibling comparison in the very same publication.*

---

866. Kety et al., 1975, p. 154.
867. *Ibid.*, p. 155.
868. p = .094; Kety et al., 1976, p. 418, Table 2a.

Table 7.3. The 1975 Danish-American Adoptees' Family Study's "Compelling Evidence" in Favor of Genetics: "Schizophrenia Spectrum" Diagnoses Among Index vs. Control Biological Paternal Half-Siblings

	Schizophrenia Spectrum as Defined by Kety et al.[1]					
	Number of Half-Siblings	Diagnosed B1, B2, or B3	Diagnosed D1, D2, or D3	Diagnosed C	TOTAL SPECTRUM (INCLUDING C) [2]	TOTAL SPECTRUM (EXCLUDING C) [3]
INDEX GROUP Biological Paternal Half-Siblings	63	8	6	4	18	14
CONTROL GROUP Biological Paternal Half-Siblings	64	1	1	9	11	2
Probability*		.015	.055 (ns)	(ns)	.094 (ns)	.001

*Adapted from Kety et al., 1976, p. 418. * Fisher's Exact Test, one-tailed.*

(ns) = Not statistically significant at the .05 level.
B1 = Chronic schizophrenia, B2 = Acute schizophrenia, B3 = Borderline schizophrenia
D1 = Uncertain chronic schizophrenia, D2 = Uncertain acute schizophrenia, D3 = Uncertain borderline schizophrenia
C = Schizoid or inadequate personality.

[1] A table showing that C (schizoid and inadequate personality) was included in the spectrum is found in Kety et al., 1975, p. 154. On page 155, they explained why it was included.
[2] These totals, in addition to C diagnoses among the biological paternal half-sibs, were not provided in the Kety et al. 1975 study. They were first published, in a table without comment, in Kety et al., 1976, p. 418.
[3] This distribution was put forward in Kety, 1974, p. 961, and Kety et al., 1975, p. 156 in support of the genetic position.

Thus, Kety et al. removed category C from their paternal half-sibling comparison *even though it was a schizophrenia spectrum disorder*. Lidz and Blatt called this a "post hoc change in criteria," and correctly called the procedure invalid.[869] The significant difference Kety cited as "compelling evidence" for the genetic transmission of schizophrenia, therefore, turned out to be statistically *non-significant* — a fact which most leading psychiatry and psychology textbooks published since the mid-1970s have failed to mention.[870]

Problems of the Interview Method

Due to death or unavailability, Kety and colleagues were able to interview only 72% of the identified biological relatives, and 48% of the identified adoptive relatives.[871] Although such occurrences are common in studies of this type, they

---

869. Lidz & Blatt, 1983, p. 430.
870. Joseph, 2000b.

counted all 347 identified biological relatives (whether interviewed or not) in the study's statistical calculations, deciding that 347 was "the most unbiased and conservative denominator."[872] This created a problem of how to count relatives who had received a record-based SSD diagnosis in 1968, but who were unavailable for interview. Unfortunately, as Lidz and Blatt pointed out, Kety and colleagues counted these dead or unavailable relatives in an inconsistent manner.[873] For example, they diagnosed one of the three 1968 D (uncertain) relatives who had either died, emigrated, or had refused an interview with a 1975 SSD, whereas the other two were not diagnosed. However, they counted *all* 1968 B and D diagnoses in their 1975 biological paternal half-sibling comparison (see Table 7.3).

The case of control adoptee C9's biological father highlights the investigators' inconsistent and sometimes arbitrary manner of counting family members. Kety et al. diagnosed this person B1 in 1968 on the basis of records, but he had died before he could be interviewed. In 1975 they did not count his B1 diagnosis in their statistical calculations, yet counted him as a non-SSD relative. Adding to the confusion, they counted him as a screened control record-based diagnosis in one table, but not in another table.[874] Although Kety and colleagues' 1975 calculations made it appear as though B1 biological relatives were significantly concentrated in the index group (5 to 0), the difference would have been statistically non-significant (5 to 1) had they decided to include this control relative.[875] Fortunately, this issue has been resolved because in 1988, Kety began counting C9's biological father as a Copenhagen B1 control biological relative.[876]

Of the 364 alive and accessible biological and adoptive relatives, 12 refused to be interviewed. However, the investigators believed they had "adequate information" to be able to diagnose them with an SSD. As they explained it, "even though the individual persistently refused to give an interview, Dr. Jacobsen nevertheless obtained considerable information in the process."[877] A description of how they may have obtained "considerable information" was provided in a 1974

871. See Kety et al., 1975, p. 151.
872. *Ibid.*, p. 153.
873. Lidz & Blatt, 1983.
874. See Table 4b on pages 160-161 in Kety et al., 1975, versus Table 3 on page 154.
875. Index 5/173 vs. control 1/174, p = .10, Fisher's Exact Test, one-tailed.
876. Ingraham & Kety, 1988, p. 122; Kety et al., 1994, p. 452.
877. Kety et al., 1975, p. 150.

paper by Paikin, Kety, Rosenthal, Wender, and others, who described some of the problems encountered in the interview process:

> In those cases where the psychiatrist was not invited into the house, the face-to-face contact varied between a few minutes to twenty minutes. In general, it was found that if the subject was seen for less than five minutes, the amount of information gained was not sufficient for a judgment to be made as to whether the subject was psychiatrically inadequate or not . . .

> It must be stated that much of the information obtained was somewhat superficial; it was necessary to give some weight to a general impression of the subject with particular emphasis being placed on disturbances in emotional contact, language use and thought processes. The diagnostic presentation has, therefore, been restricted to three broad categories: outside the schizophrenia spectrum, suspected schizophrenia spectrum and schizophrenia.[878]

It appears that the investigators diagnosed people with schizophrenia or another SSD on the basis of a *five minute doorstep conversation*. Although Paikin and associates described the interview process of a different Danish-American schizophrenia adoption study, it is possible that several relatives in Kety's Adoptees' Family study were diagnosed in a similar fashion.

Real or fabricated interviews? Kety and colleagues' use of fabricated "pseudo-interviews" was first brought to public attention in 1984. According to Lewontin and colleagues, who had been in correspondence with one of the psychiatrists conducting the interviews,

> In several cases, when relatives were dead or unavailable, the psychiatrist "prepared a so-called pseudo interview from the existing hospital records." That is, the psychiatrist filled out the interview form in the way in which he guessed the relative would have answered.[879]

In other words, Danish researchers *made up* interviews for several relatives. Lewontin and colleagues discussed index adoptee S11's biological mother, who had received a C diagnosis in the 1968 study. By 1975 her "interview" diagnosis had been changed to D3. But in fact, this woman "had committed suicide long before the psychiatrist had attempted to locate her, and so — from the original hospital records — she was 'pseudo-interviewed.'"[880] She had been hospitalized on two different occasions for manic-depression, a diagnosis Kety et al. viewed as genetically unrelated to schizophrenia.[881]

---

878. Paikin et al., 1974, pp. 308-310.
879. Lewontin et al., 1984, p. 225.
880. *Ibid.*, p. 224.

Kety and colleagues' use of pseudo-interviews seems to explain the changing diagnostic status of control adoptee C9's biological father. As we recall, they diagnosed him B1 in 1968 on the basis of records, but gave him no diagnosis in 1975 while still counting him as a control biological relative. His diagnostic status may have been changed on the basis of a posthumous "pseudo-interview."

Kendler and Gruenberg, the authors of a 1984 independent reanalysis of Kety's 1975 results, provided another description of the pseudo-interviews:

> Based on an extensive review of hospital records, detailed pseudointerviews were constructed for all of the index adoptees. These pseudointerviews contained more detailed information on the index adoptees than had been available to Kety and co-workers when they made their initial diagnoses. However, although they contained a detailed account of the psychiatric illness, the hospital records did not contain all the information normally present in sections of the real interviews dealing with such factors as personal history or living environment. This difference in information content as well as other differences in format made it impossible to be "blind" to whether an adoptee interview was a real interview with a control adoptee or a pseudointerview with an index adoptee.[882]

Thus, in addition to relatives, the original index adoptees were also "pseudo-interviewed." Moreover, it appears that Kety and colleagues' "blind" diagnoses may not have been as blind as they implied. Kendler and Gruenberg could tell the difference between a "real" and a "pseudo" interview, and it is possible that Kety, Rosenthal, and Wender could also tell the difference. Because they knew from the 1968 records that there were more index than control biological relative SSDs (13 index vs. 3 control), their recognition of a pseudo-interviewed relative meant that this person was much more likely to be an index relative. Thus, another bias may have been introduced into the diagnostic process.

To summarize, the interview process was plagued by many difficulties. Only 64% of biological and adoptive relatives were actually interviewed, suggesting that the study's subtitle, "A Preliminary Report Based on Psychiatric Interviews," was misleading. Kety's decision to use "pseudo-interviews" in place of a prior decision on how he would count deceased or unavailable 1968 adoptees and relatives is a serious methodological error. He should have discussed and

---

881. Kety et al., 1976.
882. Kendler & Gruenberg, 1984, p. 556.

justified his use of pseudo-interviews in the methods section of all studies where they were used.

We recall that the investigators, after completing the Copenhagen study in 1975, extended their work to encompass "the rest of Denmark." They called this half of the investigation the "Provincial" study. Because the Provincial final results were not published until 1994, most early critics of the Adoptees' Family studies focused on the Copenhagen half of the study. Previously, Kety et al. had published the preliminary Provincial record-based results in 1978.[883] They began the interview process in 1980, reporting preliminary results in 1988.[884] In 1992 they first published diagnoses for the individual family members. This was followed by their 1994 final report, which will be our starting point. [885]

Kety and colleagues selected their Provincial adoptees from a total of 8,944 Danish individuals given up for adoption between 1924 and 1947. These adoptions took place in the Danish provinces outside of the greater Copenhagen area. Like the Copenhagen study, they checked the names of their Provincial adoptees against the Psychiatric Register to determine how many had been admitted to a psychiatric facility. When the records indicated that an adoptee's symptoms were compatible with a diagnosis of schizophrenia, a summary of the institutional records was translated into English and was edited to remove family-related information, and was sent to Kety, Rosenthal, and Wender in the United States, who performed blind consensus diagnoses. This process originally produced 41 index adoptees diagnosed B1, B2, or B3. Later, the investigators removed the eight B2 adoptees.

The researchers established a matched control group, which in 1994 was reported to consist of 24 adoptees and their 121 biological and 55 adoptive relatives. The final index group consisted of 29 B1 (now called "chronic schizophrenia") and 4 B3 (now called "latent schizophrenia") adoptees. The chronic schizophrenia index adoptees had 171 biological and 71 adoptive relatives.

---

883. Kety, Rosenthal, & Wender, 1978; Kety, Rosenthal, Wender, Schulsinger, & Jacobsen, 1978.
884. Ingraham & Kety, 1988.
885. Kety & Ingraham, 1992.

Kety et al. began interviewing relatives in 1980. Although they interviewed nearly 90% of the available relatives, this represented only 63% of all identified relatives.[886] They continued to use the global diagnostic method despite the publication of DSM-III and its "operationalized" diagnostic system.[887] Kety and colleagues diagnosed relatives with two SSDs only: chronic schizophrenia and latent schizophrenia. They diagnosed 8 (4.7%) index biological relatives with chronic schizophrenia, versus zero diagnoses among control biological relatives. In addition, they diagnosed 14 (8.2%) index biological relatives with latent schizophrenia, versus 3 (2.5%) among control biological relatives. Both comparisons reached statistical significance. The combined SSDs (chronic + latent) totaled: index, 13% (22/171) versus control, 2.5% (3/121), leading Kety and colleagues to conclude,

> This study and its confirmation of previous results in the Copenhagen Study speak for a syndrome that can be reliably recognized in which genetic factors play a significant etiologic role. These findings provide important and necessary support for the assumption often made in family studies; observed familial clustering in schizophrenia is an expression of shared genetic factors.[888]

Latent Schizophrenia in the Provincial Study

I have argued here briefly and in detail elsewhere against counting latent or borderline schizophrenia in schizophrenia kinship research.[889] However, the investigators retained it the Provincial study. Although they diagnosed 14 index biological relatives with latent schizophrenia, only 4 were first-degree relatives. The remaining 10 diagnoses were made on half-siblings. Among controls, Kety et al. diagnosed one first-degree and two second-degree biological relatives with latent schizophrenia. Thus, the latent schizophrenia rate among the first-degree biological relatives of chronic schizophrenia index versus control adoptees was not statistically significant.[890] But this is only the beginning of our story.

In 1983, Kety reported that the Provincial control group consisted of 42 adoptees.[891] However, by 1994 there were only 24 control adoptees because Kety

---

886. Kety et al., 1994, p. 445.
887. APA, 1980.
888. Kety et al., 1994, p. 442.
889. Joseph, 2000a.
890. 4/82 index vs. 1/59 control, p = .30, Fisher's Exact Test, one-tailed.
891. Kety, 1983b.

had removed 18 through a screening process quite different from the one he had used in the Copenhagen study. In the Provincial study, Kety decided that all 13 interviewed control adoptees diagnosed with a "serious or confounding mental illness" (primarily non-SSD affective disorders[892]) *should be removed from the study.*[893] It might have been permissible for Kety to remove a control adoptee diagnosed with schizophrenia, but from the genetic perspective non-SSD affective disordered (e.g., depression, manic-depression) controls would not "confound" the study's results.

In contrast, Kety and colleagues retained affective disordered control adoptees in their 1975 study, and even placed them in their "screened control" group. In the 1975 study they had described their screened control group as follows:

> Nine of these we called normal, one was diagnosed neurotic, *one as affective disorder*, and 12 were called personality disorders other than schizoid or inadequate personality. These 23 controls we designate as "screened" to indicate that there was no suggestion of schizophrenic disorder among them [emphasis added].[894]

Apparently, a diagnosis certifying Copenhagen control adoptees as "screened" was enough to remove them from the Provincial control group.[895]

In 1992, Kety and Ingraham discussed the Provincial control group:

> In the Provincial sample, we have found latent schizophrenia in 5 of 37 (13.5%) biological relatives of probands screened out of the study due to a diagnosis of major affective disorder, a rate comparable to that in the biological relatives of the schizophrenic adoptees. It is possible that more precise diagnostic criteria could characterize a specifically schizophrenia-related syndrome.[896]

In other words, Kety excluded affective disordered controls because he diagnosed their biological relatives with SSDs at rates comparable to index adoptees, thereby *assuming* the genetic basis of schizophrenia in the process of

---

892. Kety & Ingraham, 1992.
893. Kety et al., 1994, p. 446.
894. Kety et al., 1975, p. 155.
895. This did not prevent Kety from combining the 1975 Copenhagen screened controls (with an affective disordered adoptee) with Provincial screened controls (affective disordered adoptees removed) to constitute the 47 National sample "Control Adoptees With No History of Major Mental Illness" (Kety et al., 1994, p. 452).
896. Kety & Ingraham, 1992, p. 250.

investigating it. Although diagnosed with a major affective disorder, according to Kety these controls may have been misdiagnosed "schizophrenics" and therefore had to be removed from the study.

Had Kety decided to retain these control adoptees, the index-control latent schizophrenia difference for *all* biological relatives (first- and second-degree) would have been statistically non-significant.[897] This finding was confirmed by Kendler and Diehl (after viewing the final Provincial data while in preparation), who reported that *before* Kety's reduction of his control group, "latent and uncertain schizophrenia was not found to be significantly more common in the biologic relatives of the schizophrenia adoptees than in those of the control adoptees (6.5% vs. 5.5%, respectively)."[898]

In their subsequent independent reanalysis of the 1994 data, Kendler and colleagues blindly re-diagnosed Kety's interviewed Provincial adoptees and relatives. Of the 37 control adoptees, they made no psychiatric diagnosis on 25. Of the remaining 12 controls, they diagnosed 6 with major depression, 5 with anxiety disorder, and 1 with schizotypal personality disorder.[899] Thus, only one control adoptee (schizotypal personality disorder) could be suspected of having an SSD, and none was diagnosed with chronic schizophrenia.

Kety et al. decided to remove five additional Provincial control adoptees on the grounds that they refused (or were unable) to be interviewed, even though they *retained* eleven non-interviewed control adoptees in their 1975 Copenhagen control group. Kety et al. did not disclose whether they found SSDs among these controls' biological relatives, nor did they indicate whether they were aware of these relatives' diagnostic status.

The investigators initiated their Provincial study in 1975, produced their first publication in 1978, began interviewing relatives in 1980 . . . but first reported that they had removed control adoptees *in 1992*. Thus, 14 years elapsed between the control group's ascertainment and its announced reduction in size from 42 to 24.

Lidz and Blatt argued in 1983 that there is little difference between B3, D3, and (non-SSD) C, and this was confirmed in the 1994 study. In 1980, the third edition of the DSM (DSM-III) was published. Its authors moved away from the older global diagnostic method and towards an "operationalized" system based

---

897. 14/171 index vs. 8/158 control, p = .18, Fisher's Exact Test, one-tailed.
898. Kendler & Diehl, 1993, p. 265.
899. Kendler, Gruenberg, & Kinney, 1994, p. 458.

on standardized criteria and categorical diagnoses. A new DSM category, "schizotypal personality disorder" (SPD), was created from the differentiating symptoms of Kety and colleagues' Copenhagen B3, D3, and C relatives.[900] According to DSM-III diagnostic criteria, a person (failing to meet criteria for schizophrenia) should be diagnosed with SPD if they manifest at least four of the following eight symptoms: (1) "magical thinking," (2) "ideas of reference," (3) "social isolation, e.g., no close friends of confidants," (4) "recurrent illusions," (5) "odd speech (without loosening of associations or incoherence), e.g., speech that is digressive, vague, overelaborate, circumstantial, metaphorical," (6) "inadequate rapport in face-to face interaction due to constricted or inappropriate affect, e.g., aloof, cold," (7) "suspiciousness or paranoid ideation," (8) "undue social anxiety or hypersensitivity to real or imagined criticism."[901]

Kety sometimes described SPD as comparable to latent schizophrenia. For example, "The components of the spectrum have assumed different names in DSM-III: schizotypal personality disorder takes the place of our latent or uncertain schizophrenia, from which its characteristics were derived."[902] In Kety and Ingraham's 1992 update, they used the terms "schizotypal personality disorder" and "latent schizophrenia" interchangeably.

Schizotypal personality disorder carries the distinction of being, in the words of psychiatric investigators Gunderson and Siever, "the first diagnostic category introduced into standard diagnostic usage that is built on a genetic rationale and explicitly gives familial relationship primacy as a validating criterion."[903] The SPD architects viewed the diagnosis as "merely a subdivision of what has for years been referred to as Schizoid Personality Disorder."[904] By 1983, however, Kety had removed schizoid and inadequate personality (category C) from the spectrum: "There was . . . no justification for believing that schizoid and inadequate personality, as we had diagnosed them in the interview study, were related to schizophrenia, and were therefore excluded from the subsequent analyses."[905] According to the DSM-III, the only symptom differentiating SPD from schizoid personality was the former's "eccentricities of communication or behavior."[906] And in 1978, Kety and associates recognized that "it is doubtful

---

900. Spitzer & Endicott, 1979.
901. APA, 1980, p. 313.
902. Kety, 1983a, p. 724.
903. Gunderson & Siever, 1985, p. 532.
904. Spitzer & Endicott, 1979, p. 98.
905. Kety, 1983a, p. 723.

that we could demonstrate a significant differentiation between" latent schizo-phrenia, uncertain schizophrenia, and schizoid/inadequate personality.[907] Thus Kety and colleagues' remarkable admission that they could not differentiate between a "schizophrenia spectrum disorder" and a "non-spectrum disorder," even though one aspect of their study's validity depended upon their ability to reliably make such a distinction.

In 1994, Kety et al. provided a table listing the "course an symptoms" of Provincial relatives diagnosed with latent schizophrenia.[908] This table showed how few symptoms were necessary to diagnose someone with latent schizo-phrenia, and how often SSD latent schizophrenia was essentially indistin-guishable from non-SSD schizoid personality.[909] According to my calculations, only 5 of 17 latent schizophrenia relatives met the criteria for the supposedly comparable DSM-III schizotypal personality disorder. Moreover, Kety et al. diagnosed two relatives on the basis of *one* major symptom, diagnosing a member of family #420 on the sole criterion of "inappropriate or constricted affect," and another, from family #936, on the basis of "odd or digressive speech." A diag-nosed member of family #871 further illustrates the lack of any meaningful dis-tinction between latent schizophrenia and schizoid personality. The investigators diagnosed this person with latent schizophrenia on the basis of (1) "no close friends/seclusive/withdrawn," and (2) "suspicious/paranoid ideation." However, they described *schizoid* personality as "most frequently reflect[ing] severe introversion, suspicious or referential symptoms in otherwise psychiatri-cally unremarkable individuals."[910] In other words, the difference between SSD latent schizophrenia and non-SSD schizoid personality was known only to the practitioners of the "global diagnostic system."

How much further could Kety and associates have strayed from the teachings of the schizophrenia concept's inventor, Eugen Bleuler, who believed that schizophrenia should be diagnosed conservatively. "Only a few isolated psy-chotic symptoms can be utilized in recognizing the disease," wrote Bleuler, "and these too, have a very high diagnostic threshold value."[911] On the other extreme,

---

906. APA, 1980, p. 310.
907. Kety, Rosenthal, & Wender, 1978, p. 220.
908. Kety et al., 1994, p. 447.
909. *Ibid.*, p. 447.
910. *Ibid.*, p. 445.
911. E. Bleuler, 1950, p. 294.

Kety diagnosed someone with an SSD manifesting the single symptom of "inappropriate or constricted affect."

In Kendler and colleagues' 1994 reanalysis, the SPD rate among all Provincial index biological relatives versus all control biological relatives was not statistically significant.[912] Limiting the comparison to first-degree relatives only, the difference remained non-significant.[913]

The evidence suggests that, as in 1968 and 1975, Kety and colleagues should not have counted "latent schizophrenia" as schizophrenia. Their results confirmed Lidz and Blatt's 1983 observation that latent schizophrenia describes non-psychotic people considered odd and reclusive, who are not easily distinguished from people diagnosed with non-SSD schizoid personality.

Chronic (B1) Schizophrenia in the Provincial Study

Although B3 or latent schizophrenia is unworthy of being counted as schizophrenia, Kety et al. diagnosed eight Provincial index biological relatives with chronic (B1) schizophrenia, versus zero control biological relatives. As always, these figures require closer examination. Of the eight relatives they diagnosed with chronic schizophrenia, two were second-degree relatives (maternal half-sibs).[914] The remaining six diagnoses were given to one biological mother, two biological fathers, and three biological full-siblings. Regarding the siblings, the investigators wrote that, whereas the 1975 Copenhagen study suffered from a lack of biological full-siblings "that prevented a meaningful estimate of disease prevalence in first-degree relatives" (although Kety et al. studied 131 biological parents, only one of whom they diagnosed B1), in the Provincial sample,

> there were a substantial number of full siblings (24 biological index siblings compared with three in the Copenhagen Sample), probably the result of differences in lifestyle from that of the large city. Schizophrenia was found in three (12.5%) of the 24 full siblings, nearly six times the rate in the half-siblings.[915]

However, a chronic schizophrenia rate of 3/24 (12.5%) among these biological full-siblings is misleading, because *all three were reared in the same family.*[916]

---

912. Index 10/140 vs. control 5/162, p = .09, Fisher's Exact Test, one-tailed.
913. Index 7/51 vs. control 3/60, p = .10, Fisher's Exact Test, one-tailed.
914. Kety et al., 1994, p. 448.
915. *Ibid.*, pp. 447-448.
916. *Ibid.*, p. 448, Figure 1.

And what a family this was! Index biological family #069 consisted of the adoptee's biological mother, father, five full-siblings, and three paternal half-siblings. As always, the researchers provided no information on how or where these children were reared.

In addition to adoptee #069's three biological full-siblings diagnosed by Kety et al. with chronic schizophrenia, one of the two remaining full-sibs was diagnosed with non-SSD "bipolar illness." Of the three paternal half-siblings in family #069, they diagnosed two with non-SSD schizoid personality. Because the Provincial sample was drawn from the small towns and rural communities of Denmark, one can envision this family living together in the Danish countryside, possibly on a farm. The parents (neither of whom received a psychiatric diagnosis) were perhaps hard working people tending to their land, but from a psychosocial perspective they raised some pretty screwed-up kids. What kind of a household could this have been in which, of the eight children, three were diagnosed with SSD chronic schizophrenia, two with non-SSD schizoid personality, and one with non-SSD bipolar illness? And what were the circumstances surrounding the abandonment of a child (the index adoptee) by parents who raised eight biological children together? Unfortunately, we can only speculate due to the investigators' failure to provide information about any of these people's *environments*.

Of the remaining 20 index biological full-siblings, who were born to six different sets of biological parents, *none* was diagnosed with chronic schizophrenia.[917] This suggests that family environment, rather than genetic background, best explains the clustering of chronic schizophrenia among three children reared in the same family. This underscores Benjamin's and Mosher's (and to a certain degree Kety's) understanding that counting diagnoses among individual relatives, rather than among families, is misleading because the common environment shared by family members violates the assumption of the independence of individual observations.

Kety and colleagues gave their three remaining first-degree biological relative chronic schizophrenia diagnoses to one mother and two fathers. One of these was the biological father of index adoptee #139, whose diagnosis suggests that the investigators did not require distinguishing symptoms of schizophrenia in order to make a chronic schizophrenia diagnosis. According to a table listing

---

917. Although Kety et al. counted 24 index biological full-sibs, the pedigree chart on page 448 of their 1994 paper indicates that there were 25.

the course and symptoms of the chronic schizophrenia adoptees, adoptee #139 manifested the following five symptoms: (1) "insidious onset," (2) "schizoid fea-tures observed premorbidly," (3) "chronic course observed," (4) "withdrawal from social interaction," and (5) "flat affect."[918] Other symptoms *not* manifested by this adoptee included "autistic behavior," "poverty of thought/speech," "loose associations," "suspicious/ideas of reference," "delusions," "auditory hallucina-tions," and "other hallucinations." This adoptee, therefore, presented no uniquely psychotic symptoms, or even one symptom that would have differentiated him from a person diagnosed with latent schizophrenia or schizoid personality. Nor would he have been diagnosed with chronic schizophrenia by the criteria of DSM I through DSM-IV-TR. This individual's chronic schizophrenia diagnosis is therefore questionable; in fact, it calls into question the validity of every chronic schizophrenia diagnosis in the entire Adoptees' Family series.

Of the original 42 index adoptees (29 of whom Kety et al. diagnosed with chronic schizophrenia), Kendler and colleagues, in their independent reanalysis, diagnosed only 19 with chronic schizophrenia.[919] Although it is unclear how many of these overlapped with Kety and colleagues' 29 diagnosed adoptees, Kendler et al. diagnosed at least 10 of the latter with conditions other than chronic schizophrenia. Among the 28 index first-degree biological relatives of chronic schizophrenia adoptees, Kendler et al. diagnosed 2 with chronic schizo-phrenia. Among the 60 first-degree biological relatives of controls, they diag-nosed 1 case.[920]

* * *

The Provincial study, as we have seen, was plagued by serious method-ological problems and bias. Moreover, how many relatives did Kety and col-leagues diagnose after a five minute doorstep interview? How many did they diagnose by pseudo-interview? How many index adoptive parents had been admitted to a mental hospital? How many deceased hospital-diagnosed relatives did Kety et al. exclude from statistical calculations? In short, as Lewontin and associates argued in 1984 when discussing the Provincial study in its preliminary phase, "there is no reason to suppose that the more recent work is free of the invalidating flaws we have outlined above."[921] Most importantly, Kety and col-

918. *Ibid.*, p. 446.
919. Kendler, Gruenberg, & Kinney, 1994, p. 458.
920. *Ibid.*, p. 460; p = .24, Fisher's Exact Test, one-tailed.
921. Lewontin et al., 1984, p. 225.

leagues' Provincial study results are confounded by the selective placement of adoptees, meaning that adoption studies' critical "no selective placement" assumption was violated. Therefore, like the 1968 and 1975 investigations, the Provincial study's massive flaws and biases invalidate its authors conclusions in favor of genetic influences on schizophrenia.

The "National Sample"

In 1994, Kety and associates calculated the results of what they called the "National sample," meaning the combined diagnoses of the Copenhagen and Provincial studies. They reported highly significant differences for chronic and latent schizophrenia rates among all index biological relatives versus controls.[922] However, as they acknowledged, the Provincial study constituted a "replication of the Copenhagen study in the rest of Denmark."[923] By its very nature, a replication attempt must stand on its own merit; otherwise, two statistically non-significant results could be combined into one large significant result, or a non-significant replication could be added to the significant original study and thus lead to a completely different conclusion. As others have written, the "independence of samples is essential to the interpretation of replication research."[924]

ROSENTHAL AND COLLEAGUES' DANISH-AMERICAN ADOPTEES STUDY

As we recall, the Adoptees method differs from Kety's Adoptees' Family method because it begins with parents, and then diagnoses their adopted-away biological offspring (see Figure 7.2). David Rosenthal and colleagues' Adoptees study began with a group of 5,483 Danish adoptees placed between 1924 and 1947, and from that group identified about 10,000 of their biological parents. These parents' names were then checked against the Psychiatric Register of the Institute of Human Genetics. Case reports were assembled in Denmark and were sent to Rosenthal, Wender, and Kety in the United States. When they arrived at a consensus B1, B2, or B3 diagnosis, that person became an index case. The investigators then retrieved information on the child this index parent had

---

922. Kety et al. 1994, p. 452.
923. *Ibid.*, p. 442.
924. Gottfredson & Hirschi, 1990, p. 55.

given up for adoption, which was sent to the examiners in Copenhagen. Of the 69 index adoptees identified in the 1968 study, 30 could not be interviewed for reasons such as refusal, death, and emigration. This left a total of 39 index adoptees, who were interviewed by two Danish psychiatrists involved in the study (J. Welner and F. Schulsinger). The investigators established a control group consisting of the adopted-away biological offspring of 47 people with no history of admission to a psychiatric hospital. These adoptees were matched with the index group on the basis of sex, age, age at transfer to the adopting family, and socioeconomic status (see Figure 7.2).

Rosenthal based his study on four assumptions: (1) "That heredity was an important contributor to schizophrenia," (2) "That this inherited factor manifested itself in the behavior or personality of persons who were not frankly schizophrenic," (3) "That these manifestations could be detected by tests or in one or two interviews," and (4) "That we would know which questions to ask or which tests to use to detect these manifestations."[925] Although he found no statistically significant results in the 1968 study, in a subsequent textbook Rosenthal concluded, "The data provide strong evidence indeed that heredity is a salient factor in the etiology of schizophrenic disorders."[926] By 1971, Rosenthal and colleagues were reporting statistically significant results on the basis of an expanded sample of 76 index and 67 control adoptees. Of these, they diagnosed 24 index and 12 control adoptees with a schizophrenia spectrum disorder (SSD), a difference just reaching statistical significance.[927]

The Inclusion of Manic-Depressive Subjects

Even though Rosenthal viewed schizophrenia and manic-depression (currently known as "bipolar disorder") as "genetically distinct and different disorders,"[928] and that in 1976, Kety, Rosenthal, et al. wrote that "manic-depressive illness was never thought to be in the schizophrenia spectrum by us,"[929] one or all of the judges diagnosed 11 of the 69 (16%) index biological parents with

---

925. Rosenthal et al., 1968, p. 380.
926. Rosenthal, 1970b, p. 129.
927. Rosenthal et al., 1971, p. 310.
928. Rosenthal, 1971a, p. 124.
929. Kety et al., 1976, p. 417. This is a curious claim because Rosenthal's 1971 results listed "manic-depressive psychosis" as one of the diagnoses "that we are tentatively including in the 'schizophrenia spectrum'" (Rosenthal et al., 1971, p. 309).

manic-depressive disorder.[930] Why did this "genetically distinct and different disorder" qualify a parent as an index case? Rosenthal offered two explanations:

> These [manic-depressive] cases were included for two reasons. *The first was one of expediency.* There were periods when we simply did not have enough schizophrenic parents processed and the staff in Copenhagen had no subjects to examine. The second and more important reason derived from this question: What if we should find differences between our Index and Control groups. . . . If we had a comparison pathology group, we might be able to learn something. . . . about the possible genetic relationship between schizophrenia and manic-depressive psychosis [emphasis added].[931]

Thus, Rosenthal included an admittedly non-related diagnosis in a schizophrenia genetic study *for reasons of expediency.* Imagine a study on the genetics of heart disease that added cancer patients because the staff "had no subjects to examine." Of course, it would have been fine to study people diagnosed with manic-depression *in addition to* people diagnosed with schizophrenia, but it was erroneous to *count* them as "schizophrenia." According to Alvin Pam, the inclusion of manic-depressive subjects rendered this study "invalid on its face."[932]

Although Rosenthal maintained that learning about the relationship between schizophrenia and manic-depression was the most important reason for including the latter disorder, the evidence suggests that "expediency" was a more important factor. Had Rosenthal been able to obtain enough SSD index parents to complete his sample, he would have had no need to look outside of the spectrum for additional cases. At this point, observed Lidz and colleagues, "the study could no longer properly be termed a study of adopted-away offspring of schizophrenic parents."[933] It is also difficult to reject the idea that manic depression was placed "tentatively" in the spectrum because, as we will see, the index/control SSD diagnosis difference would not have reached statistical significance without it.

---

930. Rosenthal et al., 1968, p. 382.
931. *Ibid.*, p. 382.
932. Pam, 1995, p. 31.
933. Lidz et al., 1981, p. 1064.

Selective Placement

The problem of selective placement I discussed in relation to the Adoptees' Family studies applies to all studies using adoptees placed in mid-20th century Denmark, and is especially problematic in Rosenthal's Adoptees study.

Because about 87% of the parents had their first psychiatric hospitalization well after their child was born, Rosenthal argued that his study was only "minimally complicated" by parents who may have been psychotic before the birth of their child.[934] Although the 87% figure might appear to have minimized the possibility of selective placement on the basis of a parent's mental health status (in direct contrast to Heston's 1966 study, where *all* adopted children were born to institutionalized women diagnosed with schizophrenia), the Danish adoption agencies, as we have seen, checked the psychiatric records of the biological *families* of potential adoptees, not just their parents. Therefore, Rosenthal needed to demonstrate that the existence of psychiatric disorders among index biological family members did not create conditions leading to the placement of index adoptees into more psychologically harmful adoptive families. That such placements likely occurred is because, *from either a genetic or environmental perspective*, the biological family of a person diagnosed with schizophrenia — or a person later diagnosed with schizophrenia — would be expected to have more psychiatrically diagnosed members than a control biological family. This means that in societies (such as Denmark, with its Psychiatric Register) where adoption agencies use the psychiatric status of an adoptee's biological family as a placement criterion, the Adoptees method is confounded by selective placement and, therefore, constitutes an invalid research model.

In 1994, Kety and colleagues wrote,

> The possibility of knowledge on the part of adoptive parents, adoptees, and mental health personnel regarding the presence of mental illness in the biological parents and the effect such knowledge could have on the occurrence, perception, and diagnosis of mental illness in the adoptee. . . . could be of some significance in studies where the adoptee sample represented children born of mentally ill mothers.[935]

---

934. Rosenthal et al., 1971, pp. 308-309.
935. Kety et al., 1994, p. 453.

Thus, Kety acknowledged that the results of Adoptees method schizophrenia studies could be influenced by the knowledge of an adoptee's biological
background.

Diagnostic Procedure

Welner and Schulsinger interviewed and tested subjects in Copenhagen
and provided, in lieu of a consensus SSD diagnosis, a preliminary "thumbnail
diagnostic formulation" for each. All 1968 and 1971 diagnoses, statistical comparisons, and conclusions in favor of genetics were based on these thumbnail formulations. By the end of the interview process, Rosenthal would claim that "none of
the examiners knew if the subject before him was an index or control case,"[936]
although four years later it would develop that in "perhaps a few cases" they did
know the group status of the subject.[937] According to Lewontin et al. (who had
been in personal correspondence with several of the collaborators), these
thumbnail formulations did not specify whether an individual should or should
not be counted as an SSD, a decision that eventually was made "in a manner and
by parties unknown."[938] Some examples of thumbnail diagnoses from the 1971
study include "possible paranoid borderline," "paranoid character," "almost
pseudoneurotic borderline," "moderately schizoid," "pronounced preschizo-
phrenic diathesis," and "conceivably paranoid borderline."[939] In statistical calculations, the researchers counted all these diagnoses as "schizophrenia."

Assumptions

According to the investigators, the purpose of their study was to test the
components of an "assumed diathesis" (predisposition) of schizophrenia, and as
we recall from the study's four assumptions, the first was that "heredity was an
important contributor to schizophrenia." In 1983, Kety reaffirmed this position:

> It is important . . . to explain what sometimes appear to be differences in the
> prevalence of one or another diagnosis between our "adoptee" studies
> (Rosenthal et al., 1968, 1971) and the present studies on families (Kety et al.,
> 1968, 1975). In the former studies *the emphasis was on the components in the diathe-*

---

936. Rosenthal et al., 1971, p. 308.
937. Rosenthal, 1975b, p. 21.
938. Lewontin et al., 1984, p. 226.
939. Rosenthal et al., 1971, pp. 309-310.

sis for schizophrenia, and we adopted a low threshold for their notation. Diagnosis was a secondary consideration. . . . In the family studies, on the other hand, because of the hypothesis being tested, the emphasis was placed on diagnosis [emphasis added].[940]

Thus, the Adoptees study assessed the components of an *assumed* predisposition for schizophrenia, which explains that study's broader definition of schizophrenia. This is a perfectly acceptable explanation, provided that the investigators draw no conclusions in favor of an *already assumed* genetic predisposition for schizophrenia. However, in 1971 the investigators concluded that "the evidence supports the theory that heredity plays a significant role in the etiology of schizophrenia spectrum disorders."[941] This conclusion in not valid, as a goal of the study was to investigate the components of an assumed genetic basis of schizophrenia. The study thus revolved around the investigators' circular argument in which they assumed that which they concluded, and concluded that which they assumed.

Kety also claimed in 1983 that "we did not make independent and consensus diagnoses in the 'adoptee' studies but merely cited the investigator's diagnostic impressions."[942] This was true in the 1968 and 1971 studies but, as we will see, the researchers reported consensus diagnoses in a 1978 article of which Rosenthal and Wender were co-authors. Although by 1983 Kety seemed to downplay the importance of Rosenthal's 1968 and 1971 studies, they have been cited in countless textbooks and review articles as providing important evidence in favor of the genetic basis of schizophrenia.

The Failure to Find Statistically Significant Index-Control Differences

Although in 1971 the investigators claimed statistically significant findings, Lidz and Blatt found statistically non-significant results simply by removing the non-SSD manic-depressive index parents and their biological children.[943] And clearly, further reanalysis is in order. Another parental diagnostic category, "B/D/M/A," cannot be included because the judges could not agree if these parents should be diagnosed as either (B) "process schizophrenia," (D) "doubtful schizophrenia," (M) "manic-depressive psychosis," or (A) "not in the schizophrenia

---

940. Kety, 1983a, p. 724.
941. Rosenthal et al., 1971, p. 310.
942. Kety, 1983a, p. 724.
943. p = .075; see Lidz et al., 1981, p. 1066.

spectrum."[944] We must also exclude "paranoids," which the researchers never counted as an SSD. Furthermore, we have already seen that the investigators eventually removed category C from the spectrum.[945] Interestingly, in 1975 Rosenthal defended the inclusion of category C in his study while at the same time (and in the same book) arguing that it should *not* be counted in Kety's studies:

> In the Kety study, which we call the Family study, the findings strongly favor the view that diagnostic categories such as uncertain schizophrenia and borderline schizophrenia are genetically related to classical process schizophrenia, but schizoid personality is not. Thus, we now talk about a hard spectrum and a soft spectrum. . . . Undaunted, I want to show you why I think the soft spectrum, which includes a number of syndromes that we call schizoid, is indeed genetically related to process schizophrenia.[946]

In their jointly authored publications of the late 1970s, Kety and Rosenthal wrote that although the evidence suggested that category C was unrelated to chronic schizophrenia, the evidence was not conclusive.[947] This allowed Kety to remove C diagnoses from certain comparisons in his study, and Rosenthal to retain them in his, and allowed both to claim statistical significance for the comparisons they considered most important. Contrary to Rosenthal's reasoning, however, a "spectrum" diagnosis cannot be appropriate for one study but not the other (both of which, I should stress, were co-authored by Rosenthal). Category C is either genetically related to schizophrenia, or it is not.

Additionally, we must remove B2-diagnosed parents and adoptees because the Danish-American investigators removed B2 from the spectrum in 1978.[948] This means that Rosenthal's 1971 results, based on the Danish-American investigators' own post-1983 definition of the schizophrenia spectrum as consisting only of chronic schizophrenia and latent schizophrenia, were not statistically significant (12.5% index vs. 7.5% control).[949] Therefore, apart from methodological problems, Rosenthal and colleagues' "thumbnail diagnosis" comparison

---

944. Rosenthal et al., 1971, p. 309.
945. Kety, 1983a; Kety et al., 1994.
946. Rosenthal, 1975a, p. 201.
947. For example, see Kety, Rosenthal, Wender, Schulsinger, & Jacobsen, 1978.
948. Kety, Rosenthal, & Wender, 1978
949. Index = 6/48 (12.5%), control = 5/67 (7.5%), p = .35, Fisher's Exact Test, one-tailed.

was not statistically significant, further illustrating that their study provided no evidence in support of genetics.

Later Reanalyses

Haier, Rosenthal, and Wender, 1978. Two reanalyses of Rosenthal's data appeared in subsequent years. The first, published in 1978, was co-authored by Haier, Rosenthal, and Wender.[950] This study reported an analysis of MMPI scores ("Minnesota Multiphasic Personality Inventory") for 128 adoptees from the 1971 study (64 index, 64 control). All index adoptees' biological parents had been hospitalized for "schizophrenia," so apparently Haier et al. excluded the manic-depressive index parents and their adopted-away children.

The investigators made several comparisons of index and control MMPI profiles, all of which failed to differentiate the two groups. They produced a table of MMPI clinical scale raw scores in which neither the index nor the control group had a significant elevation on Scale 8 (Schizophrenia). Moreover, there were no significant differences on two MMPI-derived psychosis scales — "Eichman Schizophrenia Signs" and "Peterson Signs." Finally, the researchers devised a formula that produced statistically significant results by comparing differences between index and control adoptees receiving an SSD *and* having an elevated score on *any* MMPI clinical scale.[951] Using this formula, they were able to show that 66% of index adoptees, versus 25% of controls, met these criteria. The difference is statistically significant, leading Haier et al. to conclude that their study "continued to support the genetic hypothesis." Still, they recognized that "on the basis of the MMPI data, there are no overall personality differences between the index and control groups."[952]

Unfortunately, there was no indication that the investigators agreed on their procedures and comparisons *before* they collected and analyzed their data. At first, Haier, Rosenthal, and Wender described a process in which they found no differences between their index and control adoptees. Finally, they devised a comparison providing statistically significant results in the genetic direction, and based their conclusions on *this* comparison. "The authors," wrote Boyle,

---

950. Haier et al., 1978.
951. T score (≥ 70).
952. Haier et al., 1978, p. 175.

"were by now arbitrarily searching for anything which would discriminate index and control groups."[953]

This study provided the first published report of index and control consensus schizophrenia spectrum diagnoses. As we have seen, Rosenthal reported only thumbnail diagnoses in his 1968 and 1971 studies. Although Rosenthal's results were preliminary, most reviewers have focused on them and ignored (or were unaware of) the consensus SSD diagnoses reported in 1978 by Haier and associates (reproduced in Table 7.4), who found no significant difference (33% vs. 25%) between index and control adoptees on the basis of an investigator-defined schizophrenia spectrum consisting of chronic schizophrenia, borderline schizophrenia, and "schizophrenic personality" (the latter included schizoid and paranoid personalities).[954] Haier et al. failed to highlight these *negative* results, discussing them mainly in the context of adoptees' MMPI profiles.

Table 7.4. Index and Control Adoptees Receiving Consensus "Schizophrenia Spectrum" (SSD) Diagnoses as Defined by Rosenthal et al.			
Diagnosis	Index Adoptees	Control Adoptees	Probability*
Chronic Schizophrenia	3	0	.12 (ns)
Borderline Schizophrenia	10	7	.30 (ns)
Schizophrenic Personality	8	9	(ns)
Non-SSD Diagnoses	21	21	
No Diagnosis	22	27	
Totals	64	64	
Schizophrenia Spectrum (SSD)	21 (33%)	16 (25%)	.22 (ns)

Source: Adapted from Haier, Rosenthal, & Wender (1978, p. 174).
* *Fisher's Exact Test, one tailed. (ns) = not statistically significant at the .05 level.*
Diagnoses in bold type indicate schizophrenia spectrum diagnoses (SSDs) as defined by Haier et al.

In 1970 Rosenthal had written that "through psychological tests and intensive psychiatric interview, we hope to be able to discriminate the two groups [index and control] with respect to a number of psychological and behavioral characteristics."[955] As the 1978 results show, Rosenthal was *unable* to discriminate the two groups on the basis of tests and interviews.

Lowing, Mirsky, and Pereira, 1983. A final reanalysis, this time using DSM-III diagnostic criteria, was published by Lowing, Mirsky, and Pereira in 1983. Here, the investigators reduced the index and control groups to 39

---

953. Boyle, 1990, p. 153.
954. Haier et al., 1978, p. 174, Table 3.
955. Rosenthal, 1970a, p. 255.

adoptees each by including only those adoptees whose biological parents had been diagnosed B1, B2, B3, or "mixed schizophrenia symptoms."[956] Two raters blindly reviewed the available information on these adoptees, making SSD diagnoses which included chronic schizophrenia, schizotypal personality disorder, paranoid personality disorder, borderline personality disorder, and "mixed spectrum disorder," which was defined as overlapping "at least two of the five spectrum disorders but did not meet the full criteria for any one of the five."[957] Adoptees whose diagnoses fell into this broadly defined spectrum were called "disordered subjects." Lowing et al. provided no justification for broadening the spectrum, nor did they explain why they included the offspring of three parents diagnosed B2 (a diagnosis which Kety et al. had removed from the spectrum in 1978) in the study.

The investigators diagnosed 15 index adoptees as "disordered subjects," versus 5 among controls, a statistically significant difference. When they narrowed their comparison to chronic schizophrenia, schizotypal personality disorder, and schizoid personality, the difference remained significant. If the comparison is limited to chronic schizophrenia and schizotypal personality disorder, however, the results are not statistically significant.[958] In fact, Lowing et al. diagnosed only one adoptee with chronic B1 schizophrenia.

Ingraham and Kety wrote in 1988,

> Using DSM-III criteria, Lowing et al. found three times as many schizophrenia spectrum disorders in the index group as in the control group. These authors specified DSM-III schizophrenia, schizotypal personality disorder, and schizoid personality disorder as the schizophrenia spectrum.[959]

But because he had removed schizoid personality in 1978, Kety himself *did not* define the spectrum this broadly. Thus, he should have concluded that Lowing and colleagues found no statistically significant differences between Rosenthal's index and control adoptee groups.

According to Lowing et al., "On the basis of the findings reported here of the significantly greater risk for schizophrenia spectrum disorders among the index subjects, we conclude that the hypothesis has been strongly sup-

---

956. Lowing et al., 1983, p. 1168.
957. *Ibid.*, p. 1169.
958. Index 7/39 vs. control 3/39, p = .16, Fisher's Exact Test, one-tailed.
959. Ingraham & Kety, 1988, p. 125.

ported."[960] But the fact remains that they identified only one case of DSM-III schizophrenia among the adoptees, and even with the questionable inclusion of DSM-III schizotypal personality disorder, they found no statistically significant results. Their study, in fact, confirmed the *negative* results of the previous studies.

Additional Problems with Rosenthal's Study

There are several other problems with Rosenthal's study which, in the interest of brevity, I chose not to examine. These include his failure to find index schizophrenia rates significantly higher than the general population expectation as a prerequisite for being able to generalize results to the non-adoptee population; his questionable practice of adding two samples together (1968 + 1971) into one total, as opposed to treating them as separate samples; his failure to describe his adoptees' rearing environments; his use of adoptees who spent months or years with their biological mothers before being transferred to their adoptive families; his use of homosexuality as a diagnostic criterion for the B1, B2, and B3 diagnoses;[961] and that being diagnosed as a "pervert" qualified as an SSD.[962] Rosenthal's Adoptees study was methodologically unsound to the extreme, and his bias is seen at every step and in every conclusion. This bias also blinded him to the confounding influence of selective placement.

THE FINNISH ADOPTIVE STUDY OF SCHIZOPHRENIA[963]

Pekka Tienari has made the study of schizophrenia his life's work, beginning with his well-known twin study, and culminating in a longitudinal study of adoptees who had been given away by their schizophrenia spectrum diagnosed biological mothers. Tienari and colleagues are unique among schizophrenia adoption researchers because, in addition to genetics, they assess their adoptees' family rearing environments. The Finnish study is, by a large margin, the most comprehensive and methodologically sound schizophrenia adoption

---

960. Lowing et al., 1983, p. 1170.
961. Kety et al., 1968, p. 352. Also see Joseph, 2000a.
962. Rosenthal et al., 1968, p. 387. This index adoptee's complete thumbnail diagnosis read "border-line schizophrenia or pervert."
963. This section is based on an updated and revised version of a previous publication (Joseph, 1999a). See also Joseph, 2004.

study to date. Therefore, the investigators' conclusion that genetic factors influence schizophrenia must be carefully examined.

Tienari and colleagues have criticized the Danish-American investigators for assessing their adoptees' family environments "in a very limited manner."[964] And although they wrote in 1994 that the Danish-American studies have been "misinterpreted as a final proof of genetics, though they only propose suggestions and trends,"[965] in 2004 they concluded, "Earlier adoption studies have confirmed convincingly the importance of a genetic contribution in schizophrenia."[966] Tienari and colleagues believe that schizophrenia's genetic component is real but overemphasized, and that family environmental factors have been given insufficient attention. "The major goal of the Finnish Adoptive Family Study," wrote Tienari and colleagues, "is to reassess genetic contributions to schizophrenia and to add measures of the adoptive family rearing environment."[967]

The investigators' general approach has been:

(1)To present evidence that the adopted-away children of mothers diagnosed with an SSD manifest SSDs at a significantly higher rate than controls, thereby establishing the genetic component, *then*,

(2)To note that virtually all adoptees diagnosed with schizophrenia have been raised in chaotic or disturbed adoptive families. Thus, "genetically predisposed" children may be more vulnerable to receive this diagnosis when raised in such families, whereas healthy rearing environments may serve to protect them from schizophrenia. Both genetic background *and* family environment are seen as predictor variables of schizophrenia.

Method

There is no national adoption register in Finland, and the arduous task of collecting the names of hospitalized women diagnosed with schizophrenia produced 9,832 such patients between 1969 and 1972. Because this sample was too small to provide enough children adopted-away at an early age, Tienari collected a second sample, which included the entire psychiatric population of Finland, plus consecutive hospital admissions from 1960-1979. These two samples yielded

---

964. Tienari et al., 1985, p. 21.
965. Tienari et al., 1994, p. 25.
966. Tienari et al., 2004, p. 216.
967. Tienari, Sorri, et al., 1987, p. 477.

information for a total of 19,447 women diagnosed with schizophrenia or "paranoid psychosis." Through the use of local population registers, Tienari identified the adopted-away offspring of biological mothers diagnosed with these two conditions. All offspring had been placed in non-relative homes and had gone through formal adoption procedures. From this group, the researchers excluded several adoptees for various reasons, leaving a total of 190 index adoptees.

Tienari and colleagues established a matching control group on the basis of sex, age of placement, age of adoptive parents, and social status. Adoptees were eligible to be controls if their biological mother had no record of having been treated for psychosis. About one-half of the biological fathers were identified, and adoptees were excluded if these fathers were found to have been treated for psychosis. Index and control adoptees were matched independently by persons not involved in the study. By 2003, 192 control adoptees were matched against 190 index adoptees with SSD-diagnosed biological mothers.

The investigators performed joint interviews with the entire adoptive family, followed by separate interviews with the adoptive couple only. Parents and adoptees were given a battery of psychological tests. Where possible, the biological parents were also interviewed and tested. The interviewers and testers were blind to the group status of the families with whom they worked.

Tienari et al. outlined five different types of adoptive families: (1) "healthy," (2) "mildly disturbed," (3) "neurotic," (4) "rigid, syntonic," and (5) "severely disturbed."[968] Their decision to assess adoptees' rearing environments was a huge step forward compared with Heston's and the Danish-American investigators' practice of ignoring family environments (other than to record psychiatric diagnoses). Moreover, as a longitudinal study, they could observe family interaction before a child was diagnosed with schizophrenia. As Tienari wrote,

> The Finnish program will be the first attempt in schizophrenia research to combine the direct adoptive family strategy and the risk research strategy to study adoptive families and adoptees prospectively, beginning prior to the onset of illness in the offspring.[969]

---

968. Tienari, Lahti, et al., 1987, pp. 40-41.
969. Tienari, 1991, p. 465.

Results

Tienari and colleagues created a "narrow spectrum" of schizophrenia spectrum disorders, which included "schizophrenia," "schizoaffective disorder," "schizophreniform disorder," and "schizotypal personality disorder." They also made diagnoses for "broad spectrum" SSDs, which included the above diagnoses plus "paranoid personality disorder," "schizoid personality disorder," "delusional disorder," "bipolar psychosis," and "depressive psychosis." Of the 190 index adoptees whose biological mothers were diagnosed with DSM-III-R (Revised)-based SSDs, 37 (19.5%) were diagnosed with an SSD, versus 8 (4.2%) SSDs among the 192 control adoptees, a statistically significant difference.[970]

The investigators concluded in 2000 that "the genetic liability to 'typical' DSM-III-R schizophrenia is decisively confirmed. Additionally, the liability also extends to a broad spectrum of other psychotic and non-psychotic disorders."[971] However, although they, like Kety, Rosenthal, et al., based their conclusions on counting all SSDs as "schizophrenia," limiting the comparison to chronic schizophrenia, as schizophrenia was traditionally diagnosed in Finland, produces different results.[972] In their 2003 publication, Tienari et al. diagnosed DSM-III-R schizophrenia in 7 out of 137 index adoptees (5.1%) whose biological mothers also were diagnosed with DSM-III-R schizophrenia, versus 3 of 192 (1.6%) control DSM-III-R schizophrenia diagnoses.[973] The index/control difference in these chronic schizophrenia rates is not statistically significant.[974] Thus, Tienari and colleagues found no statistically significant clustering of chronic schizophrenia in their index adoptee group. In a companion study, Wahlberg and associates compared scores of a subsample of index and control adoptees on the "Thought Disorder Index" (TDI), which according to the investigators is "the most widely used test measure of thought disorder."[975] The results showed no significant differences between index and control adoptee TDI scores, and that both groups' scores were about the same as normal subjects. However, on the basis of other comparisons Wahlberg concluded that genetic factors influence some aspects of thought disorder.

---

970. Tienari et al., 2003.
971. Tienari et al., 2000, p. 433.
972. In their 2000 publication, Tienari et acknowledged that Finnish diagnosticians traditionally used "restrictive diagnostic criteria" (Tienari et al., 2000, p. 441).
973. Tienari et al., 2003.
974. 7/137 vs. 3/192, p = .065, Fisher's Exact Test, one-tailed.
975. Wahlberg et al., 2000, p. 128. Also see Wahlberg et al., 1997, 2001.

Because index adoptees who grew up in seriously disturbed adoptive families were diagnosed with SSDs more often than control adoptees reared in these types of adoptive families, Tienari and colleagues concluded that both genes *and* adoptive family rearing environment are "predictor variables" for schizophrenia. Thus, as they wrote in 2004,

> In adoptees at high genetic risk of schizophrenia, but not in those of low genetic risk, adoptive-family ratings were a significant predictor of schizophrenia-spectrum disorders in adoptees at long-term follow-up.[976]

The correlation between adoptive family disturbance and the index SSD rate was so great that, at least through 1987, Tienari could report that "all adoptees who had been diagnosed either as schizophrenic or paranoid had been reared in seriously disturbed adoptive families."[977]

Selective Placement

Tienari has recognized that a "major assumption" of adoption studies "is that genetic background and the rearing environment are not correlated, in terms of factors relevant to the development of the specific psychopathology in question."[978] Here we will review evidence that, like the Oregon and Danish studies, the perceived genetic background of many adoptees *was* correlated with their rearing environment in ways that could have made them more vulnerable to being diagnosed with schizophrenia or an SSD for non-genetic reasons.

Like Denmark and Oregon, Finland had a long history of eugenics-inspired legislation aimed at curbing the reproduction of people labeled mentally retarded and mentally ill.[979] Eugenic ideas took root in Finland during the 1920s, and a government commission was created in 1926 to look into the desirability of promoting the sterilization people seen as "mentally retarded," "mentally ill," or epileptic. In 1935 the Finnish parliament passed the Sterilization Act, which allowed the compulsory sterilization and castration of "idiots," "imbeciles," and the "insane," which included people diagnosed with schizophrenia and manic-depression.[980] The Finnish National Board of Health made the decision of

976. Tienari et al., 2004, p. 216.
977. Tienari, Sorri, et al., 1987, p. 482.
978. Tienari, 1992, p. 53.
979. The discussion of eugenics in Finland is based largely on the historical research of Hietala, 1996.
980. Hietala, 1996, p. 232.

whether or not to sterilize.[981] The law permitted compulsory sterilization if there was reason to believe that a person's condition could be genetically transmitted to his or her children. The widespread support for this law is evidenced by the fact that only 14 out of 200 members of Finnish Parliament voted against it. The Castration Act was passed in 1950, permitting the compulsory castration of criminals, the mentally retarded, and the "permanently mentally ill." It wasn't until the Abortion Act of 1970 that compulsory eugenic sterilization was legally abolished in Finland.

Tienari's adoptees were born between 1927 and 1979, and therefore most were placed when eugenic ideas were widespread in Finland and sterilization for eugenic purposes was permitted by law. A child born to a person with a family history of schizophrenia was seen as someone who should "never have been born," and as someone who represented a threat to the purity of the Finnish gene pool. Those who did make their way into the adoption process were seen as carriers of the "hereditary taint of mental illness."

Significantly, up to one-third of Tienari's index adoptees were placed *after* their mother was diagnosed as psychotic.[982] Because two-thirds were placed before their mothers' psychosis, Tienari argued that in "most cases . . . it is not likely that information about psychosis could have influenced the placement,"[983] as it undoubtedly did in Heston's study. Although Tienari considered two-thirds of his index adoptees to be "most cases," this still leaves many index adoptees placed when their mother's psychiatric status was known by the adoption agency, and probably also by prospective adoptive parents.

The confounding influence of the adoptive parents' knowledge of a child's biological heritage is evidenced in one of the few case histories presented by the research team. In 1987 Tienari et al. briefly discussed an adoptee named "P," who was raised in a "severely disturbed" family. They did not list P's psychiatric status, but her case history suggests that she was not diagnosed with a psychotic disorder. More important is the apparent impact of her biological family's psychiatric history on her rearing environment:

---

981. Hemminki et al., 1997, p. 1877.
982. Tienari, Lahti, et al., 1987, p. 44.
983. *Ibid.*, p. 45.

The adoptive father of P, a girl of 19, was a primary school teacher; the mother was a kindergarten teacher. They had been married during the war in 1943 after hardly any contact except by correspondence. The father had threatened to commit suicide unless the mother married him. The parents had been emotionally distant from each other throughout their marriage, devoting themselves to P and four older, biologic children. Before P's adoption, the youngest biologic child had started school, and the marriage seemed about to break down. The adoption of P was a conscious effort to save the marriage. Ever since her infancy, P had been the bond between her adoptive parents, understanding both of them and solving their conflicts, while both parents had leaned upon and parentified her. *At the same time, however, both parents attributed to P's biologic background some of the potential madness they had explicitly feared in themselves. They spoke of their strenuous efforts to protect her from becoming crazy, and their relief that they had been able to preserve her sanity for so long* [emphasis added].[984]

There are several points worth making about P's adoptive family. First, her parents do not appear to have been typical candidates for the adoption process, and their turbulent history may have prevented them from being permitted to adopt an "untainted" child. It is also possible that this couple intentionally sought out a potentially psychotic child in order to help them resolve their own "explicit fears" of going mad. Second, the fact that they made "strenuous efforts to protect her from becoming crazy" implies that P's biological family history of schizophrenia had become a central preoccupation of the family. From a family systems perspective, we might say that P's biological family history was a major factor in her becoming "triangled" into her adoptive parents' relationship. A control adoptee, free from the perceived predisposition to madness, would not face this type of potentially "crazy-making" family environment.

A third point is that P's parents *expected* her to "become crazy." For them, it was a question of when, not if. Given the prevailing attitudes in Finland, one might ask the same question posed by Kringlen in 1987 in relation to Heston's study: What type of parents would want to adopt such a child? As we have seen, up to one-third of Tienari's index adoptees were born to mothers already diagnosed with a psychotic disorder. He did not state how many of the remaining two-thirds had a known history of mental disorder in their biological families, and we have already seen that the Adoptees method is problematic in societies (such as Denmark) where genetic ideas were strong and psychiatric records

---

984. Tienari, Lahti, et al., 1987, p. 43.

were available. If adoptees' biological families were similarly checked in Finland, this would constitute a serious problem for Tienari's study, as well. As we saw in Rosenthal's study, on either environmental or genetic grounds, a psychotic or pre-psychotic person would be expected to have more psychiatrically diagnosed biological relatives than a "normal" person.

At an earlier point, Tienari recognized that adoptive parents' knowledge of the biological mother's psychosis could influence his results. In 1975, he wrote that separation from the biological mother

> ought to have been complete, in the sense that neither the child nor his adoptive parents should have had any contact with the child's biological mother, nor should they even have been aware of her psychosis. The group meeting these criteria was still too small.[985]

This passage demonstrates both Tienari's understanding of the impact of the child's and adoptive parents' "awareness," and the likely reason he was compelled to violate his standard. Like Kety, Tienari found it necessary to change his criteria when confronted with the problem of small sample sizes. Critics could sympathize with researchers who have invested enormous amounts of time and effort in their search for the causes of schizophrenia. The same critics, however, would be abdicating their responsibility should they also fail to point out the potentially confounding and invalidating nature of such changes in criteria.

Genes and Environment

Tienari and colleagues have failed to see the larger picture presented by their data. Instead, they have concluded in favor of genetics without first assessing whether environmental factors influenced the results upon which their genetic conclusions were based. We have seen that adoption researchers must assume that agency policies did not lead to the placement of their index adoptees into inferior rearing environments. If this assumption is untrue, a higher schizophrenia or SSD rate among index adoptees versus controls could be the result of their having been systematically placed into these types of environments. A way of illustrating this is to look at *favism* (G6PD deficiency), which requires both genetic and environmental factors, and which, unlike schizophrenia, is a real disease. Gottesman and Shields described the favism gene-environment interaction as follows:

---

985. Tienari, 1975, pp. 34-35.

Favism, a hemolytic anemia that follows the eating of fava or broadbeans, provides a textbook example of a genotype X environment interaction. Only those persons with the particular X-linked G6PD enzyme variant develop favism, and then, only after eating the bean. Both the gene and the bean are necessary for the disease to appear, neither alone is sufficient, and the disease is *both* a genetic and an environmental one.[986]

Elsewhere, Gottesman pointed out that a favism adoption study looking at genetically vulnerable children adopted by residents of a "nonbean" community could erroneously conclude that favism lacks a genetic component.[987] Both genetic predisposition and the consumption of fava beans are "predictor variables" of favism — with one or none, there is no favism; with both, there is.

If the principles of this "textbook example" are applied to Tienari's data, a problem becomes apparent. Tienari has concluded that genetic background and dysfunctional rearing environments are predictors of schizophrenia. However, although "genes and beans" predict favism virtually 100% of the time, Tienari and his colleagues diagnose SSDs in index adoptees raised in severely disturbed homes only 15-18% of the time.[988] One would think that adoptees exposed to two major theorized predictor variables of schizophrenia would have a much higher rate. It follows that the investigators cannot account for most of the prediction variability of schizophrenia. What explains the fact that roughly 80% of the "at risk" adopted-away biological offspring of people diagnosed with an SSD, who were also raised in severely disturbed adoptive homes, were *not* diagnosed by Tienari et al. with an SSD? Clearly, it is possible that a large and unknown factor or set of factors, related to both the biological background (though not necessarily itself biological) and the adoptive family environment of the child, is wholly responsible for an adoptee being diagnosed with a schizophrenia spectrum disorder. Thus, a failure to gauge the interaction of all aspects of the Finnish adoption process, including selective placement, has led Tienari to the unwarranted conclusion that genetic factors are a cause of schizophrenia.

As Tienari has acknowledged, it is difficult to determine the direction of causality in adoptive parent and adoptee dysfunction. About this, Kendler has written,

---

986. Gottesman & Shields, 1976b, p. 447.
987. Gottesman, 1978.
988. For example, Tienari et al. (1991, p. 49, Table 2) reported that only 8 (17.8%) of the 45 index adoptees raised in "severely disturbed" adoptive homes were diagnosed as "spectrum psychotic."

Most of the adoptive families [in Tienari's study] were evaluated when their adoptees were older than 15. It may have been the disturbed adoptees who affected their adoptive families, not the other way around. Only prospective studies examining families when the adoptees are quite young can hope to disentangle the direction of causality of disturbances in the adoptees and the adoptive family.[989]

This is an interesting argument when we realize that Kendler defends the validity of the twin method in schizophrenia research because he believes that family environments do not contribute to the causes of schizophrenia (see Chapter 3). Although when speaking of *adoptees*, Kendler recognizes that family members can contribute to each other's "disturbances," he fails to recognize that parental behavior can lead to disturbances in *twins*. He takes differing positions on this question depending on whether he is talking about twins or adoptees, and, as always, his position supports the genetic argument.

Age of Transfer (Abandonment)

An additional flaw in Tienari's study is that some adoptees were placed as late as four-years-old.[990] Although the investigators have presented evidence that adoptee mental health ratings are not associated with the age at adoption,[991] the trauma of parental abandonment and placement in temporary homes or institutions was greater for these late-placement adoptees. It appears that a fair number of adoptees spent considerable time with their biological parents or in foster homes, although the researchers provided little information on this period in their early lives.[992] In 1987 Tienari recognized that children adopted away at age four "were doubtful as to their suitability for a study discriminating between genetics and environment,"[993] and 12 years earlier he had written that a child's separation from the biological mother should have occurred "in no case later than the age of 3."[994] Although the number of late-placed adoptees apparently was not large, in adoption studies every case is

---

989. Kendler, 1986, pp. 36-37.
990. Tienari, Sorri, et al., 1987, p. 479.
991. Tienari et al., 1985.
992. Breggin, 1991; Jackson, 2003; Jacobs, 1994; Lehtonen, 1994.
993. Tienari, Sorri, et al., 1987, p. 478.
994. Tienari, 1975, p. 34.

important because it can potentially make the difference between statistically significant and non-significant results.

Attachment theorists point to the negative psychological impact of a rupture in the parent-child bond, even in cases of less-than-nurturing parents. One could argue that a child separated from his or her mother at or near birth would not be psychologically damaged in the process, but a child ripped from its caregivers at age one, two, or three is truly an "abandoned child." Matching controls on the basis of age-of-placement or separation cannot eliminate this problem.

Conclusion

Tienari is the creator of the most well-planned and executed schizophrenia adoption study. However, statistically significant diagnostic differences which might exist between his index and control adoptees are confounded by environmental factors such as selective placement and late placement. In addition, like the Danish-American studies, his finding of significantly higher rates of schizophrenia among his index adoptees depends on the questionable use of a "schizophrenia spectrum of disorders." Tienari and colleagues have concluded that disturbed family rearing environments can predict which children will receive a schizophrenia diagnosis. However, the evidence does not support their additional conclusion in favor of genetics.

SCHIZOPHRENIA ADOPTION RESEARCH: SUMMARY AND CONCLUSIONS

Earlier, I mentioned two broad categories of problems in schizophrenia adoption research: selective placement, and unsound methodology and bias. In these concluding comments I will briefly summarize my argument around these issues.

If we look at schizophrenia adoption research in the context of the social and political environments in which it was performed, we can conclude the following: *The great majority of adoptees studied in schizophrenia adoption research were given up for adoption at a time when the compulsory sterilization of "schizophrenics" for eugenic purposes was permitted by law in the country or state in which their adoptions took place (Denmark, Finland, Oregon).* Leaving aside all other problems, the evidence suggesting the occurrence of the selective placement of adoptees in these studies is reason enough to reject any conclusions in favor of genetic factors, until convincing evidence is produced that placements were not influenced by the genetic stigmatization of index adoptees on the basis of their perceived biological background. Unfortunately, adoption researchers rarely discuss selective placement, other than to briefly dismiss its impact on their results.

There is a parallel between adoption researchers' denial that selection factors influenced their results, and twin researchers' handling of the equal environment assumption. Most twin researchers now recognize that identical twins share a more similar environment than fraternals. Yet they claim that, after all, it doesn't really matter. The claims of some adoption researchers are similar. Although Mednick recognized that the Danish adoption agencies had a "policy of selective placement," the Danish-American schizophrenia adoption researchers acted as if it didn't matter. One of my goals in this chapter has been to show that, in fact, it *did* matter.

If genetic inferences from these studies are invalidated by selective placement, the Oregon and Danish-American studies are equally invalidated by unsound methodology and bias. Investigators such as Kety, Rosenthal, and Wender, were intent on confirming their strong genetic views. As seen clearly in their published works, they changed definitions, comparisons, and ways of counting to ensure that they would find what they were looking for. It is not a matter of fraud, but rather of how conclusions drawn from ostensible scientific experiments are transformed into a statement of the investigators' *beliefs*. Genetic research has a long history of these types of conclusions, going all the way back to Galton. If the early schizophrenia adoption studies helped "break a paradigm," then the paradigm they helped gain acceptance must be reevaluated as soon as possible.

Tienari's study cannot and should not be dismissed as easily as its predecessors. He and his colleagues were the first to make diagnoses on the basis of standardized criteria, and they made a conscious effort to assess their adoptees' rearing environments. They also put in place safeguards not used in the previous studies. However, the four main errors of the Finnish investigators have been, (1) their failure to recognize the influence of selective placement on their results; (2) their inclusion of late-adopted children; (3) their unwarranted acceptance of a broad definition of schizophrenia; and (4) their failure to seriously consider completely non-genetic explanations of their results.

In conclusion, the results from family, twin, *and* adoption studies provide little scientifically acceptable evidence that genes influence the appearance of a set of behaviors given the name "schizophrenia." It is much more likely that "schizophrenic" behavior is the way that some people respond to having experienced "seriously disturbed families," and seriously disturbing social and political environments. "All symptoms of schizophrenia," wrote psychologist Bertram Karon, "may be understood as manifestations of chronic terror and defenses against terror."[995] According to Szasz,

> Every "mental" symptom is a veiled outcry of anguish. Against what? Against oppression, or what the patient experiences as oppression. The oppressed speak in a million tongues. . . . What of the psychiatrist or of others who wish to help such a person? Should they amplify the dissent and help the oppressed shout it aloud? Or should they strangle the cry and reoppress the fugitive slave? This is the psychiatric therapist's moral dilemma.[996]

Calling this response a "disease" or a "genetic disorder" delays the discovery of environmental factors and perpetuates the belief that people diagnosed with schizophrenia are a problem for society. In reality, society is a problem for them.

---

995. Karon, 1999, p. 3. See also Karon & VandenBos, 1981.
996. Szasz, 1968, p. 52.

CHAPTER 8. IS CRIME IN THE GENES? A CRITICAL REVIEW OF TWIN AND ADOPTION STUDIES OF CRIMINAL AND ANTISOCIAL BEHAVIOR

> Crime is a medical problem in at least half of its extent; and I look forward to the day when the physician will be looked on as the criminologist of the country, and when our police courts and our police administration will be put largely under medical control. . . .We have the humiliating spectacle of the defective criminal insane of our community . . . because we have never taken the steps to stop this at its fountain head by preventing the reproduction of defectives.
>
> — American physician J. N. Hurty in 1909.[997]

> Expunge the criminal, but leave the environment untouched: this has always been the message of the biological or genetic determinist point of view.
>
> — Biologist and historian Garland E. Allen in 2001.[998]

In today's "everything is genetic" atmosphere, even traits usually thought of as being the product of environmental influences are subject to the claims of hereditarians. It should therefore come as no surprise that, increasingly, criminal behavior is seen as the result of defective genes.[999] In this chapter I focus on twin and adoption studies of "criminal," "antisocial," and "psychopathic" behavior as examples of how genetic research is used in the study of human behavioral differences.[1000] Topics beyond the scope of this chapter include the "XYY" controversy of the 1960s and 1970s, and molecular genetic research. Suffice it to say that, apart from claims made about Brunner and colleagues' 1993 study of one Dutch family, no genes have been found to cause criminal or antisocial behavior.[1001]

---

997. Discussion in Sharp, 1909, pp. 1900-01.
998. Allen, 2001, p. 210.
999. This chapter is based on a revised and updated version of a previous publication in *The Journal of Mind and Behavior* (Joseph, 2001c).
1000. The terms "antisocial" and "psychopath" are political and moral judgments that have been transformed into "mental illness." Were it not for the fact that it would make for more difficult reading, I would place these terms in quotation marks every time they are used.
1001. Brunner et al., 1993; Rowe, 2002; Wassermann & Wachbroit, 2001. For a review of the "XYY fiasco," see Hubbard & Wald, 1993.

When people read about an alleged genetic link to crime, they are apt to conclude that "crime is genetic." This leads to a belief that criminality is the result of genetic makeup — and not of racism, poverty, and other oppressive social conditions — thereby influencing public attitudes toward ethnic groups having a relatively high conviction/incarceration rate.

As a prelude to a discussion of twin and adoption studies, it is worthwhile looking into the historical background of the "genetics of criminality" question. This is the subject of the following section.

SOME EARLY VIEWS OF THE GENETICS OF CRIMINALITY

Like other types of socially disapproved behavior, the inherited basis of criminality was widely accepted long before adoption, twin, or even family studies had been performed. In the 19th century, Italian physician Cesare Lombroso argued that criminals could be identified by their physical features and that they represented, to use Gould's description in *The Mismeasure of Man*, "evolutionary throwbacks in our midst."[1002] According to Lombroso,

> Many of the characteristics presented by savage races are very often found among born criminals. Such, for example, are: the slight development of the pilar system; low cranial capacity; retreating forehead; highly developed frontal sinuses . . . the thickness of the bones of the skull; enormous development of the maxillaries and the zygomata . . . greater pigmentation of the skin; tufted and crispy hair; and large ears. To these we may add the lemurine appendix; anomalies of the ear; dental diastemata; great agility; relative insensitivity to pain; great visual acuteness; ability to recover quickly from wounds; blunted affection; precocity as to sensual pleasures . . . laziness; absence of remorse; impulsiveness; physiopsychic excitability; and especially improvidence, which sometimes appears as courage and again as recklessness changing to cowardice.[1003]

For the eugenicist Charles Davenport, criminals represented "our . . . apelike ancestors" from "animalistic strains," who should be rooted out of the American breeding stock:

1002. Gould, 1981, p. 124.
1003. Lombroso, 1968, pp. 365-366.

The acts of taking and keeping loose articles, of tearing away obstructions to get at something desired, of picking valuables out of holes and pockets, of assaulting a neighbor who has something desirable or who has caused pain or who is in the way, of deserting a family and other relatives, of promiscuous sexual relations — these are crimes for a twentieth century citizen but they are the normal acts of our remote, ape-like ancestors. . . . Imbecility and "criminalistic" tendency can be traced back to the darkness of remote generations in a way that forces us to conclude that these traits have come to us directly from our animal ancestry and have never been got rid of. . . . If we are to build up in America a society worthy of the species *man* then we must take such steps as will prevent the increase or even the perpetuation of animalistic strains.[1004]

Today, the publicly stated views of Lombroso and Davenport are an embarrassment to genetic researchers, but the suggestion of criminals as "evolutionary throwbacks" remains in some quarters.

The role of genetic factors in crime was widely accepted during the late 19th and early 20th centuries.[1005] As an American physician wrote in 1914, "That a criminal father should beget a child pre-destined to criminality is a foregone conclusion. The father exerts a hereditary influence equal to all the previous ancestors in the paternal line."[1006] A physician at Sing Sing prison in New York held this opinion:

My own observations, which have been practically unlimited along lines of information connected with the male offender, have led me to believe, in the last few years, that criminal character depends in the first instance on heredity. . . . So the born criminal is the product, mind and body, of the forces of heredity. Not only his body, but his mind is deeply impressed with the character of the parentage. And few indeed are the criminals who come to our prison at Sing Sing with minds that were at birth *tabula rasa*, whose mental powers at birth were not already thickly sown with seeds of crime.[1007]

There were frequent calls for the sterilization of criminals on eugenic grounds during this period. Vasectomy was proposed in the late nineteenth century by A. J. Ochsner as a "humane" alternative to castration.[1008] Ochsner jus-

---

1004. Davenport, 1911, pp. 262-263. See Allen, 2001, for a critical discussion of Davenport's views on crime.
1005. See Fink, 1938, for a review.
1006. Hall, 1914, p. 87.
1007. Irvine, 1903, p. 750.

tified this procedure on the grounds that it was "demonstrated beyond a doubt that that a very large proportion of all criminals, degenerates and perverts have come from parents similarly afflicted."[1009] He believed that a large-scale sterilization of criminals "would do away with hereditary criminals from the father's side," and recommended the same treatment for "chronic inebriates, imbeciles, perverts and paupers."[1010]

In 1907, Indiana became the first of many US states to pass a law permitting compulsory eugenic sterilization. The law sanctioned sterilization "to prevent procreation of confirmed criminals, idiots, imbeciles, and rapists" residing in a state institution, and who had been judged as "unimprovable" by a panel of physicians.[1011] A prison physician, Harry Sharp, had performed vasectomies on inmates in Indiana since 1899. According to Sharp, "There is no disputing the fact that mental as well as physical defects are transmitted to the offspring. . . . The decidedly defective individual is very easily recognized, as the mental abnormality is usually accompanied with prominent physical defects, described by Lombroso."[1012] In a 1909 discussion of Sharp's article in the pages of the *Journal of the American Medical Association*, Dr. J. N. Hurty, who had recently visited the estate of a wealthy family, added the following comments:

> I was standing near the man who was in charge of the beautiful collies at the kennels; one of them (a female) came up to me, and she looked so pleasant that she seemed to me to have a laugh on her face. I patted her on the head, and she was duly grateful for the attention. I asked him, "Do you have any vicious dogs here?" He said, "Do you suppose that we would breed from vicious animals? *If a vicious animal appears here we kill it*; we have nothing to do with them at all; and the result is that we have no biting animals, but only those amenable to instruction." *Why cannot we apply this to the human family?* [emphasis added].[1013]

While the call to kill (and thereby, as imagined, eliminate) "vicious" strains in the "human family" was an extreme view even for that time, a glance at the bibliography of a 1938 review of the period's literature on the causes of crime demonstrates the common concern over the procreation of "criminals" and

---

1008. Ochsner, 1899.
1009. *Ibid.*, p. 867.
1010. *Ibid.*, p. 868.
1011. Quoted in Reilly, 1991, pp. 46-47.
1012. Sharp, 1909, pp. 1897-1898.
1013. *Ibid.*, p. 1901.

"defectives."[1014] The consensus was that society could greatly reduce criminality and antisocial behavior by preventing the reproduction of people manifesting these behaviors.

The period 1877-1919 saw the publication of several histories of American "degenerate families,"[1015] the most well-known being Richard Dugdale's *The Jukes*, and Henry Goddard's *The Kallikak Family*. For many years these "studies" were cited as proof that criminality and "feeble-mindedness" were hereditary conditions. For Lombroso, Dugdale's book provided "the most striking proof of the heredity of crime."[1016] According to Davenport, Dugdale showed that criminal family pedigrees could be the result of "a single focal point" of bad heredity, which in this case was "traced back to Max [Juke] living in a lonely mountain valley."[1017] Davenport created pedigrees of his own and concluded, "The foregoing cases are samples of scores that have been collected and serve as fair representations of the kind of blood that goes into the making of thousands of criminals in this country."[1018]

Today, the Jukes and Kallikaks "studies" are largely discredited because, among other reasons, it is widely recognized that poverty, criminality, and illiteracy can run in families for social and environmental reasons. That being said, I should point out that Dugdale and Goddard recognized the importance of environmental conditions. According to Dugdale :

> Where the organization is structurally modified, as in idiocy and insanity, or organically weak as in many diseases, the heredity is the preponderating factor in determining the career; but it is, even then, capable of marked modification for better or for worse by the character of the environment. . . . Where the conduct depends on the knowledge of moral obligation (excluding idiocy and insanity), the environment has more influence than the heredity. . . . For instance, where hereditary kleptomania exists, if the environment should be such as to become an exciting cause, the individual will be an incorrigible thief; but if, on the contrary, he be protected from temptation, that individual may lead an honest life, with some chances in favor of the entailment stopping there.[1019]

1014. Fink, 1938.
1015. See Rafter, 1988.
1016. Lombroso, 1968, p. 161.
1017. Davenport, 1911, p. 183.
1018. *Ibid.*, p. 92.

And Goddard wrote that "there is every reason to conclude that criminals are made and not born," while adding that the "best material out of which to make criminals, and perhaps the material from which they are most frequently made, is feeble-mindedness."[1020]

Few modern proponents of a hereditary basis for criminal behavior claim that there are "genes for crime." Rather, they argue that people inherit predispositions for personalities that make them more likely to commit crime. According to Goldsmith and Gottesman,

> Notions such as "genes for crime" are nonsense, but the following notion is reasonable: There may be *partially* genetically influenced *predispositions* for basic behavioral tendencies, such as impulsivity, that in certain experiential contexts make the *probability* of committing certain crimes higher than for individuals who possess lesser degrees of such behavioral tendencies.[1021]

Still, despite disclaimers of this type, the results of genetic research lead many to believe that criminal behavior is caused by faulty genes.

DEFINING CRIMINAL AND ANTISOCIAL BEHAVIOR

Any study of the genetics of criminal and antisocial behavior is confronted with the problem of how to define these concepts. Not surprisingly, political and moral considerations come into play. As eugenicist Paul Popenoe pointed out in an article written during the Great Depression, "A few years ago the man who had a bag of gold in his safe was a thrifty and praiseworthy citizen; today he is a criminal."[1022] Criminality, observed critics Hubbard and Wald, is a social construct that depends on context. "Killing," they wrote, "can be heroism or murder," and "taking someone's property can be confiscation or theft."[1023] Even Lombroso would have likely applauded the deeds of his "criminal men," had

---

1019. Dugdale, 1910, p. 65. Commenting on Dugdale's book, Fink (1938, p. 179) wrote, "Perhaps no one book in the field of criminology in America has lent itself to such partisan interpretation as has Richard Dugdale's *The Jukes*, published in 1877. Unread, misread, or willfully distorted, it has been used by hereditarians and environmentalists alike to assert and supposedly to prove their respective positions."
1020. Goddard, 1927, p. 54.
1021. Goldsmith & Gottesman, 1996, p. 9.
1022. Popenoe, 1936, p. 388.
1023. Hubbard & Wald, 1993, p. 105.

their "savagery" only been directed at the opponents of the Italian army's colonial campaigns.

Another problem is that being a registered or convicted criminal is dependent on one's being apprehended and charged. According to Rutter, "Many surveys have shown that, at some time, almost all boys commit acts that fall outside the law and which could have led to prosecution if they had been caught."[1024] Surveys in Scandinavian countries such as Norway, Sweden, and Denmark, where some of the most frequently cited twin and adoption studies were performed, show that most young men, when answering anonymously, admitted to committing one or more criminal offenses at some point.[1025] If these surveys are representative of the population, one could conclude that most Scandinavian males are genetically predisposed to commit criminal acts! One might argue that genetic studies of registered or convicted criminality look for genetic factors in the apprehension for, rather than the commission of, criminal acts. According to sociologist Troy Duster, "if one looks at the record in 250 years of US history, no white man ever committed the crime of rape on a black woman in twelve southern states."[1026] Another critic wrote that an illegal act one day can become legal the next, and that the reporting of certain types of crime varies widely. He also noted that wealthy people usually have better legal representation than the poor. His conclusion: "In using any sort of recorded crime figures, researchers are taking on something that is very messy indeed, and that is one reason why confusion will follow."[1027]

In this chapter I analyze the most frequently cited twin and adoption studies of criminal and antisocial behavior. We have seen that family studies can at best demonstrate "scientifically" what society already knows — that the relatives of criminals are more likely to be criminals than are the relatives of non-criminals. According to adoption researchers Brennan, Mednick, and Gabrielli, "In terms of genetics, very little can be learned from . . . family data alone," because "The parents have a major influence on the child's environment as well as on his/her genetic makeup; family studies cannot disentangle these hereditary and environmental influences."[1028]

1024. Rutter, 1996, p. 2.
1025. Hurwitz & Christiansen, 1983, pp. 42-43.
1026. Duster, 1990, p. 97.
1027. P. Taylor, 1996, p. 137.
1028. Brennan et al., 1991, p. 232.

CRIMINAL TWINS: BLOOD BROTHERS OR PARTNERS IN CRIME?

The Twin Method and the Study of Criminality

The German investigator Johannes Lange published the first criminal twin study in 1929, only a few years after the twin method had been developed. He identified 13 identical and 17 fraternal pairs where at least one twin had been imprisoned. Because Lange (who was often referred to as "Kraepelin's favorite student"[1029]) found that ten identicals but only two fraternals had a co-twin who was also imprisoned, he concluded that genetic factors "play a predominant part."[1030]

In 1934, Rosanoff et al. published their study of 97 twin pairs, finding male concordance rates of 22/33 identical (67%) and 3/23 same-sex fraternal (13%). They concluded that criminal behavior is caused by "pre-germinal [genetic] rather than germinal" factors.[1031]

Two criminal twin studies were published in Nazi Germany in 1936. Friedrich Stumpfl, who produced the most influential criminal biology research during the Nazi era,[1032] studied 18 identical and 19 fraternal pairs, finding concordance rates of 65% and 37% respectively.[1033] While opposing the blanket sterilization of criminals, Stumpfl called for extending the sterilization law to include all "incorrigible recidivists."[1034] Kranz found concordance rates of 66% identical (21/32) and 54% same-sex fraternal (23/43). This comparison is not statistically significant, which did not prevent Kranz from calling for the sterilization of criminals in the interest or racial hygiene (see Chapter 2).

Following the publication of a handful of small studies between 1937 and 1976, two major Scandinavian criminal twin studies were published in the 1970s by Christiansen, and Dalgard and Kringlen. These investigations were based on national registers and studied an entire population of twins unselected for criminality.

Dalgard and Kringlen studied 49 male identical and 89 same-sex male fraternal pairs, finding no statistically significant concordance rate difference using either a "broad" or "strict" definition of crime. The identical twin pairwise rate

1029. Wetzell, 2000, p. 157.
1030. Lange, 1930, p. 46.
1031. Rosanoff, Handy, & Rosanoff, 1934, p. 929.
1032. Wetzell, 2000.
1033. Cited in Christiansen, 1977b, p. 72.
1034. Quoted in Wetzell, 2000, p. 277.

for broadly defined crime was 22%, and 26% for strictly defined crime. On the basis of these results and of the more similar environments experienced by identical twins, the authors concluded, "These findings support the view that *hereditary factors are of no significant importance in the etiology of common crime.*"[1035]

Christiansen studied 85 identical and 147 same-sex fraternal pairs (and an additional 196 pairs of opposite-sex fraternals), reporting pairwise concordance rates of 33% identical (28/85) and 12% same-sex fraternal (17/147).[1036] Christiansen's sample included all twins born between 1881 and 1910 in the Danish islands east of the Little Belt in which both twins were alive past the age of 15. Their names then were checked against Danish police records. Christiansen's final results were published by Cloninger and Gottesman in 1987, who reported a pairwise identical rate of 48% (56/116), and a same-sex fraternal rate of 28% (56/202).[1037]

In 1995, Lyons and associates published a study of pairs obtained from the Vietnam Era Twin Registry (VETS). They performed diagnostic interviews by telephone, and concluded that several antisocial traits were "significantly heritable."[1038] However, their self-report data was subject to bias. For example, most respondents in a companion study denied any "early criminal behavior" such as "swip[ing] things from stores or from other children or steal[ing] from your parents or from someone else."[1039] As a critic asked, "Who hasn't swiped something before the age of 15?"[1040] In fact, two-thirds of the respondents denied *all* antisocial behaviors. The very title of Lyons's 1996 investigation, "A Twin Study of Self-Reported Criminal Behavior," illustrates an important problem with studies of this type. Unless answering anonymously, people tend to refrain from telling other people (especially strangers) about their past and present criminal and antisocial behavior.

Other twin studies of antisocial behavior have been performed in the last 20 years, which I will not review here.[1041] As should by now be clear, assessing the validity of the underlying assumptions of the twin method is the crucial question at hand.

1035. Dalgard & Kringlen, 1976, p. 231. Emphasis in original.
1036. P < .0001. Christiansen, 1977a.
1037. Cloninger & Gottesman, 1987.
1038. Lyons et al., 1995.
1039. Lyons, 1996, p. 63.
1040. M. Daly, discussion in Lyons, 1996, p. 71.
1041. e.g., Eley et al., 1999; Rowe, 1986.

Twins Reared Apart

Although there are several individual case studies of *reared-apart* criminal twins in the literature, there has been no systematic study of reared-apart criminal twins, apart from a 1990 study of antisocial behavior by William Grove and his MISTRA colleagues.[1042] Although Grove et al. claimed to have found "significant heritability" for childhood and adult antisocial behavior, we saw in Chapter 4 that reared-apart twin studies are plagued by methodological problems and the failure to control for the common environmental influences shared by reared-apart twins.

An example of the problems with case histories of allegedly reared-apart twins is evidenced in a 1952 report of 18-year-old Mexican-American identical twins Esther and Elvira. Although the girls were separated at nine months, Elvira soon returned to her mother's home, while Esther was sent to live with another relative. For eight years following their separation the twins lived "in neighboring houses, perhaps a couple of hundred feet apart."[1043] Their mother reported that the girls played together, had the same type of clothing and toys, and "were aware of their twin relationship."[1044] Although the girls became geographically separated at age nine, they continued to maintain contact with each other, including yearly visits in the summertime. As an adolescent Elvira was beaten by her "brutal" stepfather, who threw her out of the house. She was subsequently sent to a correctional institution because of her delinquency. Around this time, Esther returned to her mother's home determined to be reunited with her twin: "Her one desire was to be with her twin, just as the latter's was to be with her. . . . the bond between the girls was a close one."[1045] Soon, Esther was sent to a correctional school for delinquency different from the one to which Elvira had been sent. Esther escaped from this school, while Elvira was on release from her school "very much upset, and determined to find Esther." The girls were picked up and sent to the same school where they "proved devoted to each other."[1046] At the age of 18, Esther, who unlike Elvira, "had not basked in warm mother love throughout her formative years,"[1047] was found dead of a morphine overdose in the hallway of a big city hotel.

1042. Grove et al., 1990.
1043. Schwesinger, 1952, p. 40.
1044. *Ibid.*, p. 40.
1045. *Ibid.*, p. 42.
1046. *Ibid.*, p. 42.
1047. *Ibid.*, p. 46.

Clearly, these "reared-apart" twins were raised in different branches of the same family and had considerable contact and a strong twinship bond. Although genetically-oriented reviewers such as Adrian Raine considered the handful of reported reared-apart concordant pairs as "clear evidence for the role of genetic factors," Susan Farber viewed Esther and Elvira's case as typical of the "dubious separation" found in many stories of reared-apart twins. [1048] She concluded, "There is no substantive evidence in the twin reared-apart data to support the claim that genetic determination is significant in this area [crime], and there is much to support the idea that environment is potent."[1049]

Does Greater Identical Twin Concordance Point to a Genetic Predisposition to Crime?

According to Dalgard and Kringlen, the twin method requires that the "environmental conditions are in general similar for MZ and DZ pairs." However, they concluded, "This assumption is obviously not true."[1050] Later, they wrote that the EEA is "an assumption which today cannot be accepted."[1051] Dalgard and Kringlen found that 86% of their identical pairs had felt an extreme or strong interdependence, while only 36% of fraternals felt this way.[1052] When they grouped twins on this basis, there was no significant concordance rate difference among pairs with an "extreme or strong" level of interdependence (identicals: 6/26, or 23.2%, vs. fraternals: 3/14, or 21.4%).

According to Christiansen,

1048. Raine, 1993, p. 59; Farber, 1981, p. 51.
1049. *Ibid.*, p. 229.
1050. Dalgard & Kringlen, 1976, p. 214.
1051. *Ibid.*, p. 223.
1052. *Ibid.*, p. 224. In a discussion of the Dalgard & Kringlen study, Cloninger & Gottesman (1987, pp. 98-99) wrote, "As expected, a slightly greater proportion of MZ twins than of DZ twins were psychologically close (84% of 31 vs. 74% of 54)." However, they were looking at the concordance figures of twins' level of "intra-pair interdependence" in Dalgard & Kringlen's Table 12 (1976, p. 224), which compared concordance rates among a select group. When Dalgard & Kringlen reported information on their *entire* sample of interviewed pairs (Dalgard & Kringlen, 1976, p. 224, Table 11), we find that 42 of 49 identical pairs (86%) had an "extremely strong" or "strong" level of closeness, which was true for only 32 of the 89 fraternal pairs (36%). Seven identical pairs (14%) were as close as ordinary siblings, which was true for 57 (64%) of fraternals.

The fundamental assumption underlying conclusions about heredity and environment that have usually been drawn from criminological twin studies is that the relevant environment of the two twins is (and has been) equally similar or equally different, regardless of zygosity. Stated in another way: intrapair environmental variations must be the same for MZ co-twins as they are for DZ co-twins. *It is open to serious doubts whether this condition is fulfilled with respect to social behavior* [emphasis added].[1053]

Christiansen wrote that "it is generally accepted that the experienced environment of MZ twins is more similar than that of DZ twins,"[1054] citing several studies as evidence. Thus, the authors of two large population-based Scandinavian criminal twin studies had serious doubts about the validity of their research method.

Behavior geneticist Gregory Carey confirmed the doubts of Dalgard, Kringlen, and Christiansen. After analyzing the Danish criminal twin study data, Carey concluded in 1992,

The assumptions of the traditional twin method may be violated for phenotypes related to externalizing antisocial behavior. . . . If MZ twins influence each other more than do DZ twins — a hypothesis that cannot be rejected in this analysis — genetic effects for criminal liability may actually be small.[1055]

Carey, a proponent of the twin method in general, believed that the equal environment assumption was likely invalid for criminal and antisocial behavior.

The following case illustrates the likely effect of the twinship bond on identical twin concordance rates. In 1951 British investigator Lorna Wheelan chronicled the lives of a pair of reared-together British identical twins whose behavior differed greatly. The "patient" and his twin brother (whom I will call "Twin A" and "Twin B") had been raised in an abusive and alcoholic family environment. Twin A, who was later diagnosed as an "aggressive psychopath" and was convicted of larceny, was the acknowledged "leader of the twins" from an early age.[1056] Wheelan investigated the twins when they were 27-years-old, and described their divergent personalities as follows:

1053. Christiansen, 1977a, p. 93.
1054. *Ibid.*, p. 94.
1055. Carey, 1992, p. 21.
1056. Wheelan, 1951, p. 134.

> The patient [Twin A] has no friends, and he soon tires of acquaintances whom he makes easily; he is described as being cold-hearted, selfish and unpredictable; he never heeds advice; borrows money, is dishonest, and [is] a shiftless worker; he shows no affection or consideration for anyone — even his four children, for whom he has never accepted responsibility. His brother [Twin B] is steady, stable, modest, less quick tempered and more ambitious. In contrast to his brother's agnosticism, he recently joined the R. C. Church.[1057]

We see that Twin B, upstanding citizen that he was, was nonetheless convicted as an accessory to a jewelry theft. His brother Twin A, who was living in a different part of the country, had asked him to pawn stolen jewelry for him. Twin B agreed and was subsequently arrested and convicted for the deed. Thus, vastly different identical twin brothers were concordant for criminality on the basis of the criteria used in most family, twin, and adoption studies. We should recall that Twin A was the leader of the pair, and it was probably difficult for Twin B to turn down his request in spite of his likely distaste at being involved in illegal activities. Like most of the twins recorded in the various studies, concordance for criminal behavior in this case appears to be the result of common influence and environment rather than common blood.

Contrary to genetic predictions, there is a marked difference between same-sex and opposite-sex fraternal twin concordance rates. Rosanoff found an adult criminality concordance rate for same-sex fraternals of 18% (5/28), but only 3% (1/32) for his opposite sex fraternal pairs. Kranz reported a same-sex fraternal rate of 23/43 (54%), and an opposite-sex fraternal rate of 7/50 (14%). Stumpfl reported rates of 7/19 (37%) and 2/28 (7%). Large differences were reported in the final publication of Christiansen's data. The rates were same-sex fraternal 56/202 (28%), and opposite-sex fraternal 14/228 (6%).[1058] Eley et al. also reported higher same-sex versus opposite-sex correlations in 1999. Dalgard and Kringlen believed that the higher rate among same-sex fraternal twins "emphasizes the significance of environmental factors."[1059] This statement is true, but requires clarification: The difference emphasizes the role of environmental factors *affecting concordance rates*, which suggests that identical-fraternal concordance rate differences are also affected (or explained entirely) by environmental factors.

1057. *Ibid.*, p. 136.
1058. Reported in Cloninger & Gottesman, 1987, p. 99.
1059. Dalgard & Kringlen, 1976, p. 217.

Behavior geneticist David Rowe attempted to test the validity of the equal environment assumption in his 1983 study of teenage twins, and concluded that the assumption is valid.[1060] According to Rowe, the EEA is supported because delinquency among his twins was not predicted by their level of association, and because identicals did not commit delinquent acts together more frequently than fraternals. Not surprisingly, however, these findings are problematic for reasons that include: (1) The study depended on mailed responses from twins, who were asked to self-report delinquent acts. It is unlikely that these teenage respondents were willing to honestly report antisocial (and sometimes criminal) acts. (2) Only 50% of the twins returned questionnaires, which likely biased the sample in the direction of better behaving pairs. (3) A statistically significant association between shared activities and fraternal males' delinquent behavior *was* found. However, as Rowe described it, he inspected the data, removed an outlier, and declared the association statistically non-significant.[1061] (4) Rowe's test involved only a few environmental variables, while many others went unchecked. The shared activities questionnaire contained a nine-item Likert scale. Rowe calculated mean scores from the Likert scale responses, which potentially obscured important associations between answers on the ends of the scale and concordance.[1062]

Conclusion

The genetic basis of criminal or antisocial behavior, or any other type of behavior, is not established by a greater concordance rate or correlation among identical versus fraternal twins (see Chapter 3). Like schizophrenia, several researchers have turned to adoption studies in an attempt to disentangle possible genetic and environmental influences on crime. It is to these studies that we now turn.

1060. Rowe, 1983.

1061. *Ibid.*, p. 478.

1062. The practice of deriving mean scores from ordinal scales is potentially misleading because differences between responses are not equivalent. According to Bradley & Schaefer (1998, pp. 118-119), "It is frequently forgotten that arithmetic is not meaningful on these scales. . . . a mean is not an appropriate statistic to calculate for the Likert scale or, in fact, for any ordinal data. Mathematically, the use of a mean requires a scale that is unique up to positive linear transformation, that is, at least an interval scale. . . . The appropriate statistic to use for this data is a frequency count."

According to adoption researchers Sarnoff Mednick and Elizabeth Kandel, family and twin studies suffer from possible environmental contamination: "To address this problem, adoption studies have been utilized. These are natural experiments in which the effects of genetic and rearing influences may be separated to a relatively high degree."[1063] Like schizophrenia, Scandinavian adoption studies are frequently cited in support of a genetic predisposition for criminal behavior. Not surprisingly, however, these studies suffer from many of the invalidating flaws we saw in Chapter 7.

There have been five adoption studies of criminality, antisocial personality, or psychopathy: two minor North American studies (Crowe, Cadoret) and one minor Danish study (Schulsinger),[1064] plus two major Scandinavian investigations (Bohman and colleagues in Sweden, Mednick and colleagues in Denmark). In none of these studies, however, did the researchers find evidence in support of genetic influences on *violent* crime. They claimed only to have found a genetic component for "petty" or "property" offenses, or for a vaguely defined notion of antisocial or psychopathic behavior.

Here, I analyze the Scandinavian studies in order to determine whether they provide evidence in support of genetic influences on crime. The studies of Crowe, Cadoret, and Schulsinger contain numerous invalidating flaws which have been documented in detail elsewhere.[1065] Regarding Crowe's study, a pair of critics wrote that it "is so far from the minimum standards of scientific adequacy that it deserves only minimal comment."[1066]

Bohman and Colleagues' Swedish Study

In 1978, Swedish investigator Michael Bohman reported rates of criminality and alcoholism in a large group of Swedish male and female adoptees given up by their biological parent(s) and placed into nonrelative adoptive homes during the first three years of life. He checked the names of these adoptees and their biological and adoptive parents against the records of the Swedish

---

1063. Mednick & Kandel, 1988, p. 103.
1064. Cadoret, 1978; Crowe, 1972, 1974; Schulsinger, 1977.
1065. Gottfredson & Hirschi, 1990; Joseph, 2001c; Walters & White, 1989.
1066. Gottfredson & Hirschi, 1990, pp. 59-60.

Criminal Register and of the Excise Board, which maintained records on people fined for intemperance.

Bohman sought to determine whether alcoholism and criminality rates were elevated among adoptees who had a criminal or alcoholic biological parent. He found that the adopted-away sons of biological parents with a criminal record were themselves criminal at rates comparable to the expected population rate. He also performed a control study of 50 male and 50 female adoptees whose biological fathers "were among those with the most serious criminal records. . . . Most of these men had been sentenced to long terms in prison."[1067] Only 8% (4/50) of the male adopted-away biological children of these men, and 8% of controls (4/50), had criminal records. Bohman concluded, "The results suggest that there is a genetic determinant for alcoholism but not for criminality (defined as repeated offenses with long prison sentences)." Thus Bohman found no evidence of a genetic predisposition for criminality, while stating that the results "must be regarded as preliminary."[1068]

In 1982, Bohman and colleagues reported the criminality rate of 862 Swedish men from the same adoptee cohort as the 1978 study. They reanalyzed the data and argued that, although there was no association between criminality in the biological parents and their adopted-away sons, genetic influences were detected if the type of offense and the use of alcohol were controlled for. The investigators claimed that "nonalcoholic petty criminals had an excess of biologic parents with histories of petty crime but not alcohol abuse."[1069] They concluded in favor of a genetic influence for petty offenses — a claim that cannot go unchallenged — but *not* for violent crime.

Critique. Although in 1982 Bohman et al. studied 862 male adoptees, in 1978 Bohman had studied 1,125 male adoptees. Because the 1982 study was essentially a reanalysis of the 1978 data, one must ask why the male adoptee group was reduced from 1,125 to 862, a drop of 23.4%. The investigators offered an explanation:

1067. Bohman, 1978, p. 272.
1068. *Ibid.*, p. 276.
1069. Bohman et al., 1982, p. 1233.

> The subjects included all 862 men born out of wedlock in Sweden from 1930 through 1949 who were adopted at an early age by nonrelatives. Other subjects included by Bohman in preliminary analyses (Bohman, 1978) were excluded because of incomplete data, late placement, or intrafamilial adoption. The age at adoption was less than 3 years in all cases and 8 months on the average. The adoptees ranged in age from 23 to 43 years at the time of last information.[1070]

The reduction of the male adoptee group is puzzling because, as Bohman reported in his 1978 paper, all 1,125 male adoptees had been placed with nonrelatives before the age of three, and were checked against the Criminal Register. Thus, there is no indication that the 1982 adoptee group was subject to more stringent criteria than the 1978 group, and it is unclear whether the investigators were aware of the Criminal Register status of the 263 adoptees (and their relatives) who they removed from the study.

The 1982 study reported a 13% criminality rate for the adopted-away biological sons of parents with a criminal record (with or without alcohol abuse). This figure is comparable to the Swedish male population risk of 11%.[1071] Contrary to genetic expectations, the fact that an adoptee had a criminally-registered biological parent did not lead to an elevated crime rate when compared to the population rate. The researchers presented several pages of complex statistical procedures which, they claimed, suggested a genetic predisposition for petty offenses. In addition to the problems I have mentioned, the investigators failed to pay sufficient attention to a pair of important issues: (1) they did not indicate that the rate of index petty criminality was significantly higher than the general population expectation, and, (2) the selective placement of adoptees.

Even when using a control group, adoption researchers must demonstrate that the index group rate of the trait in question is significantly higher than the general population expectation, or more accurately, than the non-adoptee population expectation.[1072] For example, Bohman stated that the general population risk for registered male criminality in Sweden was 11%. Suppose that Bohman, in his 1978 study, had found that 11% of his male adoptees with seriously offending biological fathers had criminal records, whereas only 2% of the control adoptees had a criminal record — a statistically significant difference. Contrary to the beliefs of most adoption researchers, we can draw no valid conclusions in favor

---

1070. *Ibid.*, p. 1234.
1071. *Ibid.*, p. 1235.
1072. Boyle, 1990.

of genetics from these figures because the experimental group adoptees had roughly the same criminality rate as one would expect in a randomly selected group of non-adopted Swedish males. The 2% rate among controls, however, might indicate that they experienced environments less conducive to producing criminality compared to the experimental group. In Bohman and colleagues' 1982 reanalysis, there is no indication that any group of adoptees had a rate of petty criminality higher than the population rate, which could have been determined through the records of the Criminal Register. The researchers compared adoptee groups to each other, but not to the all-important population rate.

Regarding selective placement, it is unlikely that the biological relatives of registered criminals were placed into the same types of environments as adoptees lacking such a family history. Adoptees in the Swedish study were born between 1930 and 1949, a period coinciding with a widespread belief in the inherited nature of most mental abnormalities, including antisocial behavior.[1073] The world's first "racial biology" institute, the Swedish *Uppsala Institute for Race Biology*, was established in 1922. The opening of the Institute is said to have inspired the creation of similar institutions in Germany.[1074] Sweden passed its first eugenic sterilization law in 1934, and another law was passed in 1941 permitting eugenic sterilization for those demonstrating "an anti social way of life."[1075] The existence of such laws suggests that children with a criminal family background, easily checked through the use of registers, would not have been attractive candidates for adoption because of their perceived "hereditary taint." According to Bohman, Swedish children "thought to be at high risk for heritable disorders were unlikely to be considered eligible for adoption."[1076] But what about children thought to be at *moderate* genetic risk? They might well have been included in the adoption process, even if they were considered less desirable adoptees. According to Bohman, Swedish "adoptee[s] and the adoptive parents were never informed [by the adoption authorities] about the identity or behavior of the biological parents,"[1077] although in an earlier paper Bohman indicated that a child's biological background *was* a factor influencing the adoption process. In a discussion of a subgroup of children from his 1971 study, Bohman wrote,

1073. Broberg & Tydén, 1996. It is also known that there were close contacts between Swedish and Nazi-era German scientists and eugenicists (Weingart, 1999).
1074. Broberg & Tydén, 1996, p. 87.
1075. *Ibid.*, p. 108.
1076. Bohman et al., 1982, p. 1234.
1077. *Ibid.*, p. 1234.

> This group may be regarded as a negative selection from the primary series, in that many of the children were considered at birth, or while at the infants' home, to be difficult to place on account of retarded development, *poor heredity*, or somatic complications. *It often took longer to place these children in their ultimate home environment than it did the other children in this study;* more than half of them spent over nine months at an institution before being placed [emphasis added].[1078]

It appears that a potential adoptee's "poor heredity" was a factor in the adoption process. This could have been manifested in two ways: (1) prospective adoptive parents were informed of the biological background of the child, or (2) the authorities were reluctant to place children with "poor heredity" into the homes of "good" adoptive families. In any case, Bohman indicated that these children were placed later and spent more time in an institution than other children.

In Bohman's 1978 control study, the biological male children of fathers and mothers registered for criminality *but not for alcohol abuse* were recorded alcohol abusers at rates significantly higher than controls.[1079] Among these adopted-away biological offspring of registered criminal fathers, 9 of 50 (18%) were registered alcohol abusers, versus 2 of 50 controls (4%).[1080] Among the offspring of registered mothers, the rate of alcohol abuse is 9 of 48 (19%), versus 2 of 48 controls (4%).[1081] Of course, one could argue that these figures suggest a genetic relationship between criminality and alcohol abuse, but a far more plausible explanation is that the children of criminally registered parents were placed into more psychologically harmful environments than were control adoptees, leading to significantly more alcoholism.

According to Bohman and Sigvardsson, "children were selected and placed . . . according to the social and occupational status of their biological parents,"[1082] and in 1978 Bohman wrote that adoptees with antisocial or alcoholic biological parents were placed on average two to three months later than controls. He found that "later placement is associated with selective factors that contributed independently to poorer social adjustment later in life and hence to

---

1078. Bohman, 1971, p. 6.
1079. Bohman, 1978, p. 274, Table 6.
1080. p = .026, Fisher's Exact Test, one-tailed.
1081. p = .025, Fisher's Exact Test, one-tailed.
1082. Bohman & Sigvardsson, 1980, p. 348.

an increased risk of appearing in the registers."[1083] If true, placement policies might account for any possible significant elevation in so-called petty offenses.

* * *

Bohman and colleagues' 1982 study was essentially an after-the-fact reanalysis of Bohman's 1978 data, from which the authors drew a different set of conclusions more in line with genetic thinking on the subject. Ironically, Kety, Rosenthal, and Wender used the *opposite* approach in their Danish schizophrenia adoption studies. For the most part, they found no statistically significant elevation of any one diagnostic category (which is especially true if the comparisons are limited to first-degree relatives). In a 1988 article, Kety wrote,

> At the prototypical end of the spectrum, chronic schizophrenia is found exclusively in the biological relatives of chronic schizophrenia patients where it occurs at a low prevalence (approximately 3%), whereas the prevalence in the biological relatives in the normal controls is negligible. The same is true for uncertain chronic schizophrenia. Latent or borderline schizophrenia was found at a 4-5% prevalence in the biological index relatives and 1% to 1.5% in the biological relatives of controls. This is also true where the symptoms are less distinct and the diagnosis is designated uncertain. *Since neither in chronic nor in latent schizophrenia the results for the definite or uncertain diagnoses are statistically different, it appears justified to combine them* [emphasis added].[1084]

Thus, Kety needed to *combine* his spectrum disorders into one total to find statistically significant results — and his investigation is frequently cited as producing the most compelling data in favor of the genetics of schizophrenia. Contrast this to Bohman's 1982 reanalysis, where the definition of crime was *narrowed* in order to find a significant genetic effect. And to my knowledge, neither Kety nor Bohman published criteria for inclusion or exclusion *before* the appearance of their studies. Thus, in both cases the investigators performed a post-data collection reanalyses — using opposite approaches — in order to find a statistically significant genetic effect for the trait under study. And in both studies, the researchers could very easily have concluded that they found no genetic influences on schizophrenia or criminality.

1083. Bohman, 1978, p. 275.
1084. Ingraham & Kety, 1988, pp. 121-123.

To summarize, Bohman and colleagues found no evidence for genetic influences on violent crime. Their conclusion in favor of genetic influences on petty offenses cannot be accepted due to methodological problems and potential environmental confounds.

Mednick and Colleagues' Danish Study

Mednick and colleagues' 1984 Danish Adoptees' Family criminality study is perhaps the most well-known and most frequently cited study in support of a genetic basis for criminal behavior.[1085] The investigators utilized a register of 14,427 Danish adoptees placed between 1924 and 1947, which had been compiled by Kety and associates for their schizophrenia studies. They identified 13,194 adoptees (6,129 male, 7,065 female). Like most genetic studies of criminality and antisocial behavior, Mednick concentrated on the results among male subjects.

The researchers determined criminal status by checking the names of adoptees against the records of the Danish Police Record Office and a separate criminal record, the *Personalia Blad*. They identified adoptees having a record of court conviction, and then checked the conviction records of their biological and adoptive relatives. According to Mednick and colleagues,

> The size of the population permits segregation of subgroups of adoptees with combinations of convicted and non-convicted biological and adoptive parents in a design analogous to the cross-fostering model used in behavior genetics. If neither the biological nor the adoptive parents are convicted, 13.5 percent of the sons are convicted. If the adoptive parents are convicted, and the biological parents are not, this figure rises only to 14.7 percent. However, if the adoptive parents are not convicted and the biological parents are, 20.0 percent of the sons are convicted. If the adoptive parents as well are convicted, 24.5 percent of the sons are convicted. These data favor an assumption of a partial genetic etiology.[1086]

They also performed a "sibling analysis," which compared rates among split-up half- and full-sibling pairs placed into different adoptive homes. Genetic theory predicts higher concordance in full-sibling pairs because they are more similar genetically. Of the 126 male-male *half-sibling* pairs, 31 had at least one convicted member and four pairs were concordant (concordance rate = 4/31, 12.9%). Of the 40 male-male *full-sibling* pairs, 15 had at least one convicted member and

---

1085. Mednick et al., 1984.
1086. *Ibid.*, p. 892.

three pairs were concordant (concordance rate = 3/15, 20%). Although the full-versus half-sibling difference was not statistically significant, Mednick and colleagues concluded, "The numbers are small but indicate that as the degree of genetic relationship increases, the level of concordance increases."[1087]

The investigators wrote that although the evidence pointed to a genetic predisposition for property offenses, "This was not true with respect to violent crimes."[1088] This finding was elaborated upon in a subsequent article:

> A significant relationship exists between parents' convictions and property offending. A significant relationship does not exist for violent offending. . . .
> Genetic factors predispose to property offending but not to violent offending. *If a biological predisposition to violence does exist, then it must be a result of other postconception factors* [emphasis added].[1089]

Danish adoptees at greater risk for registered criminality. It appears that Danish adoptees in general were significantly more susceptible to criminal conviction than non-adoptees. According to Mednick, 84.1% of the identified male adoptees had no conviction record, meaning that the adoptee conviction rate was about 16%.[1090] They stated that the general population rate for male conviction in Denmark was about 9%,[1091] while criminologists Hurwitz and Christiansen put the figure at 8%.[1092] As we have seen, Mednick found a conviction rate of 13.5% for male adoptees whose biological and adoptive parents had no record of conviction,[1093] which means that about 827 such adoptees were registered criminals (13.5% of 6,129). However, applying the 8% Danish male conviction rate to 6,129 hypothetical members of the Danish male general population, we would expect to find only 490 registered criminals (8% of 6,129). Thus, Danish adoptees placed during the era under study were far more likely to be convicted when compared to the general population prevalence, and thereby constituted a distinct population with regard to criminal conviction.

Moreover, an 8% Danish population prevalence is probably too high. First, this rate is slightly inflated by the inclusion of adoptees, who have a higher rate. Second, it is likely that a fair percentage of the non-adopted convicted criminals

---

1087. *Ibid.*, p. 893.
1088. *Ibid.*, p. 891.
1089. Mednick et al., 1988, p. 29.
1090. Mednick et al., 1984, Table 1.
1091. Hutchings & Mednick, 1975.
1092. Hurwitz & Christiansen, 1983, p. 41.
1093. Mednick et al., 1984, p. 892.

had a parent with a conviction record, yet Mednick found a 13.5% conviction rate among adoptees whose adoptive (and biological) parents were *not* convicted. Although the Danish figures are not available, it is probable that the conviction rate among non-adoptees whose parents had no conviction record was well below 8%.[1094] Thus, Danish *adoptees* placed between 1924 and 1947 having non-convicted biological and adoptive parents were at roughly twice the risk of being convicted versus *non-adoptees* with non-convicted biological parents.[1095] Therefore, with regards to criminality, we cannot generalize findings among this Danish adoption cohort to the non-adoptee population, Danish or otherwise. Mednick and colleagues' findings, therefore, *apply only to the population of Danish adoptees.*

A controversy in the schizophrenia twin study literature is relevant to this discussion. We recall that in his devastating critique of these studies, Don Jackson theorized that the unique aspects of the identical twinship might lead to higher rates of schizophrenia. If true, identical twin schizophrenia concordance rates would have been inflated, and identical-fraternal differences would not be generalizable to the single-born (non-twin) population. In response, Rosenthal and others argued that identical twins were no more likely to receive a schizophrenia diagnosis than members of the single-born population, and that the schizophrenia twin studies were therefore valid.[1096] As we have seen, Danish *adoptees* were about twice as likely to be registered criminals than non-adoptees, meaning that the Rosenthal's argument in defense of schizophrenia twin studies restricts Mednick's generalizations to the Danish adoptee population.

Selective placement. Another problem is that, as we saw in the Swedish study, it is likely that the biological children of convicted criminals were placed into inferior rearing environments when compared to adoptees lacking this biological background.[1097] In Chapter 7 we saw that the biological background of children put up for adoption in mid-20th century Denmark was an important factor in the placement process. I should add a few points to this discussion as

---

1094. In another paper by the investigators (Kandel et al., 1988), which looked at a consecutive series of Danish children born between 1936 and 1938 who were *not* adopted away, 39% of the sons of "severely sanctioned" criminals were themselves recorded criminals by 1972, while 7% of the sons of non-registered fathers had been registered (p. 225).
1095. The higher rate among adoptees is consistent with the results of a 1997 publication from the investigators which found a correlation between early maternal rejection and adult violence (Raine et al., 1997).
1096. For a more detailed discussion of this point, see Joseph, 2001b, pp. 41-43.
1097. Kamin, 1985.

they relate to criminality, which in Denmark was also viewed as the result of bad heredity.

Denmark passed a law in 1935 allowing the compulsory sterilization of mentally "abnormal" people on eugenic grounds. In the same year, a panel of leading Danish medical authorities reviewed the results of the 1929 law, writing,

> [The psychopaths] are often — to a larger extent than for example, the mentally retarded — asocial or antisocial (criminal); and their erotic activity and inventiveness, considered together with their fertility — often extramarital — is considerable. . . . With respect to hereditary tainted progeny the psychopaths are comparable to the more well-defined mental diseases, even though the pattern of inheritance is still unknown.[1098]

As this document suggests, the children of criminals and "psychopaths" were regarded as "hereditary tainted progeny." Historian Bent Hansen noted that "the leading [Danish] medical experts in 1935 were ready to go very far in their pursuit of eugenic goals and social control of the marginal groups of society." He pointed out that "there were no adverse reactions" to the 1935 document, and that "the medical world seemed to agree with the conclusions."[1099]

Between 1940 and 1945, Denmark was under Nazi occupation, which represents about one-fifth of the period during which Mednick's adoptees were placed. The war years saw a dramatic rise in the Danish conviction rate, and Mednick included these convictions in the study.[1100] This period was likely marked by an even greater emphasis on eugenics, since Denmark was under the occupation of the German government of "applied racial science." Mednick noted on several occasions that the twin studies of Stumpfl and Kranz were "tainted by their origins in Nazi Germany,"[1101] which he considered "a politically unfortunate period."[1102] However, he failed to highlight fact that about 20% of the period in which his Danish adoptees were placed, and criminals were registered, occurred during a similarly "unfortunate period" in Danish history.[1103]

Thus, the period during which adoptees were placed in Mednick's study (1924-1947) coincided with a period of Danish history in which undesirable traits were seen largely as the product of bad genes; so much so that the authorities passed laws in an attempt to prevent the alleged carriers of these genes from

---

1098. Quoted in B. Hansen, 1996, pp. 42-43.
1099. *Ibid.*, p. 43.
1100. Hurwitz & Christiansen, 1983.
1101. Mednick & Volavka, 1980, p. 92.
1102. *Ibid.*, p. 95.

reproducing. According to Mednick and Hutchings, a "potential problem with the adoption method is the possibility that the adoptive family is informed by the adoption agency of deviance in the biological family," and (as seen in Chapter 7) they noted that a social worker who had read the older adoption journals "formed the impression that serious deviance in the biological parents was routinely reported to the prospective adoptive parents unless they refused the information."[1104]

The investigators stated that 37% of the biological parents had their first conviction before their child was adopted, whereas 63% were convicted after the adoption. They produced figures showing that there was no difference between the conviction rates of adoptees placed before and after their biological parents' conviction, and concluded, "The fact that the adoptive parents were informed of the biological parents' criminality did not alter the likelihood that the adoptive son would be convicted of a crime."[1105] However, this comparison looks only at the parents and leaves out knowledge of the criminal status of *other* family members who, as we saw in Chapter 7, were of great interest to the adoption authorities.

Mednick and his colleagues failed to grasp the potential impact of the differing rearing environments of desirable and undesirable children. For them, prospective adoptive parents' knowledge of the adoptee's biological family history of criminality might cause them to expect the adoptee to commit criminal or antisocial acts, which might affect the way they perceive and rear their adoptee. But more important is that a child's biological background, under the prevailing conditions in Denmark, meant that the child was not a desirable adoptee in the eyes of the most qualified adoptive parents who, as we have seen, were routinely provided information on the criminal records of a potential adoptee's biological family. Into what kinds of adoptive families were the adopted-away biological children of families with convicted criminals eventually

---

1103. Not all researchers of the genetics of criminality considered studies published during the National Socialist period to be "tainted." For example, Fini Schulsinger wrote, "The very special eugenic ideas of the Third Reich involved some German psychiatric geneticists in the classical type of family studies on relatives of psychopathic probands. . . . Their work was carried out in the same neat way as other, respectably intended, genetic work from the famous Munich school. The results of these studies unanimously indicated that heredity plays a role in the etiology of psychopathy" (Schulsinger, 1977, pp. 112-113).

1104. Mednick & Hutchings, 1977, p. 161.

1105. Mednick et al., 1984, p. 893.

placed? Most likely, they were placed into the homes of people who, for various reasons, were not qualified to adopt the most desirable children. These homes may have been more chaotic or potentially exploitative than the others. In a 1975 control study, for example, Hutchings and Mednick compared the biological and adoptive fathers of criminal adoptees (N = 143) to the biological and adoptive fathers of adoptive (non-criminal) controls (N = 143). Among the criminal adoptees, 33 (23%) had a criminally registered *adoptive* father, whereas only 14 (9.8%) of the control adoptive fathers were registered.[1106] The comparison is statistically significant,[1107] and Hutchings and Mednick added that the difference "was also reflected in the various indices of criminality such as number of recorded cases and total length of sentence."[1108]

The rates among these adoptive fathers add to the evidence suggesting that criminally convicted adoptees experienced different types of family environments compared with non-criminal adoptees. In fact, a major aspect of the Adoptees' Family model compares the prevalence of the trait in question among index and control *adoptive* (rearing) relatives. In Rosenthal's 1970 description of how the model is used in schizophrenia, he explained that a "higher incidence among the adoptive relatives of index cases than of controls supports the view that rearing by, of, or with schizophrenics contributes to the development of the disorder."[1109] Rosenthal observed that, in the Kety et al. studies,

> among adoptive relatives, there was no appreciable difference between the two groups. . . . Genetic theory would predict such a finding. Most environmentalist theories would predict a higher incidence of such disorders among the adoptive parents of the index cases as compared with the controls, but the findings do not support this prediction.[1110]

It appears that the conviction rate difference between index and control adoptive fathers in Mednick's study is, as Rosenthal outlined, consistent with environmental predictions.

Placement factors probably influenced the previously discussed Mednick et al. "sibling analysis," which compares concordance rates among adopted-away half- and full-sibling pairs. Because the full-siblings had a more similar biological family background, they likely were placed into more similar environments than

---

1106. Hutchings & Mednick, 1975, p. 109; also discussed by Kamin, 1985.
1107. p = .002, Fisher's Exact Test, one-tailed.
1108. Hutchings & Mednick, 1975, p. 109.
1109. Rosenthal, 1970b, p. 57.
1110. *Ibid.*, p. 127.

the half-sibling pairs. As noted, the concordance rate difference between the pairs was not statistically significant. Contrary to basic statistical principles, however, Mednick and colleagues claimed that a non-significant difference (based on an admittedly "small" sample) suggested a correlation between genetic relationship and concordance for registered criminality.[1111]

In a footnote to their 1984 *Science* article, Mednick et al. wrote,

> Among males, there was a statistically significant association between adoptee criminality and the amount of time spent in the orphanage waiting for adoption. This effect, which was not true for females, may be due to institutionalization. Or it may be a function of selection bias ("less desirable" boys adopted later and also being convicted).[1112]

Thus, like Bohman, Mednick found an association between time spent in an institution and subsequent conviction, which might be due to selection bias or the effects of institutionalization. The psychological damage inflicted on children spending significant periods of time under the typically appalling and nurtureless conditions of an orphanage (particularly under the conditions of the Great Depression, foreign occupation, and war) was greater for convicted than non-convicted adoptees, and was not controlled for in this study.

Conclusion. Despite having found no evidence in support of a genetic basis for violent crime, the Mednick group typically emphasizes the theme of alleged genetic influences on criminality in general. The investigators could have given their papers titles such as, "No Genetic Basis for Violent Crime," or "Environmental Causes of Violence Must be Identified," but instead used potentially misleading titles such as "Genetic Influences in Criminal Convictions: Evidence from an Adoption Cohort," "Genetic Correlates of Criminal Behavior," and "Predisposition to Violence."[1113] Although the investigators believed they had found evidence for property (but not violent) crime, the high registered criminally rate among the population of adoptees suggests that their results cannot be generalized to the non-adoptee population. In addition, the evidence suggests that Mednick's study was confounded by selection factors in the Danish adoption process, and it is therefore unlikely that children with a criminal family background were placed into the same types of environments as children lacking such a history. [1114]

1111. Mednick et al., 1984, p. 893.
1112. *Ibid.*, p. 894.
1113. Mednick et al., 1984; Gabrielli & Mednick, 1983; Mednick et al., 1988.

DISCUSSION

We have seen that the idea of criminality as a hereditary condition is an old one, whose "scientific" basis dates back at least to Lombroso. In the United States, the sterilization of criminals on eugenic grounds was permitted in several states. Long before the publication of Lange's German twin study, the idea of "crime as destiny" was widespread in the North American population, including especially its most educated layer. Studying the genetics of criminality and anti-social behavior fell into disfavor in the years after World War II, with the revelations of Nazi genocide in the name of "racial hygiene."

Since the late 1960s, however, genetic theories for most human traits have made a comeback. So too have ideas about genetic factors in criminal and anti-social behavior, although statements by genetic proponents in this area are more cautious than for other behaviors, in part because people understand that social conditions such as racism, unemployment, and poverty contribute greatly to criminal behavior. Most genetic researchers also recognize that these and other environmental factors play a role. Why then is it necessary to study the genetics of criminality? According to the Mednick group,

> We must try to identify the specific biological mechanisms through which heritable predispositions toward criminal behavior are expressed. By identifying these mechanisms we can learn how to successfully treat and prevent criminal behavior.[1115]

The authors failed to discuss the treatments and preventive measures they had in mind, or to distinguish them from the "treatments" of previous generations. Perhaps they had in mind relatively benign measures such as early intervention programs,[1116] but they rarely discuss the possibility of improving the socioeconomic environment as a way of eliminating crime. Why focus on individuals and not the environment?

The Mednick group's call for "treatment" is even more puzzling when we realize that they found no evidence supporting a genetic basis for violent crime, the type of offense of greatest concern to the general public. Thus, they might have concluded that social conditions must be changed in order to reduce or eliminate violent crime. Furthermore, how can there be a genetic predisposition for property crime but not for violence? How can people be predisposed to steal

---

1114. See Kamin, 1985, 1986.
1115. Brennan et al., 1991, p. 243.
1116. As discussed in Gabrielli & Mednick, 1983.

but not to shoot? To write bad checks but not to rape? The explanation becomes clear when the invalidating flaws of the research are exposed, which I have attempted to do in this chapter.

But suppose that future researchers present compelling evidence in support of a genetic predisposition for criminal behavior. One could still argue that the proper environmental conditions would eliminate this behavior (even among the genetically predisposed), just as the expression of a genetic disorder such as PKU is prevented with a proper diet. While the practical significance of finding a genetic predisposition for any psychiatric condition or behavioral trait is open to debate, discovering a genetic predisposition for criminal and antisocial behavior would accomplish little more than diverting society's attention from eliminating the social conditions leading to these behaviors. But diverting attention from environments causing crime is needed by privileged economic classes and politicians seeking to absolve themselves of responsibility for the dreadful social and material conditions of a significant part of the population. Their objective, as well as the objective of those whose research they underwrite, is to locate the causes of social problems within the bodies, minds, and genes of the oppressed.

CONCLUSION

The title of a 1989 review article asked whether the evidence suggesting genetic influences on criminality is explained by "bad genes or bad research."[1117] The answer is that we have a body of methodologically flawed and environmentally confounded research. Quite erroneously, influential reviewers such as Wilson and Herrnstein have promoted the twin method as a "natural approximation to a controlled experiment for estimating the genetic involvement in a trait,"[1118] and adoption studies of criminal and antisocial behavior have been promoted as "natural experiments in which the effects of genetic and rearing influences may be separated to a relatively high degree."[1119] On paper, an adoption study might appear to be just such a natural experiment. In the real world of

---

1117. Walters & White, 1989.
1118. Wilson & Herrnstein, 1985, pp. 90–91.
1119. Mednick & Kandel, 1988, p. 103. More recent attempts to use twin and adoption data in support of genetic influences on criminal and antisocial behavior include Rhee & Waldman, 2002, and Rowe, 2002. See Joseph, 2003, for a critique of Rhee & Waldman's study.

selective placement, socioeconomic differences, and attachment disturbance, these studies typically are confounded by environmental factors — to a lesser degree than family and twin studies perhaps — but confounded nonetheless.

At bottom, criminal adoption research attempted — and failed — to confirm Galton's claim (see Chapter 2) that the adopted-away children of "wild, untamable savages" raised in "civilized" environments tend to return to a life of "contented barbarism, without a vestige of their gentle nature."

Goddard's study of the Kallikak family traced two lines of "Martin Kallikak's" descendants. One line began with the progeny of Martin and his "good Quaker wife," whereas the "defective strains" were produced by Martin's liaison with a "feebleminded tavern girl." For Goddard, a comparison of these two lines was pure science: "We have, as it were, a natural experiment with a normal branch with which to compare our defective side."[1120] Most of Goddard's contemporaries agreed that family pedigree studies (compiled by non-blinded investigators) were true "natural experiments" of the influences of heredity and environment. One day we will look upon twin studies and most adoption studies in much the same way as we now view the folly of Goddard's logic.

In summary, family, twin and adoption studies provide no scientifically acceptable evidence for the existence of a genetic predisposition for any type of "criminal," "psychopathic," or "antisocial" behavior, however it has been defined at any given time or in any given society. Finally, given (1) the potential social impact of criminal genetic research, which includes the further unwarranted stigmatization of ethnic minorities; (2) the well-known social factors leading to crime; and (3) the political aspects of deciding who is and is not labeled a criminal, it is questionable whether this type of research should even be performed.

1120. Goddard, 1927, p. 68.

CHAPTER 9. GENETICS AND IQ

> In effect, then, Galton's aim, and that of his followers, became simply an attempt to reproduce an existing set of ranks (social class) in another, the test scores, and pretend that the latter is a measure of something else. This is, and remains, the fundamental strategy of the intelligence-testing movement.
>
> — Psychologist Ken Richardson, in 2000[1121]

We now arrive at the most studied and debated aspect of the nature-nurture debate: the genetics of intelligence as allegedly measured by IQ tests. In previous chapters we saw how IQ tests were used by eugenicists and behavior geneticists to demonstrate that "intelligence" is determined largely through inheritance. Historically, IQ studies of twins and adoptees have been a central preoccupation of these fields. The issues of racial and class differences in IQ have been the subject of intense controversy for almost 100 years. Eugenically oriented psychologists' use (and creation) of IQ tests in support of a "scientific" basis for racism has been documented in several books, which include Gould's *The Mismeasure of Man*, Kamin's *The Science and Politics of IQ*, Chase's *The Legacy of Malthus*, Tucker's *The Science and Politics of Racial Research*, and Guthrie's *Even the Rat was White*.[1122] Eugenic sterilization laws in the United States, Germany, Scandinavia and elsewhere targeted "mental defectives" and the "feebleminded." A low IQ score for a German child in the late 1930s was sometimes a death sentence.[1123]

The claim that standardized IQ tests such as the Stanford-Binet and Wechsler scales measure innate intelligence (or general intelligence, represented as *g*) has been the subject of intense debate. Some of the most frequent objections to this claim are:[1124]

- The psychometric emphasis on differences in abilities is only one of many approaches to the study of human intelligence.
- General intelligence is merely the product of a mathematical formula. It has no physical reality.
- There is no consensus definition of "intelligence."

1121. K. Richardson, 2000.
1122. Gould, 1981; Kamin, 1974; Chase, 1980; Tucker, 1994; Guthrie, 1976.
1123. Aly, 1994.
1124. Some of these objections are taken from Jensen's 1980 *Bias in Mental Testing*. Jensen, of course, argued against these objections.

- IQ tests measure school learning more than innate intelligence.
- IQ tests are biased against non-Caucasians and people of lower economic classes.
- IQ tests measure only narrow abilities and ignore "real world" intelligence.
- Standardized IQ tests have been normed on non-representative populations.
- Intelligence is not normally distributed (in a bell curve), as psychometrists assume.

A large body of literature has been devoted to the "heritability of IQ" topic, with estimates ranging from 0% to 90%. In Chapter 5, we saw that heritability estimates are not appropriate for estimating the magnitude of genetic influences on a trait, and this is particularly true for IQ. Thus, the various "IQ heritability" calculations are scientifically meaningless. Of course, one might legitimately ask whether (or how much) genes influence intelligence. This question, however, cannot be answered with a heritability coefficient.

Given the size of the "genetics of IQ" literature, my intention is to briefly review a few of the important issues and to show how IQ tests were created, and continue to be used in some quarters, more as a tool of oppression than as a measurement of "intelligence."

THE PSYCHOMETRIC APPROACH TO HUMAN INTELLIGENCE

As an American Psychological Association task force acknowledged in 1996, "The psychometric approach [to cognitive ability] is the oldest and best established, but others also have much to contribute."[1125] And it is self-evident that "oldest and best established" does not equal "most useful and valid." In their 1996 book *Inequality by Design*, Claude Fischer and his colleagues compared the psychometric and information processing approaches to human intelligence. Information processing seeks to understand the way that *all* humans learn and process information. Psychometrics, on the other hand, is concerned with *differences* among people. Fischer and colleagues argued that psychometric IQ tests "are refined to magnify differences," and that the famous bell shaped curve is an invention of psychometrists who *assume* that intelligence is normally dis-

---

1125. Neisser et al., 1996, p. 80.

tributed.[1126] In fact, the test data collected by Herrnstein and Murray for *The Bell Curve* did not resemble a bell curve at all! It was only after a "good deal of statistical mashing and stretching," as Fischer et al. put it, that Herrnstein and Murray were able to create a bell shaped distribution.[1127] The Gaussian or bell shaped curve, as psychologist Julian Simon observed, "is seldom if ever observed in nature." Rather, it is the "result of scientists gradually isolating and factoring out the large elements in any situation." [1128] Even human height distribution, as Simon noted, is "skewed toward little things like babies."[1129]

Lewontin, Rose, and Kamin described the "grand illusion of psychometry" in which scales are refined to produce the desired results.[1130] The scale psychometrists choose "depends on whether one wants to make differences of scale appear large or small, and these decisions are those that psychometry arbitrarily makes."[1131] A classic example is a line graph. These days, people with retirement plans frequently log onto the Internet to check the day's stock market activity. A graph may give the initial impression that the market fluctuated widely all day. That is, until the viewer notices that the range of the graph is only 50 points. The same events captured on a graph with a range of 1000 points would elicit a much different initial reaction from the viewer. I have in front of me a psychological assessment textbook by authors who support the psychometric position. Similar to the example I just gave, they present two graphs giving "vastly different" impressions of the same data, under the title "Consumer (of graphed data) Beware."[1132] Yet, one cannot truly say which scale is more appropriate. Psychometric scales create the impression of large differences between people and groups, whereas other scales (based on the same data) would show that humans have a great deal in common. The consumer of psychometric data, therefore, should be aware of its tendency to create the appearance of large differences.

1126. Fischer et al., 1996, p. 44.
1127. *Ibid.*, p. 32.
1128. Simon, 1997, p. 203.
1129. *Ibid.*, p. 204.
1130. Lewontin et al., 1984, p. 92.
1131. *Ibid.*, p. 92.
1132. Cohen et al., 1992.

Genetic theories of IQ are based largely on the results of kinship correlations, which are derived from family, twin, and adoption studies. In previous chapters we saw that these methods are flawed instruments for detecting genetic influences on psychological traits, and in Chapter 4 we examined the invalidating flaws of reared-apart twin studies of IQ. On the other hand, factors which may invalidate genetic conclusions for psychiatric disorders and behavior are not as relevant when discussing abilities, cognitive or otherwise. For example, the identity confusion of identical twins would be much more likely to affect schizophrenia concordance rates than it would the distance twins could throw a javelin. Although in both cases important environmental similarities exist, the blurring of psychological boundaries would not be a major factor influencing twin javelin throwing correlations, as it would a schizophrenia diagnosis.

Twin Studies

Behavior geneticists generally put the reared-apart identical twin (MZA) IQ correlation (omitting Burt's figures) at about .77, although in Chapter 4 we saw that this body of research is fatally flawed. For reared-together twins, Bouchard and McGue pooled the world literature in 1981 and calculated an identical twin correlation of .86, and a fraternal correlation of .60.[1133] However, these figures are inflated by the inclusion of many poorly performed and biased studies, as well as the biases found in meta-analyses in general.[1134] Still, like other traits, both environmentalists and hereditarians would expect to find a higher identical twin correlation (see Chapter 3), which leads us back to the questions of the methods and assumptions of twin research in general.

Adoption Studies

IQ adoption studies report two seemingly contradictory findings: Although the IQs of adopted-away children tend to correlate more with their biological parents than with their adoptive parents, the typical IQ point rise for children of working-class biological parents who are reared in middle- or upper-class adoptive homes is 12-14 points. In general, hereditarian interpreters of

---

1133. Bouchard & McGue, 1981.
1134. For problems in meta-analytic research, see Begg, 1994; Joseph & Baldwin, 2000; R. Rosenthal, 1979; Simes, 1987.

adoption studies focus on the IQ correlation of adoptees and their biological parents, while environmentalists emphasize the large gains made by working-class children adopted into middle-class families. There is reason to believe, however, that the correlations are an artifact of sampling, while the IQ score gain is the statistic of value.

In his 1981 debate with hereditarian psychologist Hans Eysenck in *The Intelligence Controversy* (published in the UK as *Intelligence: The Battle for the Mind*), Kamin discussed the "restrictive variance" of adoptive families similar in SES, who have been screened by adoption agencies. Most adoptive parents are above average in IQ and provided above-average environments. Thus, the "necessary statistical consequence" of the restricted environmental variance of the adoptive families is that "child IQ correlation in adoptive families cannot be very high."[1135] Kamin used boxing as an analogy. He pointed out that, if weight classifications were not used, high correlations would be observed between boxers' weights and their won-loss records. "To avoid such a correlation," wrote Kamin,

> definite weight divisions have been established by boxing authorities. Fights can only take place between boxers of reasonably similar weight, and the correlation between weight and boxing success is consequently very low. We are suggesting that in terms of the environments provided for their children almost all adoptive parents — unlike biological parents — are in the heavyweight division. That would account for the lower parent-child IQ correlation observed in adoptive families. The correlation would presumably be much higher if parents who would provide poor environments wanted to, and were allowed to, adopt more often.[1136]

Thus, one could erroneously conclude that the lack of a correlation between boxers' win-loss records and their weights means that increased weight would not improve a boxer's chance of winning, even if weight divisions were abolished. A similar point can be made about Little League baseball age divisions, which greatly reduce the correlation between age and, for example, a player's batting average. For IQ adoption correlations to have meaning, therefore, biological and adoptive relative groups in a particular study must reflect the range of scores in the general population. Given the nature of the adoption process, however, this is very difficult to accomplish. Other reviewers have also pointed to the problem of range restriction in adoption studies.[1137]

1135. Kamin, in Eysenck vs. Kamin, 1981, p. 117.
1136. *Ibid.*, p. 117. See Stoolmiller (1998) for further evidence of range restriction in IQ adoption studies.

Another aspect of correlation versus score gain in IQ adoption studies is that, as Schiff and Lewontin demonstrated, "adopted children, even though they may correlate *individually* with their biological parents more than with their adoptive parents, are, in fact, more similar as *a group* to the adoptive parents than to their biological ones."[1138] They presented a table of hypothetical IQ data to illustrate their point. The data showed that adoptees' IQs were perfectly correlated with those of their biological parents, and there was no correlation between adoptees and their adoptive parents. Yet, adoptees *as a group* had the same mean IQ as adoptive parents as a group, and differed from their biological parents' mean group IQ by seven points. In assessing the meaning of these figures, we must keep in mind that correlation does not measure similarity, but only how traits vary together.

Recent IQ adoption investigations include a study carried out in France, and the Colorado Adoption Project (CAP) performed by leading behavior geneticists. The French study of Schiff and colleagues found 32 adoptees who had been separated near birth from their unskilled laborer parent, and were reared in the homes of families in the top 13% of the socio-professional scale. These adoptees' IQs were compared to the scores of 20 biological half-sibling controls, who were reared in the homes of their lower-class parents. The investigators found that adoptees' IQ scores averaged 14 points higher than the control group, suggesting that being reared in a professional family, as opposed to a family of unskilled workers, could raise one's IQ almost one full standard deviation (about 15 IQ points).[1139]

The Colorado Adoption Project is an ongoing longitudinal adoption study that began in 1974, when the adoptees were infants. As discussed briefly in Chapter 7, this study's design is far superior to the comparatively crude schizophrenia adoption studies. We also saw in Chapter 4 that, in sharp contrast to genetic expectations, the investigators found no personality scale correlation between birth mothers and their 245 adopted-away biological offspring. The investigators created a control group of non-adopted children matched on

---

1137. Stoolmiller, 1999. In a rejoinder to Stoolmiller, behavior genetic adoption researchers John Loehlin & Joseph Horn (2000, p. 245) acknowledged, "Stoolmiller . . . argues that restriction of range is an important consideration in interpreting the results of adoption studies of IQ, such as the Texas Adoption Project (TAP). We agree, and have regularly pointed this out in our own publications . . ."

1138. Schiff & Lewontin, 1986, p. 179.

1139. Schiff et al., 1982.

several factors. For IQ, Plomin and colleagues concluded that genetic factors are important, and that "environmental factors correlated with parents' general cognitive ability have little effect on children's cognitive ability."[1140] However, their conclusions were based on correlations. It is not likely that the CAP biological and adoptive parent IQ scores would be dramatically different, since the socioeconomic status of both adoptive and biological parents were similar. As a commentator noted, "There is an enormous degree of range restriction" in the CAP adoptive homes.[1141]

These problems, along with others I have discussed in earlier chapters such as selective placement and researcher bias, cast doubt on the claim that family, twin, and adoption studies provide important information about possible genetic influence on intelligence as measured by IQ tests. More importantly, there is no standard definition of intelligence (which the tests purport to measure), and there are many reasons to reject the claim that "intelligence" can be measured with IQ tests. Yet, several studies have shown that a superior environment can raise scores almost one standard deviation.[1142]

IQ AND GROUP DIFFERENCES

The claim that IQ score differences demonstrate the genetic inferiority of particular races and classes is almost as old as the tests themselves. In the early 20th century, the view that races and classes differed in intelligence was axiomatic in American psychology. Jensen's 1969 article in the *Harvard Educational Review* began the modern revival of this position, which was continued in Herrnstein and Murray's incendiary *The Bell Curve*, published in 1994. These authors' arguments have been decisively refuted by numerous commentators.[1143] Most

---

1140. Plomin, Fulker, et al., 1997, p. 443. In a CAP study on children's television viewing habits, Plomin et al. concluded that there is a "significant genetic influence on individual differences in children's television viewing" (Plomin, Corely, et al., 1990, p. 371).
1141. Stoolmiller, 1999, p. 395.
1142. In Jensen's 1998 book, *The g Factor*, (p. 476) he concluded, "There is simply no good evidence that social environmental factors have a large effect on IQ, particularly in adolescence and beyond, except in cases of extreme environmental deprivation." The evidence from adoption studies, however, flies in the face of this claim.
1143. Counterarguments to Jensen, and Herrnstein & Murray can be found in many books, which include Block & Dworkin, 1976; Ceci, 1996; Chase, 1980; Devlin et al., 1997; Fischer et al., 1996; Fish, 2002; Fraser, 1995; Gould, 1981; Jacoby & Glauberman, 1995; Kamin, 1974; Lewontin et al., 1984.

critics argue that, while supposedly all sides acknowledge that American blacks score 12-15 points lower than American whites on standardized IQ tests, the difference is explainable by environmental factors such as poverty and racism. Although this argument is solid even if we accept that IQ tests are unbiased and actually measure intelligence, there is good reason not to accept these positions.

Racial and Class Bias Built into IQ Tests

Intelligence testing has a long history of being used for reactionary social and political purposes. The testing method developed by Alfred Binet in France was brought to the United States and developed by eugenicists such as Henry Goddard and Lewis Terman for the purpose of quantifying what they saw as people's innate mental capacities. Indeed, in 1924 Terman would write that his fellow psychologists and their IQ tests were "the beacon light of the eugenics movement."[1144]

The historical association of IQ testing with eugenics and racism, as described by several authors, does not mean that all of the early pioneers of mental testing were driven by these motivations, or that most contemporary mental testing is performed for this purpose. Still, modern proponents of racial and class inferiority continue to use the results of genetic studies of IQ in support of their positions. It is therefore necessary to demonstrate that one of the original *purposes* of IQ tests was to prove that blacks and other minorities, southern European immigrants and Jews, and the working-class were genetically inferior to upper-class whites of Northern European descent.

Related to this point, it is critically important to understand that IQ tests can produce any result its creator desires. For those who doubt this, I call on the most well-known American defender of IQ testing, Arthur Jensen. "It is claimed," wrote Jensen, "that the psychometrist can make up a test that will yield any kind of score distribution he pleases. This is roughly true, but some types of distributions are much easier to obtain than others."[1145] Although Jensen is a leading proponent of the idea that racial differences in IQ scores are at least partly explained by genetics, he implied that psychometrists *could* create tests in which blacks and whites scored equally. Interestingly, most *choose* not to.

For the American pioneers of mental testing there was no doubt that blacks and the working class were genetically inferior to upper class whites. In

1144. Quoted in Samelson, 1979, p. 105.
1145. Jensen, 1980, p. 71.

Terman's 1916 Stanford-Binet manual, he wrote that "dullness" among blacks and Latinos "seems to be racial, or at least inherent in the family stocks from which they come."[1146] Although Terman suggested that research was needed, he believed that "when this is done there will be discovered enormously significant racial differences in general intelligence, differences which cannot be wiped out by any scheme of mental culture"[1147] Psychologist Carl Brigham wrote in his influential 1923 book *A Study of American Intelligence*, which discussed the large-scale testing of World War I army recruits, that "These army data constitute the first really significant contribution to the study of race differences in mental traits. They give us a scientific basis for our conclusions."[1148] This passage is illuminating, since Brigham admitted that he and his colleagues reached conclusions about racial differences before the "scientific" evidence had come in. Lacking a scientific basis, Brigham, who viewed "the importation of the negro" as the "most sinister development in the history of this continent,"[1149] was actually describing his *prejudices*. Brigham wrote that "Our own data from the army tests indicate clearly the intellectual superiority of the Nordic race group."[1150] Another pioneer of mental testing, Henry Goddard, lectured Princeton students in 1919 as follows:

> Now the fact is, that workman may have a ten year intelligence while you have a twenty. To demand for him such a home as you enjoy is as absurd as it would be to insist that every laborer should receive a graduate fellowship. How can there be such a thing as social equality with this wide range of mental capacity?[1151]

My purpose is to document the beliefs of the people who created the tests. The results of their tests merely confirmed — and were a product of — their pre-existing views on racial and class inequality.

When psychometrists *do* assume equality, they simply create tests to reflect this assumption. Items on standardized IQ tests do not fall from the sky; they are carefully selected (out of a large pool of potential items) to produce results desired by the test creator. In Terman and Merrill's 1937 revision of the Stanford-Binet test, they wrote that a "few tests in the trial batteries which

1146. Terman, 1916, p. 91.
1147. *Ibid.*, p. 92.
1148. Brigham, 1923, p. xx.
1149. *Ibid.*, p. xxi.
1150. *Ibid.*, p. 207.
1151. Quoted in Kamin, 1974, p. 8.

yielded largest sex differences were early eliminated as probably unfair."[1152] Because they assumed that males and females are equal in intelligence, Terman and Merrill created a test whose results reflected this assumption. Apparently, it was not "unfair" if races and classes scored differently, or if people reached conclusions about genetic inferiority on the basis of their scores.

David Wechsler, who developed the most widely used IQ tests, followed Terman in eliminating sex differences from his scales, although he had a "'sneaking suspicion' that the female of the species is . . . more intelligent that the male."[1153] Wechsler described the problem of sex differences as follows:

> In trying to arrive at an answer as to whether there are sex differences in intelligence much depends upon how one defines intelligence, and on the practical side, on the types of tests one uses in measuring it. The contemporary approach, contrary to the historical point of view, adopts a sort of null hypothesis. Unfortunately this procedure turns out to be a circular affair since the nature of the tests selected can prejudice or determine in advance what the findings will be.[1154]

As Wechsler acknowledged, the "null hypothesis" for male-female differences in intelligence is that the sexes are equal. Like his predecessors, however, another implicit assumption of his tests was that *races and classes* are unequal.[1155] This reflects little more than the beliefs and prejudices of the test creators and their backers. According to Jensen, "The practice of eliminating and counterbalancing items to minimize sex differences is based on the assumption that the sexes do not really differ in general intelligence." [1156] What Jensen failed to articulate is the assumption upon which racial and class differences are allowed to remain, which could be stated, "*The failure to remove and counterbalance items in order to eliminate racial and class differences is based on the assumption that races and classes really do differ in general intelligence.*" As Ken Richardson observed,

---

1152. Terman & Merrill, 1937, p. 34.
1153. Wechsler, 1944, p. 107.
1154. *Ibid.*, p. 144.
1155. Like Terman and Brigham, Wechsler (1958, p. 90) believed that racial differences in intelligence were real: "'Practical' handling of the problem does not, of course, imply an answer to the question of whether or not there are ethnic and cultural differences in intelligence. That such differences exist appear open to little doubt."
1156. Jensen, 1980, p. 623.

While "preferring" to see sex differences as undesirable artifacts of test composition, other differences between groups or individuals, such as different social classes or, at various times, different "races," are seen as ones "truly" existing in nature. Yet these too could be eliminated or exaggerated by exactly the same process of assumption and manipulation of test composition.[1157]

In other words, the belief that races and classes are genetically inferior is built into most standardized IQ tests.[1158] It is not simply a matter of whether individual test items are culturally biased on their face; more important is that psychometrists know that individual test items, regardless of how culturally biased they may appear, discriminate between various groups. As IQ critics Mensh and Mensh observed, "there is no distinction between crassly biased IQ-test items and those that appear to be non-biased."[1159]

The fact that the sexes, but not races and classes, were assumed equal in "native intelligence" reveals the racial and class bias of the tests. On what grounds, we might ask, did people like Terman and Wechsler anoint women with the same level of native intelligence as men? After all, during the time when they were creating their tests most women were apparently so innately deficient that they rarely worked outside of the home. Almost all of the captains of industry and most professionals, including college professors and psychologists, were men, as were most of history's great leaders, thinkers, writers, artists, military commanders, and inventors. In fact, the innate cognitive impairment of women was so widely recognized that they were not guaranteed the right to vote in the United States until 1920! (Undoubtedly, granting this right was a demonstration of the male cognitive elite's goodwill.) By now it is clear that I have slipped into facetiousness — but only to make a point. Using the standards applied for determining the relative worth and intelligence of races and classes, Terman, Goddard, Brigham and others should have viewed women as being innately inferior to men. But white upper-class women had two characteristics that set them apart from non-white races and the working class: They were of the same race and class as those for whom the tests were designed to find "scientifically" superior! Could the test creators actually decide that their own mothers, wives, and daughters were innately inferior? The pioneers of mental testing decided to create cognitive equality between men and women in order to

---

1157. K. Richardson, 1998, p. 114.
1158. Mensh & Mensh, 1991.
1159. *Ibid.*, p. 51.

eliminate a distraction from what they really wanted to show — that with regards to measured intelligence, blacks were inferior to whites, Southern Europeans and Jews were inferior to Northern Europeans, and the working-class was inferior to the capitalist class.

I recall an undergraduate university lecture by a political science professor who, I later learned, believed in the existence of racial differences in intelligence. He said that some people argued that blacks score lower than whites on IQ tests because they are treated as inferiors and are told they are stupid. In response, he pointed out that women historically have received a similar message, yet score the same as men on standardized IQ tests. The implication, of course, was that a person's or group's position in society is not reflected by lower IQ scores. My professor either had no idea about how IQ tests are constructed, or he was consciously deceiving his students. Had I known anything about the subject in those years I would have answered that women score about the same as men because the tests are *designed* to produce this result, and that blacks score lower than whites for precisely the same reason.

Even if IQ tests were a valid measure of "intelligence," separate norms for groups experiencing vastly different social environments (such as blacks and whites in the Jim Crow South!) could have been created. This was acknowledged by Wechsler in 1944, although he claimed that it was not "possible to do this at present."[1160] In fact, it was quite possible to create separate norms but it was undesirable from the standpoint of those who wanted to use the tests to provide scientific evidence for racial inequality. Had Wechsler decided to create separate norm groups he would have undermined one of the original purposes of American IQ tests, which was — literally — to demonstrate the genetic superiority of "the Nordic race group." That aspects of this "original purpose" live on is shown by the fact that contemporary IQ tests are seen as valid if they correlate well with the original Stanford-Binet. To his credit, Wechsler found it necessary to state that "our norms cannot be used for the colored population of the United States."[1161] Naturally, this did not prevent subsequent commentators from citing racial differences in Wechsler IQ scores as evidence of black people's genetic inferiority.

From the standpoint of the economically powerful classes, IQ tests helped justify the inequalities of capitalism. They could argue, as Goddard argued, that

---

1160. Wechsler, 1944, p. 107.
1161. *Ibid.*, p. 107.

genetic differences between classes precluded the egalitarian society advocated by socialists. This position is embraced by most contemporary IQ hereditarians. "The tests," wrote Mensh and Mensh in *The IQ Mythology*, "do what their construction dictates; they correlate a group's mental worth with its place in the social hierarchy."[1162]

In response to Jensen's claim that IQ tests were not designed to discriminate between social groups, Mensh and Mensh responded, "In reality — which is precisely the opposite of what Jensen claims it to be — test discrimination among individuals *within* any group is the incidental by-product of tests constructed to discriminate *between* groups."[1163] This position requires qualification. The eugenics movement, which played a major role in promoting IQ testing, was interested in individual as well as group differences. Even among the American white population, the "chronic pauper stocks" (such as the "Jukes" and the "Kallikaks") were targeted for eugenic intervention. Generally speaking, and allowing for considerable overlap, eugenicists supported IQ tests as an instrument for identifying individuals who should be prevented from reproducing; the ruling elite wanted to scientifically legitimize the class-stratified society it had created; racists and segregationists saw IQ testing as evidence supporting their causes; and psychologists such as Terman, Brigham, Goddard, and others wanted all of these things. And all the while the interests of the poor, blacks, immigrants, native Americans, and the working class were ignored. After all, they were powerless.

Nevertheless, many psychologists and psychiatrists used (and continue to use) IQ tests for benign purposes. Yet it is important to understand why the tests were created and how present-day tests, including those claiming to be "culture-fair," are based on many of the assumptions of the original Stanford-Binet, published in 1916.[1164] In this respect there are parallels between twin research and the American brand of IQ testing. Both were used (some might say developed) in order to identify people in need of eugenic intervention, which sometimes included compulsory sterilization. Yet despite contemporary society's rejection of the eugenic program, IQ testing and twin research remain with us.

1162. Mensh & Mensh, 1991, p. 30.
1163. *Ibid.*, p. 73.
1164. Rose, 1997.

The Fate of Castelike Minorities

An international perspective is often helpful in demystifying narrow national debates. Anthropologist John Ogbu documented the fate of castelike minority groups such as the Maoris in New Zealand and the Burakumin of Japan.[1165] He believed that the lower performance of minority children is the result of their status as members of a group relegated to the bottom of society, such as the case with blacks in the United States.

Expanding on this point, Fischer and colleagues documented the position of these and other groups. These include whites and aborigines in Australia, high caste and low caste in India, Jews and Arabs in Israel, English and Irish in Great Britain, and French and Flemish in Belgium. In each of these cases the latter group scores lower on standardized IQ tests. Another example is South Africa between 1950 and 1990, where Afrikaaners (Dutch origin) scored from one-half to a full standard deviation lower on IQ tests than people of English descent. By the 1970s, however, the gap had disappeared: "The convergence of Afrikaaner and English scores coincides with the rise of Afrikaaners to power in South Africa after generations of subordination to the English."[1166] Fischer and colleagues' thesis reads as follows: "A racial or ethnic group's position in society determines its measured intelligence rather than vice versa."[1167]

Fischer and colleagues also discussed Koreans in Japan, who occupy a position in Japanese society analogous to blacks in the United States or West Indians in the UK. They also score lower than members of the dominant culture on IQ tests:

> Koreans, who are of the same "racial" stock as Japanese and who in the United States do about as well academically as Americans of Japanese origin (that is, above average), are distinctly "dumb" in Japan. The explanation cannot be racial, nor even cultural in any simple way. The explanation is that Koreans, whose nation was a colony of Japan for about a half-century, have formed a lower-caste group in Japan.[1168]

Clearly, the United States is not the only country in which lower-caste groups score lower on standardized IQ tests.

1165. Ogbu, 1978.
1166. Fischer et al., 1996, p. 193.
1167. *Ibid.*, p. 173.
1168. *Ibid.*, p. 172.

CONCLUSION

The claims of behavior geneticists and others in support of important genetic influences on intelligence derive from their acceptance of many implausible assumptions. Several critics, such as Howard Taylor and Ken Richardson, have listed some of these assumptions.[1169] If we include the false assumptions underlying behavior genetic methods such as twin and adoption studies, the case for important genetic influences on intelligence collapses completely.[1170] As Richardson commented,

> I think it can be safely said that never before in any field of science have so many arbitrary assumptions been gathered together, *in full knowledge of their invalidity*, as the basis of substantive claims about the nature of people, with so many potentially dire consequences for them.[1171]

Given the racial and class bias *built into* the most widely used IQ tests, claims by the authors of *The Bell Curve* and others about the innate cognitive inferiority of ethnic minorities and the working class are preposterous, to say the least. "Nothing," wrote Lewontin, Rose, and Kamin, "demonstrates more clearly how scientific methodology and conclusions are shaped to fit ideological ends than the sorry story of the heritability of IQ."[1172]

1169. K. Richardson, 1998; Taylor, 1980.
1170. This observation applies to a 2001 brain mapping twin study by Thompson et al. claiming important genetic influences on brain structure. The authors also claimed that frontal gray matter differences are correlated with Spearman's *g* (Gray & Thompson, 2004; Thompson et al., 2001). According to Plomin & Craig, this study "shows that *g* has a biological basis" (Plomin & Craig, 2002, p 186).
1171. K. Richardson, 1998, p. 135. Emphasis in original.
1172. Lewontin et al., 1984, p. 100.

CHAPTER 10. MOLECULAR GENETIC RESEARCH IN PSYCHIATRY AND PSYCHOLOGY: AN EXERCISE IN FUTILITY?

Psychiatry as a discipline has too often been characterized by many speculations based on few facts.

—Psychiatric geneticists Edith Zerbin-Rüdin and Kenneth Kendler in 1996.[1173]

Much of what passes for scientific advance in psychiatry is, in fact, rhetorical innovation.

— *Psychiatry critic Thomas Szasz in 1964.*[1174]

The search for the genes believed to contribute to psychiatric disorders and psychological trait differences has been underway for many years. In psychiatry, researchers attempt to identify genes for schizophrenia, attention-deficit hyperactivity disorder, and other diagnoses. Other investigators attempt to identify genes for intelligence, personality, and abnormal behavior. Molecular genetic research differs from the methods discussed in the previous chapters in several respects. Most importantly, investigators must identify actual genes. With twin and adoption studies, the existence of genes was inferred from correlations of observed or reported behavior, or of test scores. To convince the public and professionals that these genes actually existed, proponents of the genetic position had merely to argue that their research was sound, and that their results pointed in the genetic direction.

There are two main reasons why we should not expect molecular genetic research in psychiatry and psychology to provide much useful information now or in the future. The first relates to the unwarranted belief that family, twin, and adoption studies have established the role of genetic factors. Most scientific articles reporting molecular genetic research begin by stating that these studies' results have established the role of genetics. For example, Tsuang and Faraone wrote in 2000, "A century of genetic epidemiologic research shows that genes play a substantial role in the etiology of schizophrenia. This is the only rea-

---

1173. Zerbin-Rüdin & Kendler, 1996, p. 332.
1174. Szasz, 1964, p. 525.

sonable conclusion we can draw from family, twin and adoption studies."[1175] More recently, M. D. Fallin and colleagues began their 2003 schizophrenia molecular genetic study by writing that "many lines of evidence, including twin, adoption, and family studies, support a strong genetic component" for schizophrenia,[1176] and Luo et al. wrote in 2004 that "evidence from family, twin, and adoption studies has indicated that both genetic and environmental factors must be involved in the etiology of schizophrenia."[1177]

The logic behind the current effort to identify genes is similar to other areas of science. For example, 19th century astronomers postulated Neptune's existence after observing irregularities in Uranus's orbit. Another astronomer subsequently calculated the position of a heavy unknown object that could account for this irregularity, leading to the discovery of Neptune in 1846. Similar logic is observed in everyday life. When approaching a crosswalk, a motorist will slow down when she sees a large vehicle in the next lane slowing down. Although she cannot see the entire crosswalk, she infers the existence of a pedestrian from the behavior of the large vehicle's driver. For molecular geneticists, family, twin, and adoption studies are analogous to Uranus's orbit and the large vehicle slowing down at a crosswalk. Based on their interpretation of the data produced by these studies, they reason that there must be a gene or genes for the condition or trait in question.

The second problem relates to the practical benefits of finding genes. Proponents of molecular genetic research argue that gene findings can be used to prevent or cure diseases influenced by identified genes. Although this is a laudable goal in the fight to cure physical diseases, does society wish to use this technology to alter "intelligence genes," or genes associated with unpleasant personalities? It might be argued that the discovery of relevant genes would be no more enlightening than the "discovery" of the original predisposition, since in both cases we might still choose to focus attention on environmental factors or triggers.

1175. Tsuang & Faraone, 2000, p. 1.
1176. Fallin et al., 2003, p. 601.
1177. Luo et al., 2004, p. 112. Another example is Brzustowicz & colleagues' claim (2004, p. 1057) that "Family, twin, and adoption studies have demonstrated that schizophrenia is predominantly a genetic disorder with high heritability . . ."

MOLECULAR GENETIC RESEARCH IN PSYCHIATRY

Schizophrenia

Researchers have been searching for "schizophrenia genes" for over 20 years. The most common methods in schizophrenia molecular genetic research are linkage and association studies. In a linkage study, researchers look for genetic markers linked with the putative disease gene among consanguineous family members. Findings are often represented as a logarithm of odds (LOD) score, which expresses the probability that the linkage occurred by chance. By convention, an LOD score higher than 3 (1000:1 odds in favor of linkage) is required in order to claim a significant linkage. Linkage studies are designed to identify areas of the chromosome where relevant genes might be located, but are unable to identify actual genes. This is the task of follow-up studies. Association studies compare the frequency of genetic markers among unrelated affected individuals and a control group.

Periodically, investigators claim to have found the location of genetic markers for schizophrenia. These studies usually are followed by retractions or failures to replicate. The most well-known example is Sherrington and colleagues' 1988 study, which was accompanied by an article in the same issue of *Nature* by Kennedy et al., who failed to replicate the findings.[1178] Another highly publicized yet subsequently withdrawn claim (this time for bipolar disorder) was published by Egeland and associates in 1987.[1179] Unfortunately, it is widely believed in the general public that genetic markers for these and other conditions have been located, when it is not the case. Furthermore, finding a marker is not the same as finding an actual gene; it merely points to an area where the gene might be located. Still, when the public hears of studies purporting to find markers they often assume that a gene has been found. In an analogy discussed by psychologist Ty Colbert, finding a marker is similar to prospectors coming across traces of gold in a river.[1180] This "marker" might indicate that a gold vein is nearby, or there could be other explanations. In any case, it would be premature to conclude that a gold vein had been discovered in the absence of other evidence. Colbert noted in this example that we at least can be sure that gold actually exists, which cannot be said about "mental illness" genes.

1178. Sherrington et al., 1988; Kennedy et al., 1988.
1179. Egeland at al., 1987.
1180. Colbert, 2001.

The numerous schizophrenia linkage studies, association studies, and genome scans have thus far failed to produce consistently replicated findings of any gene or marker. A large 2002 internationally-based schizophrenia linkage study of 382 sibling pairs by DeLisi and associates failed to replicate reports of linkage in several previous studies. The investigators acknowledged that the "most striking feature" of their results "is the failure to confirm a number of earlier claims of positive findings," and wrote further that the "present findings suggest that a critical reevaluation of the linkage approach is warranted."[1181]

Despite results of this type, the latter part of 2002 saw claims that susceptibility genes for schizophrenia had been discovered. For genetic researcher C. Robert Cloninger, the 2002 studies of Straub et al., Chumakov et al., and Stefansson et al. were a "watershed" event, and that for "the first time, specific genes have been discovered that influence susceptibility to schizophrenia . . ."[1182] Cloninger's claim, however, is more the result of wishful thinking than objective scientific evaluation. In addition to the premature nature of his conclusions, how can "schizophrenia susceptibility genes" (in reality, a mere association between "schizophrenia" and genetic markers) be found when, as a leading American biological psychiatrist and schizophrenia researcher has admitted, American psychiatry cannot even "figure out who really has schizophrenia or what schizophrenia really is."[1183] Nevertheless, genetic investigators Elkin, Kalidindi, and McGuffin echoed Cloninger in 2004 by proclaiming that "Schizophrenia genes have been found at last,"[1184] although other leading researchers have been much more cautious.[1185]

Claims to the contrary notwithstanding, as of this writing molecular genetic studies have failed to find genes for schizophrenia. Most leaders of the field now believe that many genes are involved (the polygenic theory), and have abandoned the single-gene approach. Psychiatric geneticists Tsuang and Faraone wrote in 2000, "We can now conclusively reject the idea that there is one gene of major effect that causes schizophrenia." They recommended that

---

1181. DeLisi et al., 2002, p. 808. In contrast to DeLisi & colleagues, Straub et al. (2002) found linkage between SSDs and the 6p22 gene.

1182. The Cloninger quote is from Cloninger, 2002, p. 13365. The three studies Cloninger referred to were, Chumakov et al., 2002; Stefansson et al., 2002; and Straub et al., 2002.

1183. Andreasen, 1998, p. 1659.

1184. Elkin et al., 2004, p. 107.

1185. DeLisi & Crow, 2003; Harrison & Owen, 2003; Kennedy et al., 2003.

future researchers design studies "to detect the many genes of small effect that each increase susceptibility to the disorder."[1186] The title of the article in which these comments are found is, fittingly, "The Frustrating Search for Schizophrenia Genes."

Tsuang and Faraone worried that "failures to replicate molecular genetic studies of schizophrenia might be interpreted to mean that schizophrenia genes do not exist," and cautioned that such a conclusion would be "premature" because the results from twin and adoption studies show that genes play an important role.[1187] While it is true that the failure to find schizophrenia genes does not prove that such genes do not exist, the belief that twin and adoption studies show that genes are involved is erroneous. The irony is that instead of confirming the results of schizophrenia twin and adoption research, the failure to find schizophrenia genes may lead researchers to take a much needed second look at this greatly flawed and environmentally confounded body of research. Schizophrenia genetic researcher Lynn DeLisi acknowledged in 2000 that "psychiatric genetics appears to be at a crossroads or crisis," as investigators continue to look for the "elusive gene or genes" for schizophrenia.[1188] She discussed "those researchers who entered the new [*sic*] field of psychiatric genetics [who] feared most of all that they would screen the whole genome and find nothing."[1189] This is precisely the "crisis" facing psychiatric genetics: Its adherents are looking for genes that may not even exist.

Another reason that investigators believe they will find genes is the concurrent (though not necessarily related) view of schizophrenia as a brain disease. Here again, as Peter Breggin has convincingly argued, the evidence is weak.[1190] On the other hand, real brain diseases such as Alzheimer's and Huntington's disease are put forward by biological psychiatrists as examples of what can be accomplished for schizophrenia. For example, Nancy Andreasen referred to Alzheimer's disease as "one of the current 'stars' in the molecular biology firmament."[1191] However, Alzheimer's and Huntington's are diseases of the brain, which can be seen in post-mortem examinations. These conditions, therefore,

---

1186. Tsuang & Faraone, 2000, p. 1. Other examples of researchers' rejection of the single gene hypothesis include Portin & Alanen (1997), and Moldin & Gottesman (1997).
1187. Tsuang & Faraone, 2000, p. 1.
1188. DeLisi, 2000, p. 190.
1189. *Ibid.*, p. 189.
1190. See Breggin, 1991. Also see Siebert, 1999.
1191. Andreasen, 2001, p. 121.

are not "behaviors" or "mental illnesses"; they are real brain diseases. Schizo-phrenia and most other psychiatric disorders, on the other hand, are affective states or socially disapproved behaviors given the name "disease" in the absence of convincing evidence in support of their biological bases. Why then do genetic researchers in psychiatry and psychology continue to point to Alzheimer's and Huntington's as examples of what they could find for conditions such as schizo-phrenia and attention-deficit hyperactivity disorder? The answer is that they have few other success stories to point to.

According to Thomas Szasz, when a psychiatric condition is shown to have a physical basis it leaves the domain of psychiatry and becomes a non-psy-chiatric medical concern:

> As soon as a disease thought to be mental is proven to be physical, it is removed from the domain of psychiatry and placed in that of medicine, to be treated henceforth by internists, neurologists, or neurosurgeons. This is what happened with paresis, pellagra, epilepsy, and brain tumors. It is an ironic paradox, then, that while definitive proof that mental illnesses are brain diseases would destroy psychiatry's *raison d'être* as a medical specialty distinct and separate from neurology, the claim that mental illness is a brain disease has served, and continues to serve, as the psychiatrist's most effec-tive justification for legitimacy as an independent medical discipline.[1192]

Thus, the psychiatric domain consists mainly of disorders that are claimed — but not proven — to have a biological basis. We might therefore say that psy-chiatry is where one will find *pseudo* brain diseases. The current multinational effort to find genes for schizophrenia is based on the position that it is a brain disease of genetic origin, yet there is little evidence in support of this view. The best course of action for molecular genetic investigators would be to perform their own critical reanalysis of the original twin and adoption studies, upon which their current search is based.

Attention-Deficit Hyperactivity Disorder

Another area of psychiatry where methodologically unsound research on twins and adoptees has led to the erroneous conclusion that genes must exist is attention-deficit hyperactivity disorder (ADHD). Even more than schizophrenia, ADHD exemplifies the way that socially disapproved behavior can be turned into a "disease." In Peter Breggin's words, the DSM-IV definition of ADHD is

---

1192. Szasz, 1987, p. 70.

"simply a list of behaviors that require extra attention from teachers."[1193] Like schizophrenia, the evidence supporting the brain disease theory of ADHD is weak.[1194] The genetic evidence consists of family studies, twin studies, and adoption studies far more flawed than their schizophrenia counterparts.[1195]

Although textbooks usually report that ADHD has an important genetic component, the validity of the equal environment assumption of the twin method is rarely challenged, and unsound ADHD adoption studies are often endorsed without analysis. Also like schizophrenia, the search for ADHD genes is well underway. In 2000, Faraone and Biederman claimed that molecular genetic research, while still its "infancy," has "already implicated several genes as mediating the susceptibility to ADHD."[1196] They cited several studies claiming an association between ADHD and specific genes, as well as several studies failing to replicate these findings. This undoubtedly constitutes an inconclusive body of evidence, in spite of a meta-analysis cited by the authors which found a significant association between a specific gene and ADHD.[1197] A disease-causing gene cannot be discovered by running a body of inconclusive research through a computer, as Faraone and Biederman seemed to suggest.[1198] Even in studies Faraone and Biederman cited as finding a significant association, the results are treated with caution. For example, the authors of the 1999 Faraone et al. study linking ADHD to the 7-repeat allele of the DRD4 gene found a statistically significant association, but also noted that "58% of the subjects without the 7-repeat allele had ADHD. . . . suggest[ing] that the 7-repeat allele cannot be viewed as a necessary cause of ADHD."[1199]

As of this writing, and despite concerted efforts worldwide, no genes for ADHD have been discovered. Although researchers believe that it is only a matter of time until they are able to identify genes, like schizophrenia, a more proper course would be to critically reexamine previous ADHD kinship research.

1193. Breggin, 2001, p. 203.
1194. See Breggin, (1998), and Leo & Cohen (2003) for critiques of ADHD brain studies.
1195. See Joseph, 2000c, 2000e, 2002a. Also see McMahon, 1980.
1196. Faraone & Biederman, 2000, p. 572.
1197. Faraone et al., 2001.
1198. For a critique of the Faraone et al. meta-analysis, see Pittelli, 2002b.
1199. Faraone, Biederman, et al., 1999, p. 770.

Why Search for Genes?

Tsuang and Faraone listed three potential benefits of identifying genes for schizophrenia: "pharmacogenomics," "pharmacogenetics," and the identification of high risk children.[1200] The first two words refer to producing psychotropic drugs with the aid of genetic information. Turning to the third potential benefit, despite the failure to find schizophrenia genes, some people have proposed the identification and "treatment" of so-called high risk children. But suppose, for the sake of argument, that schizophrenia really does have a genetic basis (which would then be triggered by environmental factors), and investigators are able to identify the predisposing gene or genes. The strategy of "early intervention" would still be problematic. The knowledge that a child is carrying schizophrenia genes could in itself be a life altering event, affecting the way he or she is treated by parents and the social environment. Even Andreasen admitted that "Treating young people who are not yet ill may adversely affect their self-esteem and self-image, perhaps creating a self-fulfilling prophecy that may lead them to eventually become ill."[1201] If Tienari and colleagues are correct that the predisposition manifests itself in people who grow up in disturbed homes, would this mean that predisposed children would be removed from these types of homes? And who would decide what constitutes "disturbed?" Treatments proposed by psychiatric geneticists usually include neuroleptic (also known as "anti-psychotic") drugs. These drugs, however, merely modify behavior or mood. They do not treat physical illness or "repair the brain." Moreover, neuroleptics can cause troubling adverse effects and permanent neurological damage such as tardive dyskinesia. They can also be lethal (for example, they can cause neuroleptic malignant syndrome).[1202] Furthermore, as seen in Mosher and colleagues' Soteria project, non-medicated people diagnosed with schizophrenia who enter a supportive home-like treatment program show as much improvement as others hospitalized and prescribed neuroleptics.[1203] Thus, even in the unlikely event that schizophrenia genes are found, society might still choose to focus on non-medical interventions, and on eliminating or mitigating schizophrenia-producing environments.

1200. Tsuang & Faraone, 2000.
1201. Andreasen, 2001, p. 330.
1202. See Breggin, 1991; Breggin & Cohen, 1999; Cohen, 1997; Cohen & McCubbin, 1990; Whitaker, 2002.
1203. Mosher, 2004; Mosher & Menn, 1978; Mosher et al., 1995; Whitaker, 2002.

Early intervention would also be problematic in disorders such as ADHD. As Australian commentators Yeh and colleagues wrote, even if a child is found to carry ADHD susceptibility genes, it still would not mean that the child will necessarily develop ADHD. Moreover, they argued that for "polygenic disorders," which they believe ADHD to be, "most individuals in the general population will carry one or more susceptibility alleles. Only a small percentage of individuals will carry enough susceptibility alleles to predict a high risk of developing the disorder."[1204] After citing potential stigmatization and other problems, they concluded "the importance of these issues should make us very cautious about accepting any proposals that may be made to test for ADHD susceptibility alleles in the general population."[1205]

A potentially disturbing aspect of identifying genes for psychiatric disorders is that information could be used for eugenic purposes. The identification of genes would have been warmly embraced by people like Rüdin, Luxenburger, Kallmann and other racial hygienists and eugenicists in their quest to rid the world of the "schizophrenia phenotype." As a contemporary German psychiatric genetic researcher observed, "doctors and scientists involved in the crimes of the Nazi period. . . . would undoubtedly have welcomed the technical possibilities of present-day genetics."[1206] According to genetic researchers Moldin and Gottesman, however, the discovery of genes would *reduce* stigma. They wrote that Sherrington and colleagues' subsequently withdrawn 1988 finding

> was welcomed with considerable optimism and insufficient scientific criticism not only because hopes were raised of identifying a factor that could finally provide clues to pathophysiology and ultimately new treatments, but also because localization of a locus was "confirmation" that schizophrenia was in fact "biological" and not a "psychosocial" disorder. The long history of stigma associated with schizophrenia further reinforces the desire to see schizophrenia as a genetic or medical condition.[1207]

However, the "long history" of the stigmatization of people diagnosed with schizophrenia is directly related to their being viewed as the carriers of "hereditary taint." Even behavior geneticists Plomin and Rutter acknowledged that the 20th century provided "chilling examples of compulsory sterilization an the name of eugenics and of genetic stigmatization."[1208] Although it is true that psy-

---

1204. Yeh et al., 2004, p. 15.
1205. *Ibid.*, p. 16.
1206. Propping, 1992, p. 910.
1207. Moldin & Gottesman, 1997, p. 554.

chosocial theories can also stigmatize families (or parents), there is a huge difference between the stigma of being a "bad parent," and the stigma of being seen
as a carrier of "bad genes." While psychoanalysts may have stigmatized "schizophrenogenic" parents from time to time, the stigma of being a "hereditary taint
carrier" has led to the perpetration of crimes against people diagnosed with
schizophrenia and other psychiatric or mental conditions. Thus, there is little to
be gained by finding presumed schizophrenia genes, but the potential for misuse
of this information is enormous.

THE SEARCH FOR BEHAVIORAL AND IQ GENES

Despite the poor track record of genetic research in psychiatry, many
people with psychiatric conditions suffer greatly and chronically. In the case of
schizophrenia, young people are diagnosed in the beginning of their productive
years and often behave in ways that bring great suffering to those who love and
care about them. It is therefore understandable that researchers would be sub-
jectively motivated to alleviate this suffering and to help restore people to productive lives.

The search for IQ genes and genes for normal variation in behavior is
another matter. Even if we accept the behavior genetic argument that genes
exert an important influence on behavioral and IQ differences among people, we
could still question the necessity of identifying the genes producing these differences. Yet, leading behavior geneticists are currently searching for genes believed
to influence psychological trait differences. According to Plomin and colleagues,

> Psychology is at the dawn of a new era in which molecular genetic tech-
> niques will revolutionize genetic research in psychology by identifying spe-
> cific genes that contribute to genetic variance for complex dimensions and
> disorders. The quest is to find not *the* gene for a trait, but the multiple genes
> that affect the trait in a probabilistic rather than predetermined manner.
> The breathtaking pace of molecular genetics . . . leads us to predict that psy-
> chologists will routinely use DNA markers as a tool in their research to
> identify some of the relevant genetic differences among individuals.[1209]

In 1998, Plomin and Rutter discussed and attempted to justify the search
for "behavior genes."[1210] They noted that while previous researchers looked for

1208. Plomin & Rutter, 1998, p. 1237.
1209. Plomin, DeFries, et al., 1997, p. 277.

single major genes influencing behavior, it is now believed that many genes contribute to the appearance of a trait or to differences in continuously distributed traits (such as IQ). Presumed markers for genes contributing to these traits are called *quantitative trait loci* (or QTLs). Plomin and Rutter viewed the effects of genes on behavior as "probabilistic," as opposed to "deterministic," meaning that genes increase the likelihood of the appearance of a trait, but are not the sole cause. As examples of behavioral gene discoveries, Plomin and Rutter cited studies whose authors claimed to have found genes associated with Alzheimer's disease and "novelty seeking." Why they discussed Alzheimer's disease in this context is unclear, since, as we have seen, it is best understood as a brain disease. Moreover, studies finding an association between the DRD4 gene and novelty seeking have seen many subsequent replication failures by other groups,[1211] and in studies where replication was claimed, the results are more likely due to "statistical accident" than a real association.[1212] Furthermore, it is difficult to imagine how a gene could influence the way people answer questions on personality tests. In any case, Plomin and Rutter believed that the DRD4 gene accounts for only 4% of the variance in novelty seeking, "which may prove typical for complex quantitative traits."[1213]

Regarding the potential misuse of genetic information, Plomin and others frequently argue that policy decisions are based on values, which are allegedly separate from "science." Yet behavior genetics, like all social science, is unavoidably permeated with politics and values. For example, Plomin and associates have written that "the basic message of behavioral genetics is that each of us is an individual. Recognition of, and respect for individual differences is essential to the ethic of individual worth."[1214] This is a social policy position, not a scientific finding. Plomin and colleagues' position on individuality would be rejected in Asian cultures where people are viewed more as members of their family than as individuals. There are many other ways that "each of us" can view ourselves, such as members of a class, "race," political party, ethnic group, nationality, club, etc. The danger here, and this is a constant theme throughout the entire history of human genetic research, is the presentation of political and social programs disguised as objective, apolitical science. In any case, regardless

1210. Plomin & Rutter, 1998.
1211. Barr, 2001; Bouchard & Loehlin, 2001; Prolo & Licinio, 2002.
1212. Beckwith & Alper, 2002, p. 321.
1213. Plomin & Rutter, 1998, p. 1226.
1214. Plomin, DeFries, et al., 1997, p. 279.

of public statements by contemporary behavior geneticists, there is an enormous potential for the misuse of genetic information. Those who worry about this should also be aware that hard scientific evidence is not necessary — and never has been necessary — to bring about atrocities. Eugenic sterilization was flourishing in the United States before the twin method had even been invented, and the Nazis and their psychiatric genetic collaborators performed their deeds on the basis of what we would today consider very weak evidence.

Although behavior genetic researchers continue to look for "IQ genes" (or QTLs correlated with g), the search has been unsuccessful. Plomin and colleagues published the results of their initial IQ QTL study in 1995, but were unable to find anything important: "Although several significant associations were found in an original sample, only one association was replicated cleanly as an independent sample. This finding might well be a chance result, because 100 markers were investigated."[1215] Indeed, there is a high risk for false positive results (Type I errors) in QTL studies, which can lead to spurious results.[1216] One such fishing expedition was published by Plomin and Craig in 2001.[1217] The investigators identified three markers (not actual genes) associated with general cognitive ability, whose association probability value, however, did not meet the conservative standards recommended by other genetic researchers.[1218] Nonetheless, Plomin and Craig wrote later that "the many hurdles that we set for acceptance of a quantitative trait locus association may have been to high ..."[1219] In addition, there are problems with the sampling procedures used by Plomin and Craig.[1220] By 2004, Plomin recognized that "no solid [IQ] QTL associations have yet emerged,"[1221] and that "the road ahead will be much more difficult than generally assumed ..."[1222]

Although he has failed to find QTL associations for IQ, Plomin expects to find differences along class and racial lines:

1215. *Ibid.*, p. 153. For the original study, see Plomin et al., 1995.
1216. Wahlsten, 1999.
1217. Plomin & Craig, 2001.
1218. The need to establish more conservative p values for genome-wide scans was discussed in Lander & Kruglyak, 1995, and Benjamini et al., 2001.
1219. Plomin & Craig, 2002, p. 186.
1220. Pittelli, 2002a, p. 186.
1221. Plomin & Spinath, 2004, p. 121.
1222. *Ibid.*, p. 124.

A more general concern involves group differences, such as average differences between classes and ethnic groups. As genes are found that are associated with differences among individuals within groups, the genes will inevitably be used to make comparisons between groups. . . . The societal implications of such research need to be anticipated . . . Perhaps people will become less preoccupied with average differences between groups when DNA chips make it possible to focus on individuals.[1223]

"Perhaps people will become less preoccupied" with alleged group differences . . . or perhaps they will become *more* preoccupied! In the United States, where race questions are a central obsession, these results would be explosive. The point is not whether group differences in "intelligence genes" will be found — they won't — but rather the way in which investigators minimize the potential social impact of their research.[1224] Would Plomin have the same attitude if research, for example, found that more Jews than non-Jews had QTLs associated with greed? Wouldn't there be a justifiable uproar that such research was even being done? [1225] Nevertheless, researchers such as University of California, Los Angeles neuroscientist Paul M. Thompson, who claims that "the volume of grey matter is correlated with intelligence . . . ,"[1226] supports testing grey matter differences among racial groups on the grounds that it is "harmful to simply censor all such work because this would set a terrible precedent of allowing an extrascientific agenda to constrain objective inquiry."[1227]

In an article published in the issue of *Science* containing the human genome sequencing paper (February 16, 2001), McGuffin, Riley, and Plomin could report only a few possible leads on genes for behavioral differences, acknowledging that "progress so far has been slow."[1228] In fact, progress has been so slow that McGuffin et al. created an escape hatch for themselves in the event that they find no important results. They discussed the "unsolved question of the distribution

---

1223. Plomin, 1999, p. C29.
1224. See Jonathan Beckwith's 2002 *Making Waves, Making Genes* for a discussion of how genetic researchers sometimes fail to take into account the potential misuse of their findings.
1225. This example is drawn from a similar one put forward by Breggin & Breggin in 1998, who argued against doing research on possible genetic influences on inner-city crime.
1226. Gray & Thompson, 2004, p. 473.
1227. *Ibid.*, p. 479.
1228. McGuffin et al., 2001, p. 1232. In 2002, Plomin & Craig wrote that "progress has been slow in identifying specific genes associated with most complex dimensions and disorders, probably because the effect sizes of individual genes is smaller than expected" (Plomin & Craig, 2002, p. 186).

of effect sizes of QTL; some may involve effects so small or so complicated that they will never be detected."[1229] The search for QTLs, a prominent critic recently remarked, is a "search for the proverbial needle in the haystack."[1230] Clearly, if Plomin and his associates give up looking for QTLs, they will say that genes are important but are too "complicated or small" to be found. Wouldn't it be far better to conclude that behavior genetics should take a hard second look at its core concepts, such as heritability and twin research?

Ken Richardson has provided an insightful analysis of the "QTLs for *g*" fiasco:

> It doesn't surprise me in the slightest that, empirically, this enterprise is turning out to be rather hit-and-miss. What does surprise me is that such a search should be seriously conceived in the first place. One reason for my misgivings, of course, is the idea of trying to identify genes for a character that no one can define. Another is that, even if allelic association with IQ was found, this would not mean that it was causal. Such a correlation could arise from a number of possible connections other than genes for IQ. How these are to be eliminated are not mentioned.[1231]

As Richardson pointed out, the search for IQ genes carries all of the problems and questionable assumptions of previous IQ genetic research, such as the inability to agree on what intelligence is, whether IQ tests (which guide the search for QTLs) actually measure intelligence, whether general intelligence (*g*) has a physical reality, the fact that correlation does not mean cause, and a host of other problems. Still, like the failed attempts at finding genes for psychiatric disorders, the public has been led to believe that "genes for IQ" have been found when, as Richardson reminds us, "no such thing has happened."[1232]

CONCLUSIONS

The failure to find genes for schizophrenia, ADHD, and other conditions, while not proving that these genes do not exist, is a testimony to the blind faith in the claims made by twin and adoption researchers. Rather than consider the possibility that genes do not exist, investigators such as Tsuang and colleagues

1229. *Ibid.*, p. 1249.
1230. Wahlsten, 1999, p. 607.
1231. K. Richardson, 2000, p. 72.
1232. *Ibid.*, p. 72.

discuss the "great strides" in finding genes, and that recent technology has led to the "implication of genes at several chromosomal loci."[1233] Others claim that they are "closing in on genes for manic-depressive illness and schizophrenia."[1234] More recently, Plomin wrote that according to geneticists, "we will be awash with genes associated with complex traits including behaviour in the next few years," [1235] and that the "future for genetic research in intelligence and other areas of personality looks brighter than ever in the dawn of the post-genomic era."[1236] And in a 2001 testimony to the fact that no histopathology has been shown to cause schizophrenia, Andreasen spoke of the condition being caused by an "'invisible lesion' that cannot be seen with the naked eye or under a micro-scope."[1237] (Perhaps the next step is to assert that schizophrenia is inherited through "invisible genes"?) All of this is said to avoid having to admit that they have found nothing important, and might not find anything important in the future. As Szasz once observed, "Much of what passes for scientific advance in psychiatry is, in fact, rhetorical innovation." Generally speaking, molecular genetic investigators use "rhetorical innovation" in their *publications* in order to obscure the failure to find genes in their *laboratories*. Ultimately, however, the use of language cannot eliminate the necessity of finding actual genes.

In the psychiatric genetic and molecular genetic literature it is striking how little attention is paid to people's social environment as having anything to do with the causes of "mental illness." The causes are viewed as residing at the molecular level. Andreasen has argued that it is important to understand how "proteins are created through our DNA . . . since it will ultimately explain how mental illnesses are caused, treated, and prevented."[1238] This statement epito-mizes the reductionistic view of human problems. In the past, such views were sustained because, in addition to their usefulness in absolving social conditions and political policies from causing human distress, the technology was not available to put reductionism to the test. Now that the technology *is* available, molecular geneticists in psychiatry may suffer through the failure to find genes for the dubious "diseases" they are investigating. The title of a 1992 article by psychiatric geneticist Michael Owen asked, "Will schizophrenia become a

1233. Tsuang et al., 2001, p. S18.
1234. Gershon et al., 1998, p. 233.
1235. Plomin, 2001, p. 138.
1236. Plomin & Spinath, 2004, p. 126.
1237. Andreasen, 2001, p. 209.
1238. *Ibid.*, p. 103.

graveyard for molecular geneticists?"[1239] The answer, at least as it relates to psychiatric disorders, may well turn out to be yes.

Reductionistic views can also be found among those looking for genes for "continuously distributed" traits such as personality and IQ. Although lip service is given to the environment, behavioral and performance differences are viewed as ultimately residing in the genes. Here again, the evidence suggests that nothing important will be found. Like the psychiatric geneticists, behavior geneticists have painted themselves into a corner, although as we have seen, some are already planning their escape. And escape they will, since the mistaken view that "genes are destiny" is needed by powerful and well-connected interests promoting political, professional, and business agendas.

In short, behavior genetics and psychiatric genetics may well have their Waterloo in molecular genetic research, even though this may not become apparent for quite some time. If it does, it will motivate investigators to reexamine the unsound research that inspired their search for genes in the first place.

1239. Owen, 1992.

CHAPTER 11. WHERE DO WE GO FROM HERE?

> Perhaps the most striking feature of the nature-nurture debate is the number of times it has ostensibly ended. — *Historian of Science Diane B. Paul, in 1998.*[1240]

This book has challenged positions that have been established and defended by internationally respected researchers. Naturally, it would require thousands of pages to counter all of the arguments in support the genetic theories and research I have analyzed. I therefore have covered what I view as the most important topics, which include the following points:

- The twin method is an invalid instrument for the detection of genetic influences on psychological trait differences and psychiatric conditions.

- So-called studies of twins reared apart are methodologically unsound, contain many pairs of twins who were not actually reared apart, and failed to use the proper comparison group when evaluating the meaning of MZA correlations. Therefore, these studies provide no scientifically acceptable evidence in favor of genetic influences on psychological trait differences.

- The heritability statistic is wrongly used to estimate the magnitude of genetic influences on trait differences. Its use should be discontinued, except in cases where one is interested in the results of a selective breeding program.

- There is little evidence that schizophrenia and most other psychiatric conditions have a genetic basis.

- Schizophrenia adoption studies have been (to varying degrees) methodologically unsound, and has been unable to disentangle possible genetic and environmental influences on the condition.

- There is little evidence that socially undesirable behaviors such as criminality have a genetic basis.

- Possible genetic influences on intelligence cannot be determined through the use of standardized IQ tests, twin studies, or the heritability statistic.

- There is little evidence pointing to the existence of specific genes for psychiatric disorders or human psychological trait variation.

1240. Paul, 1998, p. 82.

- Concepts such as "schizophrenia," "intelligence," "personality," and "criminality" are often difficult to define, and are controversial (even though human genetic researchers tend to ignore this). Attempts by human genetic researchers to categorize or quantify these concepts are more often means to the end of promoting various agendas than the result of scientific progress.

A running theme of this book has been that, although previous genetic theories were based on evidence considered overwhelming at the time, *today* we recognize that the evidence was actually very weak. In Chapter 1 we saw how eugenicists Charles Davenport and Harry Laughlin could write, in the absence of twin studies, adoption studies, family studies, or even family pedigrees, that many physical, mental, and moral traits "have been proved to have a hereditary basis." The early investigators' conclusions should be seen more as statements of their beliefs than as objectively drawn from the evidence, and little, unfortunately, has changed since then.

A belief in important hereditary influences in psychiatry has been around for a long time. For example, Emil Kraepelin, the father of modern psychiatry, began his chapter on the causes of "manic-depressive insanity" as follows:

> **Hereditary taint.** I could demonstrate in about 80 per cent of the cases observed in Heidelberg. Walker found it in 73.4 per cent, Saiz in 84.7, Weygandt in 90, Albrecht in 80.6 per cent, and in the forms with numerous attacks still somewhat more frequently. Taint from the side of the parents he found in 36 per cent of the cases, in the last-named forms in 45 per cent.[1241]

Although today "hereditary taint" has been replaced by "genetic predisposition,"[1242] Kraepelin was as sure of the hereditary basis of manic-depression as contemporary psychiatric geneticists are for the various disorders they investigate. This is also true of the "inheritance of intelligence" question. The point is that today, as in the past, unsupported beliefs and prejudices are passed off as scientific facts.

1241. Kraepelin, 1976, p. 165.
1242. I thank Alvin Pam for bringing this to my attention.

SOME PROPOSALS FOR THE COLLECTION OF DATA AND THE PUBLICATION OF RESEARCH

A major aspect of this book has been the reanalysis of a large body of greatly flawed research. In the process, I do not want to leave readers with the impression that flawed research is confined to human genetic investigations. In fact, it occurs in most fields and has been the subject of criticism for many years. In many areas of science there are documented cases of fraud and misconduct, and undoubtedly many more cases are never discovered.[1243] Several years of reviewing bias in human genetic research has made it clear to me that many changes must be made in the way that scientific research is undertaken and reported, and here I mention two.

(1) **The need for research registers.** We saw in Chapter 3 that Susan Farber called for a central register to house twin data. This idea should be expanded to cover all psychiatric and psychological research. Before initiating a study, researchers would be expected to submit a written description of how they will obtain participants, how they will define and measure the variables of interest, how they will perform group comparisons, and what conclusions they will draw from the possible results they obtain. An internationally-based social science central register should be established to collect and store this infor-mation. A register would create a permanent public record of the intentions and methods of researchers before data collection, analysis, and publication. A research register has been defined as "a database of research studies, either planned, active, or completed (or any combination of these), usually oriented around a common feature of the studies such as subject matter, funding source, or design."[1244]

Research registers are already used in some areas of medicine and should now be extended to the social sciences. Cooper and colleagues called for the cre-ation of "an international, standardized, computer accessible register of IRB [institutional review board] approved research projects" for psychological research studies.[1245] The most important purpose of an international social science register (ISSR) would be to record researchers' methods before the col-lection of data:

1243. For documentation of fraud in science see Altman & Hernon, 1997; Bell, 1992; Broad & Wade, 1982; Gould, 1981; LaFollette, 1992.
1244. Dickersin, 1994, p. 72.
1245. Cooper et al., 1997, p. 452.

A further advantage [of a registry] is that through access to information on stated prior hypotheses, it would be possible to identify the use of multiple comparisons (that is whether positive results were based on prior hypotheses or posterior analyses). Such information could be helpful in the publication review process.[1246]

A research register would help ensure that investigators analyze results according to their research design specifications, which would be completed and registered before data collection. Sometimes, as the research proceeds, changes are required in the design specifications. In these instances, a statement would be sent to ISSR explaining the protocol revisions. Ideally, information about research design sent to ISSR would be published in the form of a brief abstract before the study's final publication.

Registration of studies at ISSR would reduce the practice of "data dredging," that is, the purposeful post-hoc analysis/manipulation of datasets in order to locate significant relationships that support the investigator's research hypothesis. It is well known that in a large set of data probability dictates that, by chance alone, some significant correlations will be found between variables having no actual causal or associative relationship. After a study is published, the information submitted to ISSR would become part of the public record.

(2) Reduce the importance attached to researchers' conclusions. In reporting the results of genetic studies, reviewers place considerable emphasis on what researchers conclude about their data. Not surprisingly, most researchers conclude that their findings are consistent with their original hypothesis. A critic noted the "pervasive . . . manner in which scientists can deliberately or, more often, unconsciously work in such a way that their conclusions are bound to support a particular position, policy, or action."[1247] Psychologist George Albee concluded that his early belief that social scientists discover facts in order to build theories was wrong, and that

> it is more accurate to say that people, and particularly social scientists, select theories that are consistent with their personal values attitudes, and prejudices, and then go out into the world, or into the laboratory, to seek facts that validate their beliefs about the world and about human nature, neglecting or denying observations that contradict their personal prejudices.[1248]

1246. Easterbrook, 1992, p. 346.
1247. Savan, 1988, p. 26.
1248. Albee, 1982, p. 5.

Undoubtedly, there is a high degree of correlation between what human genetic researchers believe in advance that their studies will show, and what they conclude after they collect and analyze their data. However, the results of most studies are open to multiple interpretations. Thus, the original researchers' conclusions are not necessarily more valid than alternative explanations. The "Adversarial Collaboration" model proposed by Barbara Mellers and colleagues in 2001 could help resolve this problem.[1249] According to this approach, adversarial researchers ("the participants") would engage in an experiment, with a trusted third party ("the arbiter") coordinating the study under an agreed upon protocol. The participants agree at the beginning to identify results that "would change their mind, at least to some extent."[1250] These would be recorded by the arbiter. The participants and the arbiter agree in advance to publish their results, with the arbiter writing the bulk of the article. If disagreements remain, each participant would write individual comments with a pre-arranged word limit. The data should be under the control of the arbiter. Mellers et al. closed by expressing their hope that "adversarial collaboration would become the norm, not the exception" in future scientific research.[1251]

As I have argued throughout this book, the results of human genetic research are plausibly explained on the basis of environmental factors and bias. Still, the authors of secondary sources (such as textbooks) typically endorse the original researchers' conclusions, at the expense of critical analysis.[1252] Not surprisingly, genetically oriented researchers tend to conclude in favor of important genetic influences. Thus, we should be skeptical about the human genetic researchers' conclusions, since they are heavily influenced by their reductionistic views. Let them collect and publish data, but we should rely on others to arrive at less biased conclusions.

Bouchard complained in 1993 that an "entire industry has evolved up around the reanalysis of kinship data, particularly the large body of published data on identical twins reared apart."[1253] This remark captures the attitude of investigators who believe that they have a monopoly on drawing conclusions from the data they produce. The "industry" Bouchard criticized is nothing more than people pointing out methodological flaws and offering alternative interpre-

1249. Mellers et al., 2001.
1250. *Ibid.*, p. 270.
1251. *Ibid.*, p. 275.
1252. Leo & Joseph, 2002.
1253. Bouchard, 1993a, p. 43.

tations of the data. If Bouchard's conclusions in favor of genetics are correct, he and his colleagues should have little problem defending them. Similar displeasure with critics is seen in the works of the popularizers of genetic research, who often portray dissenters as "armchair critics" seeking to tear down the work of "scientists," implying that those not directly carrying out research have little right to offer alternative explanations of others' results. However, it is not only a right but a duty for others to critique research, particularly where important social policy implications are involved.

THE DANGER OF EUGENICS

Genetic researchers Garver and Garver wrote in 1994 that "a eugenic mentality has existed in the United States during the entire 20th century," although it was "more apparent in the first 45 years."[1254] From this perspective the post-World War II era's rejection of the eugenic program was a mere aberration, or an interval. Regardless of contemporary behavior geneticists and psychiatric geneticists' motivations, their work could pave the way for a rebirth of eugenics. One reason I have documented the historical relationship between genetic research and eugenics is to warn of the possibility that history can repeat itself. The United States was once the world leader in eugenics, with over 30 states passing eugenic sterilization laws.

Although human genetic researchers often remind us that nature and nurture work together and that the environment is important, there is nothing new about this.[1255] Similar statements were made by prominent eugenicists in the first half of the 20th century. For example, in 1935 Harry Laughlin wrote, "In any good eugenics exhibit there must always be a section for 'Heredity and Environment.' Such an exhibit must bring out clearly how both heredity and environment are important — each a vital factor in the development of the individual."[1256] Frederick Osborn wrote in the second (1951) edition of his *Preface to Eugenics*, "Which is more important, heredity or environment? Such a question lacks reality, because when two sets of factors are necessary to produce a given result — such as a living human being — there can be no question of the relative importance of either set of factors; both are essential to the final

1254. Garver & Garver, 1994, p. 151.
1255. See Paul, 1998.
1256. Laughlin, 1935, p. 161.

result."[1257] According to Popenoe and Johnson, in the 1933 edition of *Applied Eugenics*, "Nature and nurture are indissolubly associated in the achievements of every man and woman. . . . in daily life they are interdependent."[1258] And even Rüdin, in an exchange with a critic in 1930, stated, "Dr. Myerson has said that environment plays a great role in the production of a disease. Of course it does."[1259] Statements such as these did not prevent Laughlin, Popenoe, and Rüdin from being strident advocates of compulsory eugenic sterilization. Nor did it prevent Osborn from issuing the following chilling opinion in *Preface to Eugenics*:

> The aggregate of human suffering caused by defective genes will be enormously diminished if the destruction of the defective genes takes place *immediately after their appearance*. Human suffering will be further reduced if the destruction of the defective genes takes place by failure of reproduction rather than by the death of the individual carrying the defective genes [emphasis added].[1260]

The warning I issue has nothing to do with the motivations of contemporary human genetic researchers. However, future eugenics advocates using their research to promote eugenic goals will not be concerned with the views and caveats of today's investigators. They will use this research for their own purposes. As historian Robert Proctor observed in 1992,

> If there is a disconcerting continuity between genomics and eugenics, it is the fact that both have taken root in a climate where many people believe that the large part of human talents and disabilities are heritable through the genes.[1261]

While the danger Proctor warned of is real, some current political and social policies, such as "ending welfare as we know it" in the US, certain aspects of genetic counseling, and the continuing practice of surreptitiously sterilizing minorities and the poor, are already influenced by eugenic ideas. Eugenic sterilization was performed in Sweden until the mid-1970s, while Japan began a eugenic sterilization program in 1948 that resulted in the sterilization of over 800,000 people, and continued well into the 1990s.[1262]

1257. Osborn, 1951, p. 81.
1258. Popenoe & Johnson, 1933, p. 18.
1259. Rüdin, 1932, p. 494.
1260. Osborn, 1951, p. 26.
1261. Proctor, 1992, p. 83.
1262. Yamaguchi, 1997.

Given the lack of evidence supporting behavior genetic and psychiatric genetic positions, I would like to propose alternative approaches to the topics covered in this book.

Family studies. It may be helpful to determine if a condition or trait is familial, provided that we understand that "familial" does not mean "genetic." Family studies may be useful for identifying physical conditions showing a Mendelian pattern of inheritance, although family pedigrees can often perform this function.

Twin research. The twin method should, at long last, be relegated to its proper place alongside the discarded pseudosciences of bygone eras, such as phrenology, alchemy, and craniometry. Given the appalling history of fraud, bias, and unwarranted claims, one is tempted to place reared-apart twin studies in this category. However, a valid study using reared-apart twins is theoretically possible, although it would be carried out far differently than the studies published to date (as outlined in Chapter 4). One area of twin research that could benefit humanity is the much neglected method of studying identical twins *discordant* for various conditions, whose purpose is to discover relevant *environmental* factors.

Adoption research. Like reared-apart twin studies, it is theoretically possible to perform a valid adoption study. The problem, as always, lies in the fact that these studies are fraught with problems related to the social and psychological aspects of the human condition. The investigators in most adoption studies, and the psychiatric studies in particular, tended to see their experiments as being only minimally affected by these problems. Future adoption researchers must, at a minimum (1) determine in advance, and publish or submit to a register, the specific hypotheses, methods, definitions, and comparison groups used in their studies; (2) make a serious attempt to come to grips with problems such as selective placement and range restriction; (3) publish or place with a registry raw case history information of those under study; (4) choose as subjects only those adoptees who were placed into their adoptive homes at or shortly after birth; and (5) ensure that all interviews, diagnoses, or ratings are performed blind.

Heritability. Simply put, the heritability statistic should be discarded in human genetics, psychology, and psychiatry.

Schizophrenia. We should recognize that there is little evidence supporting a genetic basis for schizophrenia. Furthermore, the discovery of a genetic predisposition would have little meaning, because we could still choose to focus attention on discovering and eliminating environmental conditions or triggers. Schizophrenia and most other psychiatric diagnoses describe behaviors and mental states, not diseases. In fact, some critics have called for discarding the schizophrenia concept entirely. There is no qualitative difference between "schizophrenics," "manic-depressives," and people suffering from other emotional problems. These conditions merely fall on different points of a continuum marking the way that people react to psychologically damaging familial, social, and political environments.

Crime. Criminal behavior is a social problem, and what constitutes "crime" is fraught with political and moral implications. While some aspects of criminal behavior may involve psychological factors, it is not a "disorder" in need of "treatment." We could, however, say that the society producing the conditions leading to crime (such as poverty, racism, and income disparity) is in need of "treatment."

IQ Tests. In 1922 the American journalist Walter Lippmann wrote that if IQ tests become "a sort of last judgment on the child's capacity, that they reveal 'scientifically' his predestined ability, then it would be a thousand times better if all the intelligence testers and all their questionnaires were sunk without warning in the Sargasso Sea."[1263] The total experience of 100 years of IQ testing has been overwhelmingly negative. I favor the abolition of any test claiming to measure "IQ" or innate intelligence. Any positive uses or features of these abolished tests could then be incorporated into other tests lacking any claim to measure innate "intelligence."

Molecular genetic research in psychology and psychiatry. If researchers in this area choose to continue searching for genes that may not even exist, that is their right. Still, they would do much better to reexamine the body of literature upon which they have based their search. In the end, molecular genetic research in psychiatry and psychology may well prove to be a waste of time, energy, and money.

1263. Quoted in Kamin, in Eysenck vs. Kamin, 1981, p. 90.

A major theme of several recent books popularizing human genetic research has been that genetic theories about the origins of psychological trait differences are new and revolutionary. We have seen, however, that these theories are in fact quite old. We are being sold an old product in new packaging. Genetic researchers often call attention to human differences at the expense of emphasizing how much people have in common. They also tend to be interested in people's limits, as opposed to their potential for achievement and growth. The behavior genetic culture, as Richardson observed, is truly "pessimistic."[1264]

When, in the future, individuals and families come into psychotherapy seeking guidance and understanding, some have envisioned a "quick swab of a client's cheek" as a part of the psychological diagnostic procedure.[1265] However, a "Gattaca" type world will never come about, because genes *are not* destiny. "The genome is not 'the very essence' of what it means to be human," wrote Proctor, "any more than sheet music is the essence of a concert performance."[1266] Familial, social, cultural, political, and psychological environments play a crucial and dominant role in shaping who we are. Behavior genetic and psychiatric genetic research is a house of cards that falls down in the face of critical examination.

Why then have genetic theories flourished in the United States, Europe, and elsewhere for several generations? It would be a mistake to see the history of human genetic research merely as the actions of isolated individuals or groups. The "scientific racism" chronicled in Alan Chase's *The Legacy of Malthus* was financed and promoted by the wealthy elite of Western societies in order to promote their interests. Families such as the Carnegies, Rockefellers, and Harrimans financed eugenic research in the United States, Germany, Denmark, and elsewhere.[1267] Thus, in the absence of social change, scientific racism, and genetic explanations for socially disapproved behaviors and economic inequality, will continue to find fertile soil.

In contrast to the bleak hereditarian view of humans and their future, there exists a radically different perspective. Human psychological distress, to the extent that it goes beyond people's normal reactions to life events, is pri-

---

1264. K. Richardson, 1998, p. 178.
1265. Efran et al., 1998, p. 37.
1266. Proctor, 1992, pp. 83-84.
1267. See Black, 2003; Kevles, 1985; Paul, 1998.

marily the result of well-known and well-documented psychologically traumatic environments and events, and conditions such as racism, sexism, homophobia, unemployment, economic inequality, war, and social alienation. Future societies free of these conditions will see a dramatic reduction in human suffering, as well as a flourishing of ability and innovation, and any possible role of genetic influences in shaping human psychological differences will be of interest mainly to historians.

REFERENCES

Abrams, R., & Taylor, M. A. (1983). The genetics of schizophrenia: A reassessment using modern criteria. *American Journal of Psychiatry, 140,* 171-175.

Ainslie, R. C. (1985). *The psychology of twinship.* Lincoln: University of Nebraska Press.

Albee, G. W. (1982). The politics of nature and nurture. *American Journal of Community Psychology, 10,* 4-36.

Allen, G. E. (2001). The biological basis of crime: An historical and methodological study. *Historical Studies of the Physical and Biological Sciences, 31,* 183-222.

Allen, M. G., Cohen, S., & Pollin, W. (1972). Schizophrenia in veteran twins: A diagnostic review. *American Journal of Psychiatry, 128,* 939-945.

Allport, G. W. (1961). *Pattern and growth in personality.* New York: Holt, Rinehart, and Winston.

Altman, E., & Hernon, P. (Eds.). (1997). *Research misconduct: Issues, implications, and strategies.* Greenwich, CT: Ablex Publishing Corporation.

Aly, G. (1994). Medicine against the useless. In G. Aly, P. Chroust, & C. Pross (Eds.), *Cleansing the fatherland* (pp. 22-98). Baltimore: Johns Hopkins Press.

American Psychiatric Association. (1965). *Diagnostic and statistical manual of mental disorders.* Washington, DC: Author. (Originally published in 1952)

American Psychiatric Association. (1968). *Diagnostic and statistical manual of mental disorders* (2nd ed.). Washington, DC: Author.

American Psychiatric Association. (1980). *Diagnostic and statistical manual of mental disorders* (3rd ed.). Washington, DC: Author.

American Psychiatric Association. (1987). *Diagnostic and statistical manual of mental disorders* (3rd Rev. ed.). Washington, DC: Author.

American Psychiatric Association. (1994). *Diagnostic and statistical manual of mental disorders* (4th ed.). Washington, DC: Author.

American Psychiatric Association. (2000). *Diagnostic and statistical manual of mental disorders* (4th ed., text revision). Washington, DC: Author.

Andreasen, N. C. (1998). Understanding schizophrenia: A silent spring? *American Journal of Psychiatry, 155,* 1657-1659.

Andreasen, N. C. (2000). Schizophrenia: The fundamental questions. *Brain Research Reviews, 31,* 106-112.

Andreasen, N. C. (2001). *Brave new brain.* Oxford: Oxford University Press.

Anonymous. (1925). *Eugenical News, 10,* 69-71.

Arieti, S. (1974). *Interpretation of schizophrenia* (2nd ed.). New York: Basic Books.

Ash, M. G. (1998). From "positive eugenics" to behavioral genetics: Psychological twin research under Nazism and since. *Paedagogica Historica (Supplementary Series, Vol. III)*, 335-358.

Bailey, J. M., & Pillard, R. C. (1993). Reply to Lidz's "Reply to 'a genetic study of male sexual orientation' " [Letter to the editor]. *Archives of General Psychiatry, 50*, 240-241.

Barkley, R. A. (1998). *Attention-deficit hyperactivity disorder: A handbook for diagnosis and treatment (2nd ed.)*. New York: The Guilford Press.

Barr, C. L. (2001). Genetics of childhood disorders: XXII. ADHD, Part 6: The dopamine D4 receptor gene. *Journal of the American Academy of Child and Adolescent Psychiatry, 40*, 118-121.

Bateson, G., Jackson, D. D., Haley, J., & Weakland, J. (1956). Toward a theory of schizophrenia. *Behavioral Science, 1*, 251-264.

Beckwith, J. (2002). *Making genes, Making waves: A social activist in science.* Cambridge, MA: Harvard University Press.

Beckwith, J., & Alper, J. S. (2002). Genetics of human personality: Social and ethical implications. In J. Benjamin, R. Ebstein, & R. Belmaker (Eds.), *Molecular genetics and the human personality* (pp. 315-331). Washington, DC: American Psychiatric Press.

Beckwith, J., Geller, L., & Sarkar, S. (1991). IQ and heredity [Letter to the editor]. *Science, 252*, 191.

Begg, C. B. (1994). Publication bias. In H. Cooper & L. Hedges (Eds.), *The handbook of research synthesis* (pp. 399-409). New York: Russell Sage Foundation.

Bell, A. E. (1977). Heritability in retrospect. *Journal of Heredity, 68*, 297-300.

Bell, R. (1992). *Impure science: Fraud, compromise, and political influence in scientific research.* New York: John Wiley and Sons.

Benjamin, L. S. (1976). A reconsideration of the Kety and associates study of genetic factors in the transmission of schizophrenia. *American Journal of Psychiatry, 133*, 1129-1133.

Benjamini, Y., Drai, D., Elmer, G., Kafkafi, N., Goloni, I. (2001). Controlling the false discovery rate in behavior genetics research. *Behavioural Brain Research, 125*, 279-284.

Bentall, R. P. (2003). *Madness explained: Psychosis and human nature.* London: Allen Lane.

Bernstein, V. H. (1945, August 21). "Created Nazi Science of Murder. Meet 'gentle' Prof. Rudin, Theorist of 'Aryanism'" *PM Daily*, 5.

Billings, P. R., Beckwith, J., & Alper, J. S. (1992). The genetic analysis of human behavior: A new era? *Social Science and Medicine, 35*, 227-238.

Black, E. (2003). *War against the weak: Eugenics and America's campaign to create a master race.* New York: Four Walls Eight Windows.

Bleuler, E. (1950). *Dementia praecox or the group of schizophrenias*. New York: International Universities Press. (Originally published in 1911)

Bleuler, M. (1955). Research and changes in concepts in the study of schizophrenia, 1941-1950. *Bulletin of the Isaac Ray Medical Library, 3*, 1-132.

Bleuler, M. (1978). *The schizophrenic disorders: Long-term patient and family disorders*. New Haven: Yale University Press.

Block, N. J., & Dworkin, G. (Eds.). (1976). *The IQ controversy*. New York: Pantheon.

Bohman, M. (1971). A comparative study of adopted children, foster children and children in their biological environment born after undesired pregnancies. *Acta Paediatrica Scandinavica* (Suppl. 221), 1-38.

Bohman, M. (1978). Some genetic aspects of alcoholism and criminality. *Archives of General Psychiatry, 35*, 269-276.

Bohman, M., Cloninger, C. R., Sigvardsson, S., & von Knorring, A. (1982). Predisposition to petty criminality in Swedish adoptees. *Archives of General Psychiatry, 39*, 1233-1241.

Bohman, M., & Sigvardsson, S. (1980). A prospective, longitudinal study of children registered for adoption. *Acta Psychiatrica Scandinavica, 61*, 339-355.

Bouchard, T. J., Jr. (1982). Identical twins reared apart: Reanalysis or pseudo-analysis? [Book review]. *Contemporary Psychology, 27*, 190-191.

Bouchard, T. J., Jr. (1984). Twins reared together and apart: What they tell us about human diversity. In S. Fox (Ed.), *Individuality and determinism: Chemical and biological bases* (pp. 147-184). New York: Plenum Press.

Bouchard, T. J., Jr. (1993a). The genetic architecture of human intelligence. In P. Vernon (Ed.), *Biological approaches to the study of human intelligence* (pp. 33-93). Norwood, NJ: Ablex Publishing Corporation.

Bouchard, T. J., Jr. (1993b). Genetic and environmental influences on adult personality: Evaluating the evidence. In J. Hettema & I. Deary (Eds.), *Basic issues in personality* (pp. 15-44). Dordrecht, The Netherlands: Kluwer Academic Publishers.

Bouchard, T. J., Jr. (1996). Behaviour genetic studies of intelligence, yesterday and today: The long journey from plausibility to proof. *Journal of Biosocial Science, 28*, 527 555.

Bouchard, T. J., Jr. (1997a). The genetics of personality. In K. Blum & E. Noble (Eds.), *Handbook of psychiatric genetics* (pp. 273-296). Boca Raton, FL: CRC Press.

Bouchard, T. J., Jr. (1997b). IQ similarity in twins reared apart: Findings and responses to critics. In R. Sternberg & E. Grigorenko (Eds.), *Intelligence, heredity, and environment* (pp. 126-160). New York: Cambridge University Press.

Bouchard, T. J., Jr. (1997c, September/October). Whenever twain shall meet. *The Sciences, 37*, 52-57.

Bouchard, T. J., Jr. (1999). Foreword. In N. Segal, *Entwined lives*. New York: Dutton.

Bouchard, T. J., Jr. (2004). [Book review]. *Intelligence, 32,* 215-217

Bouchard, T. J., Jr., & Loehlin, J. C. (2001). Genes, evolution, and personality. *Behavior Genetics, 31,* 243-273.

Bouchard, T. J., Lykken, D. T., McGue, M., Segal, N. L., & Tellegen, A. (1990). Sources of human psychological differences: The Minnesota Study of Twins Reared Apart. *Science, 250,* 223-228.

Bouchard, T. J., Jr., Lykken, D. T., McGue, M., Segal, N. L., & Tellegen, A. (1991). Response [Letter to the editor]. *Science, 252,* 191-192.

Bouchard, T. J., Jr., Lykken, D. T., Segal, N. L., & Wilcox, K. J. (1986). Development in twins reared apart: A test of the chronogenetic hypothesis. In A. Demirjian (Ed.), *Human growth: A multidisciplinary review* (pp. 299-310). London: Taylor & Francis.

Bouchard, T. J., Jr., & McGue, M. (1981). Familial studies of intelligence: A review. *Science, 212,* 1055-1059.

Bouchard, T. J., Jr., & McGue, M. (1990). Genetic and rearing environmental influences on adult personality: An analysis of adopted twins reared apart. *Journal of Personality, 58,* 263-292.

Bouchard, T. J., Jr., McGue, M., Hur, Y., & Horn, J. M. (1998) A genetic and environmental analysis of the California Psychological Inventory using adult twins reared together and apart. *European Journal of Personality, 12,* 307-320.

Bouchard, T. J., Jr., & Pedersen, N. (1999). Twins reared apart: Nature's double experiment. In M. LaBuda & E. Grigorenko (Eds.), *On the way to individuality: Methodological issues in behavioral genetics* (pp. 71-93). Commack, NY: Nova Science.

Boyle, M. (1990). *Schizophrenia: A scientific delusion?* New York: Routledge.

Boyle, M. (1999) Diagnosis. In C. Newnes, G. Holmes, & C. Dunn (Eds.), *This is Madness: A critical look at psychiatry and the future of mental health services* (pp. 75-90). Ross-on-Wye: PCCS Books.

Boyle, M. (2002). *Schizophrenia: A scientific delusion?* (2nd ed.). Hove, UK: Routledge.

Bradburn, N. M., Rips, L. J., & Shevell, S. K. (1987). Answering autobiographical questions: The impact of memory and inference on surveys. *Science, 236,* 157-161.

Bradley, W., & Schaefer, K. (1998). *The uses and misuses of data and models.* Thousand Oaks, CA: Sage Publications.

Breggin, P. R. (1991). *Toxic psychiatry.* New York: St. Martin's Press.

Breggin, P. R. (1998). *Talking back to Ritalin.* Monroe, ME: Common Courage Press.

Breggin, P. R. (2001). Empowering social work in the era of biological psychiatry. *Ethical Human Sciences and Services, 3,* 197-206.

Breggin, P. R., & Breggin, G. R. (1998). *The war against children of color.* Monroe, ME: Common Courage Press.

Breggin, P. R., & Cohen, D. (1999). *Your drug may be your problem.* Reading, MA: Perseus.

Brennan, P., Mednick, S. A., & Gabrielli, W. F. (1991). Genetic influences and criminal behavior. In M. Tsuang, K. Kendler, & M. Lyons (Eds.), *Genetic issues in psychosocial epidemiology* (pp. 231-246). New Brunswick, NJ: Rutgers University Press.

Brigham, C. C. (1923). *A study of American intelligence.* Princeton, NJ: Princeton University Press.

Broad, W. J., & Wade, N. (1982). *Betrayers of the truth.* New York: Simon and Schuster.

Broberg, G., & Roll-Hansen, N. (Eds.). (1996). *Eugenics and the welfare state: Sterilization policy in Denmark, Sweden, Norway, and Finland.* East Lansing, MI: Michigan State University Press.

Broberg, G., & Tydén, M. (1996). Eugenics in Sweden: Efficient care. In G. Broberg & N. Roll-Hansen (Eds.), *Eugenics and the welfare state: Sterilization policy in Denmark, Sweden, Norway, and Finland* (pp. 77-149). East Lansing, MI: Michigan State University Press.

Brunner, H. G., Nelen, M., Breakefield, X. O., Ropers, H. H., & van Oost, B. A. (1993). Abnormal behavior associated with a point mutation in the structural gene for monoamine oxidase. *Science, 262,* 578-580.

Brzustowicz, L. M., Simone, J., Mohseni, P., Hayter, J. E., Hodgkinson, K. A., Chow, E. W. C., & Bassett, A. S. (2004). Linkage disequilibrium mapping of schizophrenia susceptibility to the CAPON region of chromosome 1q22. *American Journal of Human Genetics, 74,* 1057-1063.

Bucur, M. (2002). *Eugenics and modernization in interwar Romania.* Pittsburgh: University of Pittsburgh Press.

Bulmer, M. (1999). The development of Francis Galton's ideas on the mechanism of heredity. *Journal of the History of Biology, 32,* 263-292.

Burbridge, D. (2001). Francis Galton on twins, heredity and social class. *British Journal for the History of Science, 34,* 323-340.

Burleigh, M. (1994). *Death and deliverance.* Cambridge, UK: Cambridge University Press.

Byerley, W., & Coon, H. (1995). Strategies to identify genes in schizophrenia. In J. Oldham & M. Riba (Eds.), *Review of psychiatry* (Vol. 14, pp. 361-381). Washington, DC: American Psychiatric Press.

Cadoret, R. J. (1978). Psychopathology in adopted-away offspring of biologic parents with antisocial behavior. *Archives of General Psychiatry, 35,* 176-184.

Cannon, T. D., Kaprio, J., Lonnqvist, J., Huttunen, M., & Koskenvuo, M. (1998). The genetic epidemiology of schizophrenia in a Finnish twin cohort. *Archives of General Psychiatry, 55,* 67-74.

Cardno, A. G., Marshall, E. J., Coid, B., Macdonald, A. M., Ribchester, T. R., Davies, N. J., Venturi, P., Jones, L. A., Lewis, S. W., Sham, P. C., Gottesman., I. I., Farmer, A. E.,

McGuffin, P., Reveley, A. M., & Murray, R. M. (1999). Heritability estimates for psychotic disorders. *Archives of General Psychiatry, 56*, 162-168.

Cardon, L. R., & Fulker, D. W. (1993). Genetics of specific cognitive abilities. In R. Plomin & G. McClearn (Eds.), *Nature, nurture, and psychology* (pp. 99-120). Washington, DC: American Psychological Association.

Carey, G. (1992). Twin imitation for antisocial behavior: Implications for genetic and family environmental research. *Journal of Abnormal Psychology, 101*, 18-23.

Carey, G., & DiLalla, D. L. (1994). Personality and psychopathology: Genetic perspectives. *Journal of Abnormal Psychology, 103*, 32-43.

Carter, H. D. (1940). Ten years of research on twins: Contributions to the nature-nurture problem. *National Society for the Study of Education Yearbook, 39*, 235-255.

Cassill, K. (1982) *Twins: Nature's amazing mystery*. New York: Atheneum.

Cassou, B., Schiff, M., & Stewart, J. (1980). Génétique et schizophrénie: Réévaluation d'un consensus [Genetics and schizophrenia: Re-evaluation of a consensus]. *Psychiatrie de l'Enfant, 23*, 87-201.

Ceci, S. J. (1996). *On intelligence* (expanded edition). Cambridge, MA: Harvard University Press.

Chase, A. (1975, February). The great pellagra cover-up. *Psychology Today, 8 (9)*, pp. 82-86.

Chase, A. (1980). *The legacy of Malthus: The social costs of the new scientific racism*. Urbana, IL/ Chicago: University of Illinois Press. (Originally published in 1977)

Cheverud, J. M. (1990). Inheritance and the additive genetic model. *Behavioral and Brain Sciences, 13*, 124.

Christiansen, K. O. (1977a). A preliminary study of criminality among twins. In S. Mednick & K. Christiansen (Eds.), *Biosocial bases of criminal behavior* (pp. 89-108). New York: Gardner Press.

Christiansen, K. O. (1977b). A review of studies of criminality among twins. In S. Mednick K. Christiansen (Eds.), *Biosocial bases of criminal behavior* (pp. 45-88). New York: Gardner Press.

Chumakov, I., Blumenfeld, M., Guerassimenko, O., Cavarec, L., Palicio, M., Abderrahim, H., Bougueleret, L., Barry, C., Tanaka, H., La Rosa, P., et al. (2002). Genetic and physiological data implicating the new human gene G72 and the gene for D-amino acid oxidase in schizophrenia. *Proceedings of the National Academy of Sciences, 99*, 13675–13680.

Cloninger, C. R. (2002). The discovery of susceptibility genes for mental disorders. *Proceedings of the National Academy of Sciences, 99*, 13365-13367.

Cloninger, C. R., & Gottesman, I. I. (1987). Genetic and environmental factors in antisocial behavior disorders. In S. Mednick, T. Moffitt, & S. Stack (Eds.), *The causes of crime: New biological approaches* (pp. 92-109). New York: Cambridge University Press.

Cohen, D. (1997). A critique of the use of neuroleptic drugs in psychiatry. In S. Fisher & R. Greenberg (Eds.), *From placebo to panacea: Putting psychotropic drugs to the test* (pp. 173-228). New York: John Wiley & Sons.

Cohen, D., & Cohen, H. (1986). Biological theories, drug treatments, and schizophrenia: A critical assessment. *Journal of Mind and Behavior, 7,* 11-36.

Cohen, D., & McCubbin, M. (1990). The political economy of tardive dyskinesia: Asymmetries in power and responsibility. *Journal of Mind and Behavior, 11,* 465-488.

Cohen, D. J., Dibble, E., Grawe, J. M., & Pollin, W. (1973). Separating identical from fraternal twins. *Archives of General Psychiatry, 29,* 465-469.

Cohen, R. J., Swerdlik, M. E., & Smith, D. K. (1992). *Psychological testing and assessment* (2nd ed.). Mountain View, CA: Mayfield.

Colbert, T. C. (2001). *Blaming our genes.* Tustin, CA: Kevco.

Cooper, H., DeNeve, K., & Charlton, K. (1997). Finding the missing science: The fate of studies submitted for review by a human subjects committee. *Psychological Methods, 2,* 447-452.

Coryell, W., & Zimmerman, M. (1988). The heritability of schizophrenia and schizoaffective disorders. *Archives of General Psychiatry, 45,* 323-327.

Craike, W. H., & Slater, E. (1945). Folie a deux in uniovular twins reared apart. *Brain, 68,* Part III, 213-221.

Crow, J. F. (1990). How important is detecting interaction? *Behavioral and Brain Sciences, 13,* 126-127.

Crowe, R. R. (1972). The adopted offspring of women criminal offenders. *Archives of General Psychiatry, 27,* 600-603.

Crowe, R. R. (1974). An adoption study of antisocial personality. *Archives of General Psychiatry, 31,* 785-791.

Crusio, W. E. (1990). Estimating heritabilities in quantitative behavior genetics: A station passed. *Behavioral and Brain Sciences, 13,* 127-128.

Dalgard, O. S., & Kringlen, E. (1976). A Norwegian twin study of criminality. *British Journal of Criminology, 16,* 213-232.

Davenport, C. B. (1911). *Heredity in relation to eugenics.* New York: Henry Holt and Company.

Davenport, C. B. (1916). The hereditary factor in pellagra. *Eugenics Record Office Bulletin No. 16.* Cold Spring Harbor, New York.

Davenport, C. B. (1920). Heredity of constitutional mental disorders. *Psychological Bulletin, 17,* 300-310.

Davenport, C. B., & Laughlin, H. H. (1915). How to make a eugenical family study. *Eugenics Record Office Bulletin No. 13.* Cold Spring Harbor, New York.

DeGrandpre, R. (1999). *Ritalin nation.* New York: W.W. Norton.

DeLisi, L. E. (2000). Critical overview of current approaches to genetic mechanisms in schizophrenia research. *Brain Research Reviews, 31,* 187-192.

DeLisi, L. E., & Crow, T. J. (2003). Drs. DeLisi and Crow reply [Letter to the editor]. *American Journal of Psychiatry, 160,* 598-599.

DeLisi, L. E., Shaw, S. H., Crow, T. J., Shields, G., Smith, A. B., Larach, V. W., Wellman, N., Loftus, J., Nanthakumar, B., Razi, K., Stewart, J., Comazzi, M., Vita, A., Heffner, T., & Sherrington, R. (2002). A genome-wide scan for linkage to chromosomal regions in 382 sibling pairs with schizophrenia or schizoaffective disorder. *American Journal of Psychiatry, 159,* 803-812.

Detterman, D. K. (1990). Don't kill the ANOVA messenger for bearing bad interaction news. *Behavioral and Brain Sciences, 13,* 131-133.

Devlin, B., Fienberg, S. E., Resnick, D. P., & Roeder, K. (1997). *Intelligence, genes, and success: Scientists respond to The Bell Curve.* New York: Springer Verlag.

Dibble, E., Cohen, D. J., & Grawe, J. M. (1978). Methodological issues in twin research: The assumption of environmental equivalence. In *Twin research: Psychology and methodology* (pp. 245-251). New York: Alan R. Liss.

Dickersin, K. (1994). Research registers. In H. Cooper & L. Hedges (Eds.), *The handbook of research synthesis* (pp. 71-83). New York: Russell Sage Foundation.

Dugdale, R. L. (1910). *The Jukes: A study in crime, pauperism, disease, and heredity* (4th ed.). New York: G. P. Putnam's Sons. (Originally published in 1877)

Dusek, V. (1987). Bewitching science. *Science for the People, 19,* 19-22.

Duster, T. (1990). *Backdoor to eugenics.* New York: Routledge.

Easterbrook, P. J. (1992). Directory of registries of clinical trials. *Statistics in Medicine, 11,* 345-359.

Ebstein, R. P., Novick, O., Umansky, R., Priel, B., Osher, Y., Blaine, D., Bennett, E. R., Nemanov, L., Katz, M., & Belmaker, R. H. (1995). Dopamine D4 receptor (D4DR) exon III polymorphism associated with the human personality trait of novelty seeking. *Nature Genetics, 12,* 78-80.

Eckert, E. D., Bouchard, T. J., Jr., Bohlen, J., & Heston, L. L. (1986). Homosexuality in monozygotic twins reared apart. *British Journal of Psychiatry, 148,* 421-425.

Efran, J. S., Greene, M. A., & Gordon, D. E. (1998, March/April). Lessons of the new genetics. *Family Therapy Networker,* 27-32, 35-41.

Egeland, J. A., Gerhard, D. S., Pauls, D. L., Sussex, J. N., & Kidd, K. K. (1987). Bipolar affective disorders linked to DNA markers on chromosome 11. *Nature, 325,* 783-787.

Eley, T. C., Lichtenstein, P., & Stevenson, J. (1999). Sex differences in the etiology of aggressive and nonaggressive antisocial behavior: Results from two twin studies. *Child Development, 70,* 155-168.

Elkin, A., Kalidindi, S., & McGuffin, P. (2004). Have schizophrenia genes been found? *Current Opinion in Psychiatry, 17,* 107-113.

Ellinger, T. U. H. (1942). On the breeding of Aryans. *Journal of Heredity, 33,* 141-143.

Essen-Möller, E. (1941). Psychiatrische Untersuchungen an einer Serie von Zwillingen [Psychiatric investigations on a series of twins]. *Acta Psychiatrica et Neurologica* (Suppl. 23). Copenhagen: Munksgaard.

Essen-Möller, E. (1963). Twin research in psychiatry. *Acta Psychiatrica Scandinavica, 39,* 65-77.

Essen-Möller, E. (1970). Twenty-one psychiatric cases and their MZ cotwins. *Acta Geneticae Medicae et Gemellologiae, 19,* 315-317.

Eysenck, H. J., versus Kamin, L. J. (1981). *The intelligence controversy.* New York: John Wiley & Sons. (Published in the UK as *Intelligence: The battle for the mind*)

Falconer, D. S. (1965). The inheritance of liability to certain diseases, estimated from the incidence among relatives. *Annals of Human Genetics, 29,* 51-76.

Falconer, D. S., & Mackay, T. F. C. (1996). *Introduction to quantitative genetics* (4th ed.). Harlow, UK: Longman.

Fallin, M. D., Lasseter, V. K., Wolyniec, P. S., McGrath, J. A., Nestadt, G., Valle, D., Kung-Yee, L., & Pulver, A. E. (2003). Genomewide linkage scan for schizophrenia susceptibility loci among Ashkenazi Jewish families shows evidence for linkage on chromosome 10q22. *American Journal of Human Genetics, 73,* 601-611.

Fancher, R. E. (2004). The concept of race in the life and thought of Francis Galton. In A. Winston, (Ed.), *Defining difference: Race and racism in history of psychology* (pp. 49-75). Washington, DC: American Psychological Association.

Faraone, S. V., & Biederman, J. (2000). Nature, nurture, and attention deficit hyperactivity disorder. *Developmental Review, 20,* 568-581.

Faraone, S. V., Biederman, J., Weiffenbach, B., Keith, T., Chu, M. P., Weaver, A., Spencer, T. J., Wilens, T. E., Frazier, J., Cleves, M., & Sakai, J. (1999). Dopamine D4 gene 7-repeat allele and attention deficit hyperactivity disorder. *American Journal of Psychiatry, 156,* 768-770.

Faraone, S. V., Doyle, A. E., Mick, E., & Biederman, J. (2001). Meta-Analysis of the association between the 7-repeat allele of the dopamine D4 receptor gene and attention deficit hyperactivity disorder. *American Journal of Psychiatry, 158,* 1052-1057.

Faraone, S. V., & Tsuang, M. T. (1995). Methods in psychiatric genetics. In M. Tsuang, M. Tohen, & G. Zahner (Eds.), *Textbook in psychiatric epidemiology* (pp. 81-134). New York: Wiley-Liss.

Faraone, S. V., Tsuang, M. T., & Tsuang, D. W. (1999). *Genetics of mental disorders.* New York: The Guilford Press.

Farber, S. L. (1981). *Identical twins reared apart: A reanalysis.* New York: Basic Books.

Farmer, A. (2003). Ethical considerations in psychiatric genetics. In P. McGuffin, M. Owen, & I. Gottesman (Eds.), *Psychiatric genetics and genomics* (pp. 425-443). Oxford: Oxford University Press.

Feldman, M. W., & Lewontin, R. C. (1975). The heritability hang-up. *Science, 190*, 1163-1168.

Fink, A. E. (1938). *The causes of crime: Biological theories in the United States, 1800-1915*. New York: A. S. Barnes and Company.

Fischer, C. S., Hout, M., Sánchez Jankowski, M., Lucas, S. R., Swidler, A., & Voss, K. (1996). *Inequality by design*. Princeton, NJ: Princeton University Press.

Fischer, M. (1971). Psychoses in the offspring of schizophrenic monozygotic twins and their normal co-twins. *British Journal of Psychiatry, 118*, 43-51.

Fischer, M. (1973). *Genetic and environmental factors in schizophrenia*. Copenhagen: Munksgaard.

Fish, J. M. (Ed.). (2002). *Race and intelligence: Separating science from myth*. Mahwah, NJ: Lawrence Erlbaum.

Fisher, R. A. (1919). The genesis of twins. *Genetics, 4*, 489-499.

Fletcher, R. (1991). *Science, ideology, and the media: The Cyril Burt scandal*. New Brunswick, NJ: Transaction Publishers.

Frangos, E., Athanassenas, G., Tsitourides, S., Katsanou, N., & Alexandrakou, P. (1985). Prevalence of DSM-III schizophrenia among the first-degree relatives of schizophrenic probands. *Acta Psychiatrica Scandinavica, 72*, 382-386.

Franzek, E., & Beckmann, H. (1998). Different genetic background of schizophrenia spectrum diagnoses: A twin study. *American Journal of Psychiatry, 155*, 76-83.

Fraser, S. (Ed.). (1995). *The Bell Curve wars*. New York: Basic Books.

Fraser-Roberts, J. A. (1934). Twins. *Eugenics Review, 17*, 25-32.

Fuller, J. L., & Thompson, W. R. (1960). *Behavior genetics*. New York: John Wiley & Sons.

Fuller, J. L., & Thompson, W. R. (1978). *Foundations of behavior genetics*. St. Louis: C. V. Mosby.

Gabrielli, W. F., & Mednick, S. A. (1983). Genetic correlates of criminal behavior. *American Behavioral Scientist, 27*, 59-74.

Galton, F. (1875). The history of twins as a criterion of the relative powers of nature and nurture. *Journal of the Anthropological Institute of Great Britain and Ireland, 5*, 391-406.

Galton, F. (1881). *Hereditary genius*. New York: D. Appleton. (Originally published in 1869)

Galton, F. (1883). *Inquiries into human faculty and its development*. London: Macmillan and Company.

Galton, F. (1995). Hereditary talent and character. In R. Jacoby & N. Glauberman (Eds.), *The Bell Curve debate* (pp. 393-409). New York: Times Books. (Originally published in 1865)

Gardner, I. C., & Newman, H. H. (1940). Mental and physical traits of identical twins reared apart. *Journal of Heredity, 31,* 119-126.

Garver, K. L., & Garver, B. (1994). The Human Genome Project and eugenic concerns. *American Journal of Human Genetics, 54,* 148-158.

Gedda, L. (1961). *Twins in history and science.* Springfield, IL: Charles C. Thomas.

Gershon, E. S. (1997). Ernst Rüdin, a Nazi psychiatrist and geneticist [Letter to the editor]. *American Journal of Human Genetics (Neuropsychiatric Genetics), 74,* 457-458.

Gershon, E. S., Badner, J. A., Goldin, L. R., Sanders, A. R., Cravchik, A., & Detera-Wadleigh, S. D. (1998). Closing in on genes for manic-depressive illness and schizophrenia. *Neuropsychopharmacology, 18,* 233-242.

Gesell, A. L. (1942). The method of co-twin control. *Science, 95,* 446-448.

Gesell, A. L., & Thompson, H. (1929). Learning and growth in identical infant twins. *Genetic Psychology Monographs, 6,* 5-120.

Gillie, O. (1976). Crucial data was faked by eminent scientist. *The Sunday Times,* 24 October, London.

Goddard, H. H. (1927). *The Kallikak family: A study in the heredity of feeble-mindedness.* New York: The Macmillan Company. (Originally published in 1912)

Goldsmith, H. H., & Gottesman, I. I. (1996). Heritable variability and variable heritability in developmental psychopathology. In M. Lenzenweger & J. Haugaard (Eds.), *Frontiers of developmental psychopathology* (pp. 5-43). New York: Oxford University Press.

Goodman, R., & Stevenson, J. (1989). A twin study of hyperactivity — II. The etiological role of genes, family relationships and perinatal adversity. *Journal of Child Psychology and Psychiatry, 30,* 691-709.

Goodnight, C. J. (1990). On the relativity of quantitative genetic variance components. *Behavioral and Brain Sciences, 13,* 134-135.

Gosden, R. (2001). *Punishing the patient.* Melbourne: Scribe Publications.

Gottesman, I. I. (1978). Schizophrenia and genetics: Where are we? Are you sure? In L. Wynne, R. Cromwell, & S. Matthysse (Eds.), *The nature of schizophrenia: New approaches to research and treatment* (pp. 59-69). New York: John Wiley & Sons.

Gottesman, I. I. (1982). Identical twins reared apart: A reanalysis [Book review]. *American Journal of Psychology, 95,* 350-352.

Gottesman, I. I. (1991). *Schizophrenia genesis.* New York: W. H. Freeman & Company.

Gottesman, I. I. (1996). Blind men and elephants. In L. Hall (Ed.), *Genetics and mental illness: Evolving issues for research and society* (pp. 51-77). San Francisco: Freeman.

Gottesman, I. I., & Bertelsen, A. (1989). Confirming unexpressed genotypes for schizophrenia. *Archives of General Psychiatry, 46,* 867-872.

Gottesman, I. I., & Bertelsen, A. (1990). Reply to Torrey [Letter to the editor]. *Archives of General Psychiatry, 47,* 977-978.

Gottesman, I. I., & Bertelsen, A. (1996). Legacy of German psychiatric genetics: Hindsight is always 20/20. *American Journal of Medical Genetics (Neuropsychiatric Genetics), 67,* 317-322.

Gottesman, I. I., & Shields, J. (1966a). Contributions of twin studies to perspectives on schizophrenia. In B. Maher (Ed.), *Progress in experimental personality research* (Vol. 3, pp. 1-84). New York: Academic Press.

Gottesman, I. I., & Shields, J. (1966b). Schizophrenia in twins: 16 years' consecutive admissions to a psychiatric clinic. *British Journal of Psychiatry, 112,* 809-818.

Gottesman, I. I., & Shields, J. (1972). *Schizophrenia and genetics: A twin study vantage point.* New York: Academic Press.

Gottesman, I. I., & Shields, J. (1976a). A critical review of recent adoption, twin, and family studies of schizophrenia: Behavioral genetics perspectives. *Schizophrenia Bulletin, 2,* 360-401.

Gottesman, I. I., & Shields, J. (1976b). Rejoinder: Toward optimal arousal and away from original din. *Schizophrenia Bulletin, 2,* 447-453.

Gottesman, I. I., & Shields, J. (1982). *Schizophrenia: The epigenetic puzzle.* New York: Cambridge University Press.

Gottfredson, M. R., & Hirschi, T. (1990). *A general theory of crime.* Stanford, CA: Stanford University Press.

Gould, S. J. (1981). *The mismeasure of man.* New York: W. W. Norton & Company.

Gray, J. R., & Thompson, P. M. (2004). Neurobiology of intelligence: Science and ethics. *Nature Reviews Neuroscience, 5,* 471-482.

Grove, W. M., Eckert, E. D., Heston, L. L., Bouchard, T. J., Jr., Segal, N. L., & Lykken, D. T. (1990). Heritability of substance abuse and antisocial behavior: A study of monozygotic twins reared apart. *Biological Psychiatry, 27,* 1293-1304.

Gunderson, J. G., & Siever, L. J. (1985). Relatedness of schizotypal to schizophrenic disorders: Editor's introduction. *Schizophrenia Bulletin, 11,* 532-537.

Guthrie, R. V. (1976). *Even the rat was white.* New York: Harper and Row.

Gütt, A., Rüdin, E., & Ruttke, F. (1934). *Gesetz zur Verhütung erbkranken Nachwuchses* [Law for the prevention of genetically diseased offspring]. Munich: J. F. Lehmanns.

Häfner, H., & der Heiden, W. (1986). The contribution of European case registers to research on schizophrenia. *Schizophrenia Bulletin, 12,* 26-51.

Haier, R. J., Rosenthal, D., & Wender, P. H. (1978). MMPI assessment of psychopathology in the adopted-away offspring of schizophrenics. *Archives of General Psychiatry, 35,* 171-175.

Hall, W. S. (1914). The relation of crime to adolescence. *Bulletin of the American Academy of Medicine, 15,* 86-95.

Haller, M. H. (1963). *Eugenics: Hereditarian attitudes in American thought.* New Brunswick, NJ: Rutgers University Press.

Halverson, C. F. (1988). Remembering your parents: Reflections on the retrospective method. *Journal of Personality, 56,* 435-449.

Hansen, B. (1996). Something rotten in the state of Denmark: Eugenics and the ascent of the welfare state. In G. Broberg & N. Roll-Hansen (Eds.), *Eugenics and the welfare state: Sterilization policy in Denmark, Sweden, Norway, and Finland* (pp. 9-76). East Lansing, MI: Michigan State University Press.

Hansen, S. (1925). Eugenics in Denmark. *Eugenical News, 10,* 81-82.

Hardt, J., & Rutter, M. (2004). Validity of adult retrospective reports of adverse childhood experiences: Review of the evidence. *Journal of Child Psychology and Psychiatry, 45,* 260-273.

Harris, J. R. (1998). *The nurture assumption: Why children turn out the way they do.* New York: The Free Press.

Harrison, P. J., & Owen, M. J. (2003). Genes for schizophrenia: Recent findings and their pathophysiological implications. *Lancet, 361,* 417-419.

Harvald, B., & Haugue, M. (1965). Hereditary factors elucidated by twin studies. In J. Neel, M. Shaw, & W. Schull (Eds.), *Genetics and the epidemiology of chronic diseases* (pp. 61-76). Washington, DC: Public Health Service Publication No. 1163.

Hearnshaw, L. (1979). *Cyril Burt: Psychologist.* Ithaca, NY: Cornell University Press.

Hemminki, E., Rasimus, A., & Forssas, E. (1997). Sterilization in Finland: From eugenics to contraception. *Social Science and Medicine, 45,* 1875-1884.

Herrnstein, R. J., & Murray, C. (1994). *The bell curve.* New York: The Free Press.

Heston, L. L. (1966). Psychiatric disorders in foster home reared children of schizophrenic mothers. *British Journal of Psychiatry, 112,* 819-825.

Heston, L. L. (1970). The genetics of schizophrenic and schizoid disease. *Science, 167,* 249-256.

Heston, L. L. (1981/November). Family reunion [Book review]. *The Sciences, 21,* 26-28.

Heston, L. L. (1988). What about environment? In D. Dunner, E. Gershon, & J. Barrett (Eds.), *Relatives at risk for mental disorder* (pp. 205-213). New York: Raven Press.

Heston, L. L., & Denny, D. D. (1968). Interactions between early life experience and biological factors in schizophrenia. In D. Rosenthal & S. Kety (Eds.), *The transmission of schizophrenia* (pp. 363-376). New York: Pergamon Press.

Heston, L. L., Denney, D. D., & Pauly, I. B. (1966). The adult adjustment of persons institutionalized as children. *British Journal of Psychiatry, 112*, 1103-1110.

Hietala, M. (1996). From race hygiene to sterilization: The eugenics movement in Finland. In G. Broberg & N. Roll-Hansen (Eds.), *Eugenics and the welfare state: Sterilization policy in Denmark, Sweden, Norway, and Finland* (pp. 195-258). East Lansing, MI: Michigan State University Press.

Hill, D. (1983). *The politics of schizophrenia.* Lanham, MD: University Press of America.

Hirsch, J. (1970). Behavior-genetic analysis and its biosocial consequence. *Seminars in Psychiatry, 2*, 89-105.

Hirsch, J. (1990). A nemesis for heritability estimation. *Behavioral and Brain Sciences, 13*, 137-138.

Hirsch, J. (1997). Some history of heredity-vs-environment, genetic inferiority at Harvard (?), and The (incredible) Bell Curve. *Genetica, 99*, 207-224.

Hirsch, N. D. (1926). A study of natio-racial mental differences. *Genetic Psychology Monographs, 1*, 231-406.

Hirsch, N. D. (1930). *Twins: Heredity and environment.* Cambridge, MA: Harvard University Press.

Hoffer, A., & Pollin, W. (1970). Schizophrenia in the NAS-NRC panel of 15,909 veteran twin pairs. *Archives of General Psychiatry, 23*, 469-477.

Hoffman, L. W. (1991). The influence of the family environment on personality: Accounting for sibling differences. *Psychological Bulletin, 110*, 187-203.

Holden, C. (1980). Identical twins reared apart. *Science, 207*, 1323-1328.

Holmberg, D., & Holmes, J. G. (1994). Reconstruction of relationship memories: A mental models approach. In N. Schwarz & S. Sudman (Eds.), *Autobiographical memory and the validity of retrospective reports* (pp. 267-288). New York: Springer Verlag.

Holmes, S. J. (1930, September). Nature versus nurture in the development of the mind. *The Scientific Monthly, 31*, 245-252.

Holzman, P. S. (2001). Seymour S. Kety and the genetics of schizophrenia. *Neuropsychopharmacology, 25*, 299-304.

Horgan, J. (1993). Eugenics revisited. *Scientific American, 268*, 122-131.

Hubbard, R., & Wald, E. (1993). *Exploding the gene myth.* Boston: Beacon Press.

Hurwitz, S., & Christiansen, K. O. (1983). *Criminology.* Boston: George Allen and Unwin.

Hutchings, B., & Mednick, S. A. (1975). Registered criminality in the adoptive and biological parents of registered male criminal adoptees. In R. Fieve, D. Rosenthal, &

H. Brill (Eds.), *Genetic research in psychiatry* (pp. 105-116). Baltimore: The Johns Hopkins Press.

Ingraham, L. J., & Kety, S. S. (1988). Schizophrenia spectrum disorders. In M. Tsuang & J. Simpson (Eds.), *Handbook of schizophrenia, Vol. 3: Nosology, epidemiology and genetics* (pp. 117-137). New York: Elsevier Science Publishers.

Inouye, E. (1961). Similarity and dissimilarity of schizophrenia in twins. *Proceedings of The Third World Congress of Psychiatry* (Vol. 1, pp. 524-530). Montreal: University of Toronto Press.

Inouye, E. (1965). Similar and dissimilar manifestations of obsessive-compulsive neurosis in monozygotic twins. *American Journal of Psychiatry, 121,* 1171-1175.

Irvine, R. T. (1903). The congenital criminal. *Medical News, 82,* 749-752.

Jackson, D. D. (1960). A critique of the literature on the genetics of schizophrenia. In D. Jackson (Ed.), *The etiology of schizophrenia* (pp. 37-87). New York: Basic Books.

Jackson, Donald Dale. (1980, October). Reunion of identical twins, raised apart, reveals some astonishing similarities. *Smithsonian,* pp. 48-56.

Jackson, G. E. (2003). Rethinking the Finnish adoption studies of schizophrenia: A challenge to genetic determinism. *Journal of Critical Psychology, Counselling and Psychotherapy, 3,* 129-138.

Jacobs, D. (1994). Environmental failure — oppression is the only cause of psychopathology. *Journal of Mind and Behavior, 15,* 1-18.

Jacoby, R. & Glauberman, N. (Eds.). (1995). *The Bell Curve debate.* New York: Times Books.

Jacquard, A. (1983). Heritability: One word, three concepts. *Biometrics, 39,* 465-477.

Jensen, A. R. (1970). IQ's of identical twins reared apart. *Behavior Genetics, 1,* 133-148.

Jensen, A. R. (1972). How much can we boost IQ and scholastic achievement? In *Genetics and Education* (pp. 69-203). London: Methuen & Co. (Originally published in 1969)

Jensen, A. R. (1974). Kinship correlations reported by Sir Cyril Burt. *Behavior Genetics, 4,* 1-28.

Jensen, A. R. (1980). *Bias in mental testing.* New York: Free Press.

Jensen, A. R. (1998). *The g factor.* Westport, CT: Praeger.

Jinks, J. L., & Fulker, D. W. (1970). Comparison of the biometrical genetic, MAVA, and classical approaches to the analysis of human behavior. *Psychological Bulletin, 73,* 311-349.

Johnson, W., Bouchard, T. J., Jr., Krueger, R. F., McGue, M., & Gottesman, I. I. (2004). Just one *g*: Consistent results from three test batteries. *Intelligence, 32,* 95-107.

Joseph, J. (1998a). *A critical analysis of the genetic theory of schizophrenia.* Unpublished doctoral dissertation, California School of Professional Psychology, Alameda.

Joseph, J. (1998b). The equal environment assumption of the classical twin method: A critical analysis. *Journal of Mind and Behavior, 19*, 325-358.

Joseph, J. (1999a). A critique of the Finnish Adoptive Family Study of Schizophrenia. *Journal of Mind and Behavior, 20*, 133-154.

Joseph, J. (1999b). The genetic theory of schizophrenia: A critical overview. *Ethical Human Sciences and Services, 1*, 119-145.

Joseph, J. (2000a). A critique of the spectrum concept as used in the Danish-American schizophrenia adoption studies. *Ethical Human Sciences and Services, 2*, 135-160.

Joseph, J. (2000b). Inaccuracy and bias in textbooks reporting psychiatric research: The case of the schizophrenia adoption studies. *Politics and the Life Sciences, 19*, 89-99.

Joseph, J. (2000c). Not in their genes: A critical view of the genetics of attention-deficit hyperactivity disorder. *Developmental Review, 20*, 539-567.

Joseph, J. (2000d). Potential confounds in psychiatric genetic research: The case of pellagra. *New Ideas in Psychology, 18*, 83-91.

Joseph, J. (2000e). Problems in psychiatric genetic research: A reply to Faraone and Biederman. *Developmental Review, 20*, 582-593.

Joseph, J. (2001a). The Danish-American Adoptees' Family studies of Kety and associates: Do they provide evidence in support of the genetic basis of schizophrenia? *Genetic, Social, and General Psychology Monographs, 127*, 241-278.

Joseph, J. (2001b). Don Jackson's "A critique of the literature on the genetics of schizophrenia" — A reappraisal after 40 years. *Genetic, Social, and General Psychology Monographs, 127*, 27-57.

Joseph, J. (2001c). Is crime in the genes? A critical review of twin and adoption studies of criminality and antisocial behavior. *Journal of Mind and Behavior, 22*, 179-218.

Joseph, J. (2001d). Separated twins and the genetics of personality differences: A critique. *American Journal of Psychology, 114*, 1-30.

Joseph, J. (2002a). Adoption study of ADHD [Letter to the Editor]. *Journal of the American Academy of Child and Adolescent Psychiatry, 41*, 1389-1391.

Joseph, J. (2002b). Twin studies in psychiatry and psychology: Science or pseudoscience? *Psychiatric Quarterly, 73*, 71-82.

Joseph, J. (2003). Genetics and antisocial behavior. *Ethical Human Sciences and Services, 5*, 147-150.

Joseph, J. (2004). Schizophrenia and heredity: Why the emperor has no genes. In J. Read, L. Mosher, & R. Bentall (Eds.), *Models of madness: Psychological, social and biological approaches to schizophrenia* (pp. 67-83). Andover, UK: Taylor & Francis.

Joseph, J., & Baldwin, S. (2000). Four editorial proposals to improve social sciences research and publication. *International Journal of Risk and Safety in Medicine, 13*, 117-127.

Joynson, R. B. (1989). *The Burt affair*. London: RKP.

Juel-Nielsen, N. (1980). *Individual and environment: Monozygotic twins reared apart* (rev. ed.). New York: International Publishers. (Originally published in 1965)

Kallmann, F. J. (1938a). *The genetics of schizophrenia: A study of heredity and reproduction in the families of 1,087 schizophrenics*. New York: J. J. Augustin.

Kallmann, F. J. (1938b). Heredity, reproduction, and eugenic procedure in the field of schizophrenia. *Eugenical News, 13*, 105-113.

Kallmann, F. J. (1938c). Wilhelm Weinberg. *Journal of Nervous and Mental Disease, 87*, 263-264.

Kallmann, F. J. (1946). The genetic theory of schizophrenia: An analysis of 691 schizophrenic twin index families. *American Journal of Psychiatry, 103*, 309-322.

Kallmann, F. J. (1953). *Heredity in health and mental disorder*. New York: W. W. Norton.

Kallmann, F. J. (1958). The uses of genetics in psychiatry. *Journal of Mental Science, 104*, 542-552.

Kamin, L. J. (1974). *The science and politics of I.Q.* Potomac, MD: Lawrence Erlbaum Associates.

Kamin, L. J. (1985). Criminality and adoption [Letter to the editor]. *Science, 227*, 983.

Kamin, L. J. (1986). Is crime in the genes? The answer may depend on who chooses the evidence. *Scientific American, 254*, 22-27.

Kamin, L. J., & Goldberger, A. S. (2002). Twin studies in behavioral research: A skeptical view. *Theoretical Population Biology, 61*, 83-95.

Kandel, E., Mednick, S. A., Kirkegaard-Sorensen, L., Hutchings, B., Knop, J., Rosenberg, R., & Schulsinger, F. (1988). IQ as a protective factor for subjects at high risk for antisocial behavior. *Journal of Consulting and Clinical Psychology, 56*, 224-226.

Karon, B. P. (1999). The tragedy of schizophrenia. *General Psychologist, 34*, 1-12.

Karon, B. P., & VandenBos, G. R. (1981). *Psychotherapy of schizophrenia: The treatment of choice*. New York: Jason Aronson.

Keefe, R. S. E., & Harvey, P. D. (1994). *Understanding schizophrenia*. New York: The Free Press.

Kempthorne, O. (1978). Logical, epistemological and statistical aspects of nature-nurture data interpretation. *Biometrics, 34*, 1-23.

Kempthorne, O. (1990). How does one apply statistical analysis to our understanding of the development of human relationships. *Behavioral and Brain Sciences, 13*, 138-139.

Kendler, K. S. (1983). Overview: A current perspective on twin studies of schizophrenia. *American Journal of Psychiatry, 140*, 1413-1425.

Kendler, K. S. (1986). Genetics of schizophrenia. In A. Frances & R. Hales (Eds.), *American Psychiatric Association annual review* (Vol. 5, pp. 25-41). Washington, DC: American Psychiatric Press.

Kendler, K. S. (1987). The genetics of schizophrenia: A current perspective. In H. Meltzer (Ed.), *Psychopharmacology: The third generation of progress* (pp. 705-713). New York: Raven Press.

Kendler, K. S. (1988). The genetics of schizophrenia: An overview. In M. Tsuang & J. Simpson (Eds.), *Handbook of schizophrenia, Vol. 3: Nosology, epidemiology and genetics* (pp. 437-462). New York: Elsevier Science Publishers.

Kendler, K. S. (1993). Twin studies of psychiatric illness: Current status and future directions. *Archives of General Psychiatry, 50*, 905-915.

Kendler, K. S. (1995). Genetic epidemiology in psychiatry: Taking both genes and environment seriously. *Archives of General Psychiatry, 52*, 895-899.

Kendler, K. S. (1997). Reply to Gejman, Gershon, and Lerer and Segman [Letter to the editor]. *American Journal of Medical Genetics (Neuropsychiatric Genetics), 74*, 461-463.

Kendler, K. S. (2000). Schizophrenia: Genetics. In B. Sadock & V. Sadock (Eds.), *Kaplan & Sadock's comprehensive textbook of psychiatry* (7th ed., Vol. 1, pp. 1147-1158). Philadelphia: Lippincott, Williams, & Wilkins.

Kendler, K. S. (2001). Twin studies of psychiatric illness: An update. *Archives of General Psychiatry, 58*, 1005-1014.

Kendler, K. S., & Diehl, S. R. (1993). The genetics of schizophrenia: A current, genetic-epidemiologic perspective. *Schizophrenia Bulletin, 19*, 261-285.

Kendler, K., & Diehl, S. R. (1995). Schizophrenia: Genetics. In H. Kaplan & B. Sadock (Eds.), *Comprehensive textbook of psychiatry* (6th ed., Vol. 1, pp. 942-957). Baltimore: Williams & Wilkins.

Kendler, K. S., & Gruenberg, A. M. (1984). An independent analysis of the Danish adoption study of schizophrenia. *Archives of General Psychiatry, 41*, 555-564.

Kendler, K. S., Gruenberg, A. M., & Kinney, D. K. (1994). Independent diagnoses of adoptees and relatives as defined by DSM-III in the provincial and national samples of the Danish adoption study of schizophrenia. *Archives of General Psychiatry, 51*, 456-468.

Kendler, K. S., Neale, M. C., Kessler, R. C., Heath, A. C., & Eaves, L. J. (1994). Parental treatment and the equal environment assumption in twin studies of psychiatric illness. *Psychological Medicine, 24*, 579-590.

Kendler, K. S., Neale, M. C., Kessler, R. C., Heath, A. C., & Eaves, L. J. (1993). A test of the equal-environment assumption in twin studies of psychiatric illness. *Behavior Genetics, 23*, 21-27.

Kendler, K. S., & Robinette, C. D. (1983). Schizophrenia in the national academy of sciences-national research council twin registry: A 16-year update. *American Journal of Psychiatry, 140,* 1551-1563.

Kennedy, J. L., Farrer, L. A., Andreasen, N. C., Mayeux, R., & St. George-Hyslop, P. (2003). The genetics of adult-onset neuropsychiatric disease: Complexities and conundra? *Science, 302,* 822-826.

Kennedy, J. L., Giuffra, L. A., Moises, H. W., Cavalli-Sforza, L. L., Pakstis, A. J., Kidd, J. R., Castiglione, C. M., Sjogren, B., Wetterberg, L., & Kidd, K. K. (1988). Evidence against linkage of schizophrenia to markers on chromosome 5 in a northern Swedish pedigree. *Nature, 336,* 167-170.

Kety, S. S. (1959). Biochemical theories of schizophrenia, part II. *Science, 129,* 1590-1596.

Kety, S. S. (1966). Research programs of the major mental illnesses: 1. *Hospital and Community Psychiatry, 17,* 226-232.

Kety, S. S. (1970). Genetic-environmental interactions in the schizophrenic syndrome. In R. Cancro (Ed.), *The schizophrenic reactions* (pp. 233-244). New York: Brunner/Mazel.

Kety, S. S. (1974). From rationalization to reason. *American Journal of Psychiatry, 131,* 957-963.

Kety, S. S. (1976). Studies designed to disentangle genetic and environmental variables in schizophrenia: Some epistemological questions and answers. *American Journal of Psychiatry, 133,* 1134-1137.

Kety, S. S. (1978). Heredity and environment. In J. Shershow (Ed.), *Schizophrenia: Science and practice* (pp. 47-68). Cambridge, MA: Harvard University Press.

Kety, S. S. (1983a). Mental illness in the biological and adoptive relatives of schizophrenia adoptees: Findings relevant to genetic and environmental factors in etiology. *American Journal of Psychiatry, 140,* 720-727.

Kety, S. S. (1983b). Observations on genetic and environmental influences in the etiology of mental disorder from studies on adoptees and their relatives. In S. Kety, L. Rowland, R. Sidman, & S. Matthysse (Eds.), *Genetics of neurological and psychiatric disorders* (pp. 105-114). New York: Raven Press.

Kety, S. S. (1985). Schizotypal personality disorder: An operational definition of Bleuler's latent schizophrenia? *Schizophrenia Bulletin, 11,* 590-594.

Kety, S. S. (1987). The significance of genetic factors in the etiology of schizophrenia: Results from the national study of adoptees in Denmark. *Journal of Psychiatric Research, 21,* 423-429.

Kety, S. S., & Ingraham, L. J. (1992). Genetic transmission and improved diagnosis of schizophrenia from pedigrees of adoptees. *Journal of Psychiatric Research, 26,* 247-255.

Kety, S. S., Rosenthal, D., & Wender, P. H. (1978). Genetic relationships within the schizophrenia spectrum: Evidence from adoption studies. In R. Spitzer & D. Klein (Eds.), *Critical issues in psychiatric diagnosis* (pp. 213-223). New York: Raven Press.

Kety, S. S., Rosenthal, D., Wender, P. H., & Schulsinger, F. (1968). The types and prevalence of mental illness in the biological and adoptive families of adopted schizophrenics. In D. Rosenthal & S. Kety (Eds.), *The transmission of schizophrenia* (pp. 345-362). New York: Pergamon Press.

Kety, S. S., Rosenthal, D., Wender, P. H., & Schulsinger, F. (1976). Studies based on a total sample of adopted individuals and their relatives: Why they were necessary, what they demonstrated and failed to demonstrate. *Schizophrenia Bulletin, 2*, 413-427.

Kety, S. S., Rosenthal, D., Wender, P. H., Schulsinger, F., & Jacobsen, B. (1975). Mental illness in the biological and adoptive families of adopted individuals who have become schizophrenic: A preliminary report based on psychiatric interviews. In R. Fieve, D. Rosenthal, & H. Brill (Eds.), *Genetic research in psychiatry* (pp. 147-165). Baltimore: The Johns Hopkins Press.

Kety, S. S., Rosenthal, D., Wender, P. H., Schulsinger, F., & Jacobsen, B. (1978). The biologic and adoptive families of adopted individuals who became schizophrenic: Prevalence of mental illness and other characteristics. In L. Wynne, R. Cromwell, & S. Matthysse (Eds.), *The nature of schizophrenia* (pp. 25-37). New York: John Wiley & Sons.

Kety, S. S., Wender, P. H., Jacobsen, B., Ingraham, L. J., Jansson, L., Faber, B., & Kinney, D. K. (1994). Mental illness in the biological and adoptive relatives of schizophrenic adoptees: Replication of the Copenhagen study to the rest of Denmark. *Archives of General Psychiatry, 51*, 442-455.

Kevles, D. J. (1985). *In the name of eugenics.* Berkeley, CA: University of California Press.

Koch, H. L. (1966). *Twins and twin relations.* Chicago: The University of Chicago Press.

Kolb, L., & Brodie, H. K. (1982). *Modern clinical psychiatry* (10th ed.). Philadelphia: W. B. Saunders Company.

Koskenvuo, M., Langinvainio, H., Kaprio, J., Lönnqvist, J., & Tienari, P. (1984). Psychiatric hospitalization in twins. *Acta Geneticae Medicae et Gemellologiae, 33*, 321-332.

Kraepelin, E. (1976). *Manic-depressive insanity and paranoia.* New York: Arno Press. (Originally published in English in 1921)

Kranz, H. (1936). *Lebensschicksale krimineller Zwillinge* [The life destiny of criminal twins]. Berlin: Julius Springer Verlag.

Kringlen, E. (1967a). Heredity and environment in the functional psychoses: An epidemiological-clinical study. Oslo: Universitetsforlaget.

Kringlen, E. (1967b). Heredity and environment in the functional psychoses: Case histories. Oslo: Universitetsforlaget.

Kringlen, E. (1976). Twins — still our best method. *Schizophrenia Bulletin, 2*, 429-433.

Kringlen, E. (1987). Contributions of genetic studies on schizophrenia. In H. Häfner & W. Gattaz (Eds.), *Search for the causes of schizophrenia* (pp. 123-142). New York: Springer Verlag.

Kringlen, E., & Cramer, G. (1989). Offspring of monozygotic twins discordant for schizophrenia. *Archives of General Psychiatry, 46*, 873-877.

Krueger, R. F., Markon, K. E., & Bouchard, T. J., Jr. (2003). The extended genotype: The heritability of personality accounts for the heritability of recalled family environments in twins reared apart. *Journal of Personality, 71*, 809-833.

Kutchins, H., & Kirk, S. A. (1997). *Making us crazy: DSM: the psychiatric bible and the creation of mental disorders.* New York: The Free Press.

LaFollette, M. C. (1992). *Stealing into print: Fraud, plagiarism, and misconduct in scientific publishing.* Berkeley, CA: University of California Press.

Laing, R. D. (1967). *The politics of experience.* New York: Pantheon Books.

Laing, R. D. (1981). A critique of Kallmann's and Slater's genetic theory of schizophrenia. In R. Evans, *Dialogue with R. D. Laing* (pp. 97-156). New York: Praeger.

Lander, E., & Kruglyak, L. (1995). Genetic dissection of complex traits: Guidelines for interpreting and reporting linkage results. *Nature Genetics, 11*, 241-247.

Lang, J. S. (1987, April 13th). How genes shape personality. *US News and World Report* pp. 58-66.

Lange, J. (1930). *Crime and destiny.* New York: Charles Boni.

Langinvainio, H., Kaprio, J., Koskenvuo, M., & Lönnqvist, J. (1984). Finnish twins reared apart III: Personality factors. *Acta Geneticae Medicae et Gemellologiae, 33*, 259-264.

Langinvainio, H., Koskenvuo, M., Kaprio, J., & Sistonen, P. (1984). Finnish twins reared apart II: Validation of zygosity, environmental dissimilarity and weight and height. *Acta Geneticae Medicae et Gemellologiae, 33*, 251-258.

Largent, M. A. (2002). "The greatest curse of the race": Eugenic sterilization in Oregon, 1909-1983. *Oregon Historical Quarterly, 103*, 188-209.

Laughlin, H. H. (1935). The eugenics exhibit at Chicago. *Journal of Heredity, 26*, 155-162.

Lauterbach, C. E. (1925). Studies in twin resemblance. *Genetics, 10*, 525-579.

Lehtonen, J. (1994). From dualism to psychobiological interaction: A comment on the study by Tienari and his co-workers. *British Journal of Psychiatry, 164* (Suppl. 23), 27-28.

Leo, J., & Cohen, D. (2003). Broken brains or flawed studies? A critical review of ADHD neuroimaging research. *Journal of Mind and Behavior, 24*, 29-55.

Leo, J., & Joseph, J. (2002). Schizophrenia: Medical students are taught it's all in the genes, but are they hearing the whole story? *Ethical Human Sciences and Services, 4*, 17-30.

Lerer, B., & Segman, R. H. (1997). Correspondence regarding German psychiatric genetics and Ernst Rüdin [Letter to the editor]. *American Journal of Human Genetics (Neuropsychiatric Genetics)*, 74, 459-460.

Lerner, I. M. (1958). *The genetic basis of selection*. New York: John Wiley & Sons.

Lewontin, R. C. (1976). Race and intelligence. In N. Block & G. Dworkin (Eds.), *The IQ controversy* (pp. 78-92). New York: Pantheon Books. (Originally published in 1970)

Lewontin, R. C. (1974). The analysis of variance and the analysis of causes. *American Journal of Human Genetics*, 26, 400-411.

Lewontin, R. C. (1987). The irrelevance of heritability. *Science for the People*, 19, 23, 32.

Lewontin, R. C. (1991). *Biology as ideology*. New York: Harper Perennial.

Lewontin, R. C., Rose, S., & Kamin, L. J. (1984). *Not in our genes*. New York: Pantheon.

Lidz, T. (1976). Commentary on a critical review of recent adoption, twin, and family studies of schizophrenia: Behavioral genetics perspectives. *Schizophrenia Bulletin*, 2, 402-412.

Lidz, T., & Blatt, S. (1983). Critique of the Danish-American studies of the biological and adoptive relatives of adoptees who became schizophrenic. *American Journal of Psychiatry*, 140, 426-435.

Lidz, T., Blatt, S., & Cook, B. (1981). Critique of the Danish-American studies of the adopted-away offspring of schizophrenic parents. *American Journal of Psychiatry*, 138, 1063-1068.

Lifton, R. J. (1986). *The Nazi doctors*. New York: Basic Books.

Lilienfeld, S. O., Lynn, S. J., & Lohr, J. M. (2003). Science and pseudoscience in clinical psychology: Initial thoughts, reflections, and considerations. In S. Lilienfeld, S. Lynn, & J. Lohr (Eds.), *Science and pseudoscience in clinical psychology* (pp. 1-14). New York: Guilford.

Loehlin, J. C. (1981). Identical twins reared apart: A reanalysis [Book review]. *Acta Geneticae Medicae et Gemellologiae*, 30, 297-298.

Loehlin, J. C., & Horn, J. M. (2000). Stoolmiller on restriction of range in adoption studies: A comment. *Behavior Genetics*, 30, 245-247.

Loehlin, J. C., & Nichols, R. C. (1976). *Heredity, environment, & personality*. Austin: University of Texas Press.

Lombroso, C. (1968). *Crime: Its causes and remedies*. Montclair, NJ: Patterson Smith. (Originally published in 1911)

Lowing, P. A., Mirsky, A. F., & Pereira, R. (1983). The inheritance of schizophrenia spectrum disorders: A reanalysis of the Danish adoptee study data. *American Journal of Psychiatry*, 140, 1167-1171.

Ludmerer, K. M. (1972). *Genetics and American society: A historical appraisal.* Baltimore: Johns Hopkins University Press.

Luo, X., Klempan, T. A., Lappalainen, J., Rosenheck, R. A., Charney, D. S., Erdos, J., van Kammen, D. P., Kranzler, H. R., Kennedy, J. L., & Gelernter, J. (2004). NOTCH4 gene haplotype is associated with schizophrenia in African Americans. *Biological Psychiatry, 55,* 112-117.

Lush, J. L. (1945). *Animal breeding plans.* Ames, IA: Collegiate Press.

Lush, J. L. (1949). Heritability of quantitative characteristics in farm animals. *Hereditas* (Suppl.), G. Bonnier & R. Larsson, Eds., 356-375.

Luxenburger, H. (1928). Vorläufiger Bericht über psychiatrische Serienuntersuchungen an Zwillingen [Provisional report on a series of psychiatric investigations of twins]. *Zeitschrift für die Gesamte Neurologie und Psychiatrie, 116,* 297-347.

Luxenburger, H. (1931a). Möglichkeiten und Notwendigkeiten für die psychiatrisch-eugenische Praxis [Possibilities and necessities for the psychiatric-eugenic practice]. *Münchener Medizinische Wochenschrift, 78,* 753-758.

Luxenburger, H. (1931b). Psychiatrische Erbprognose und Eugenik. [Psychiatric genetic prognosis and eugenics]. *Eugenik, 1,* 117-124.

Luxenburger, H. (1931c). Psychische Hygiene und psychiatrische eugenik [Psychic hygiene and psychiatric eugenics]. *Eugenik, 2,* 49-56.

Luxenburger, H. (1934a). Paranoia und Gesetz zur Verhütung erbkranken Nachwuchses [Paranoia and the law for the prevention of hereditarily ill offspring]. *Der Erbarzt, 1,* 33-36.

Luxenburger, H. (1934b). Rassenhygienisch wichtige Probleme und Ergebnisse der Zwillingspathologie [Important racial hygienic problems and results of twin pathology]. In E. Rüdin (Ed.), *Erblehre und rassenhygiene im völkischen staat* [Genetics and racial hygiene in the völkish state] (pp. 303-316). Munich: J. F. Lehmanns.

Luxenburger, H. (1936a). Die rassenhygienische Bedeutung der Lehre von den Manifestationsschwankungen schwanfungen erblicher Krankheiten [The racial hygienic meaning of the theory of variations in the manifestation of hereditary illness]. *Der Erbarzt, 3,* 33-36.

Luxenburger, H. (1936b). Die wichtigsten neueren Ergebnisse der Empirischen Erbprognose und der Zwillingsforschung in der Psychiatrie [The most important new results of the empirical genetic prognosis and twin research in psychiatry]. *Der Erbarzt, 3,* 129-133.

Luxenburger, H. (1939). Kritische Besprechungen und Referate [Review of G. Banu L'Hygiène de la race (Racial Hygiene)]. *Archiv für Rassen- und Gesellschaftsbiologie, 33,* 433-438.

Luxenburger, H. (1940a). Erbpathologie der Schizophrenie [Genetic pathology of schizophrenia] In A. Gütt (Ed.), *Handbuch der erbkrankheiten* (Vol. 2, pp. 191-294). Leipzig: Georg Thieme.

Luxenburger, H. (1940b). Schizophrenie und manisch-depressives Irresein [Schizophrenia and manic-depressive psychosis]. *Fortschritte der Erbpathologie und Rassenhygiene, 4,* 239-259. Abstract obtained from *Psychological Abstracts (1942), 16, (4),* 168.

Lykken, D. T. (1978). The diagnosis of zygosity in twins. *Behavior Genetics, 8,* 437-473.

Lykken, D. T. (1995). *The antisocial personalities.* Hillsdale, NJ: Lawrence Erlbaum Associates.

Lykken, D. T., McGue, M., Tellegen, A., & Bouchard, T. J., Jr. (1992). Emergenesis: Genetic traits that may not run in families. *American Psychologist, 47,* 1565-1577.

Lyons, M. J. (1996). A twin study of self-reported criminal behavior. In G. Bock & J. Goode (Eds.), *Genetics of criminal and antisocial behavior* (pp. 61-75). New York: John Wiley and Sons.

Lyons, M. J., & Bar, J. L. (2001). Is there a role for twin studies in the molecular genetics era? *Harvard Review of Psychiatry, 9,* 318-323.

Lyons, M. J., Kendler, K. S., Provet, A., & Tsuang, M. T. (1991). The genetics of schizophrenia. In M. Tsuang, K. Kendler, & M. Lyons (Eds.), *Genetic issues in psychosocial epidemiology* (pp. 119-152). New Brunswick, NJ: Rutgers University Press.

Lyons, M. J., True, W. R., Eisen, S. A., Goldberg, J., Meyer, J. M., Faraone, S. V., Eaves, L. J., & Tsuang, M. T. (1995). Differential heritability of adult and juvenile antisocial traits. *Archives of General Psychiatry, 52,* 906-915.

Marchese, F. J. (1995). The place of eugenics in Arnold Gesell's maturation theory of child development. *Canadian Psychology, 36,* 89-114.

Markon, K. E., Krueger, R. F., Bouchard, T. J., Jr., & Gottesman, I. I. (2002). Normal and abnormal personality traits: Evidence for genetic and environmental relationships in the Minnesota Study of Twins Reared Apart. *Journal of Personality, 70,* 661-693.

Marshall, R. (1990). The genetics of schizophrenia: Axiom or hypothesis? In R. Bentall (Ed.), *Restructuring schizophrenia* (pp. 89-117). London: Routledge.

Martin, J. D., Blair, G. E., Dannenmaier, W. D., Jones, P. C., & Asako, M. (1981). Relationship of scores on the California Psychological Inventory to age. *Psychological Reports, 49,* 151-154.

McClearn, G. E. (1964). The inheritance of behavior. In L. Postman (Ed.), *Psychology in the making* (pp. 144-252). New York: Alfred A. Knopf.

McCourt, K., Bouchard, T. J., Jr., Lykken, D. T., Tellegen, A., & Keyes, M. (1999). Authoritarianism revisited: Genetic and environmental influences in twins reared apart and together. *Personality and Individual Differences, 27,* 985-1014.

McGue, M., & Bouchard, T. J., Jr. (1984). Adjustment of twin data for the effects of age and sex. *Behavior Genetics, 14,* 325-343.

McGue, M., & Bouchard, T. J., Jr. (1989). Genetic and environmental determinants of information processing and special mental abilities: A twin analysis. In R. Sternberg (Ed.), *Advances in the psychology of human intelligence* (Vol. 5, pp. 7-45). Hillsdale, NJ: Lawrence Erlbaum.

McGuffin, P., & Katz, R. (1990). Who believes in estimating heritability as an end in itself? *Behavioral and Brain Sciences, 13,* 141-142.

McGuffin, P., Riley, B., & Plomin, R. (2001). Towards behavioral genomics. *Science, 291,* 1232-1249.

McGuire, T. R., & Hirsch, J. (1977). General intelligence (g) and heritability (H2, h2). In I. Uzgiris & F. Weitzmann (Eds.), *The structuring of experience* (pp. 25-72). New York: Plenum Press.

McMahon, R. C. (1980). Genetic etiology in the hyperactive child syndrome: A critical review. *American Journal of Orthopsychiatry, 50,* 145-150.

Mednick, S. A. (1996). General discussion III. In G. Bock & J. Goode (Eds.), *Genetics of criminal and antisocial behavior* (pp. 129-137). New York: John Wiley and Sons.

Mednick, S. A., Brennan, P., & Kandel, E. (1988). Predisposition to violence. *Aggressive Behavior, 14,* 25-33.

Mednick, S. A., Gabrielli, W. F., & Hutchings, B. (1984). Genetic influences in criminal convictions: Evidence from an adoption cohort. *Science, 224,* 891-894.

Mednick, S. A., Gabrielli, W. F., & Hutchings, B. (1987). Genetic factors in the etiology of criminal behavior. In S. Mednick, T. Moffitt, & S. Stack (Eds.), *The causes of crime: New biological perspectives* (pp. 75-91). New York: Cambridge University Press.

Mednick, S. A., & Hutchings, B. (1977). Some considerations in the interpretation of the Danish adoption studies in relation to asocial behavior. In S. Mednick & K. Christiansen (Eds.), *Biosocial bases of criminal behavior* (pp. 159-164). New York: Gardner Press.

Mednick, S. A., & Kandel, E. S. (1988). Congenital determinants of violence. *Bulletin of the American Academy of Psychiatry and the Law, 16,* 101-109.

Mednick, S. A., & Volavka, J. (1980). Biology and crime. In N. Morris & M. Tonry (Eds.), *Crime and justice: An annual review of research* (Vol. 2, pp. 85-158). Chicago: The University of Chicago Press.

Mellers, B., Hertwig, R., & Kahneman, D. (2001). Do frequency representations eliminate conjunction effects? An exercise in adversarial collaboration. *Psychological Science, 12,* 269-275.

Mensh, E., & Mensh, H. (1991). *The IQ mythology.* Carbondale, IL: Southern Illinois Press.

Merriman, C. (1924). The intellectual resemblance of twins. *Psychological Monographs, 33,* (5), 1-58.

Mischel, W. (1968). *Personality and assessment.* New York: Wiley.

Mittler, P. (1971). *The study of twins.* Middlesex, UK.: Penguin Books.

Moldin, S. O., & Gottesman, I. I. (1997). At issue: Genes, experience, and chance in schizophrenia — positioning for the 21st century. *Schizophrenia Bulletin, 23,* 547-561.

Moore, D. S. (2001). The dependent gene: The fallacy of "nature vs. nurture". New York: Times Books.

Moos, R. H., & Moos, B. S. (1986). *Family Environment Scale manual* (2nd ed.). Palo Alto, CA: Consulting Psychologists Press.

Morris-Yates, A., Andrews, G., Howie, P., & Henderson, S. (1990). Twins: A test of the equal environments assumption. *Acta Psychiatrica Scandinavica, 81,* 322-326.

Mosher, L. R. (1975). Unpublished letter to Gottesman, dated 10/14/75.

Mosher, L. R. (2004). Non-hospital, non-drug intervention with first-episode psychosis. In J. Read, L. Mosher, & R. Bentall (Eds.), *Models of madness: Psychological, social and biological approaches to schizophrenia* (pp. 349-364). Andover, UK: Taylor & Francis.

Mosher, L. R., & Menn, A. Z. (1978). Community residential treatment for schizophrenia: Two-year follow up. *Hospital and Community Psychiatry, 29,* 715-723.

Mosher, L. R., Vallone, R., & Menn, A. (1995). The treatment of acute psychosis without neuroleptics: Six-week psychopathology outcome data from the Soteria project. *International Journal of Social Psychiatry, 41,* 157-173.

Muller, H. J. (1925). Mental traits and heredity. *Journal of Heredity, 16,* 433-448.

Müller-Hill, B. (1998). *Murderous science.* Plainview, NY: Cold Spring Harbor Laboratory Press. (Original English version published in 1988)

Munk-Jørgensen, P. (1985). The schizophrenia diagnosis in Denmark. *Acta Psychiatrica Scandinavica, 72,* 266-273.

Murray, R. M., & Reveley, A. M. (1986). Genetic aspects of schizophrenia: Overview. In A. Kerr & P. Snaith (Eds.), *Contemporary issues in schizophrenia* (pp. 261-267). London: Gaskell.

Myerson, A. (1925). *The inheritance of mental diseases.* Baltimore: Williams & Wilkins.

Myerson, A. (1932). Discussion in E. Rüdin, "The significance of eugenics and genetics for mental hygiene." In F. Williams (Ed.), *Proceedings of the First International Congress on Mental Hygiene* (pp. 471-495). New York: International Committee for Mental Hygiene.

Neale, J. M., & Oltmanns, T. F. (1980). *Schizophrenia.* New York: John Wiley & Sons.

Neel, J. V., & Schull, W. J. (1954). *Human heredity.* Chicago: University of Chicago Press.

Neisser, U., Boodoo, G., Bouchard, T. J., Jr., Boykin, A. W., Brody, N., Ceci, S. J., Halpern, D. F., Loehlin, J. C., Perloff, R., Sternberg, R. J., Urbina, S. (1996). Intelligence: Knowns and unknowns. *American Psychologist, 51*, 77-101.

Newman, D. L., Tellegen, A., & Bouchard, T. J., Jr. (1998). Individual differences in adult ego development: Sources of influences in twins reared apart. *Journal of Personality and Social Psychology, 74*, 985-995.

Newman, H. H. (1917). *The biology of twins (mammals).* Chicago: University of Chicago Press.

Newman, H. H. (1932). *Evolution, genetics, and eugenics* (3rd ed.). Chicago: University of Chicago Press.

Newman, H. H., Freeman, F. N., & Holzinger, K. J. (1937). *Twins: A study of heredity and environment.* Chicago: The University of Chicago Press.

Ochsner, A. J. (1899). Surgical treatment of habitual criminals. *Journal of the American Medical Association, 32*, 867-868.

Ogbu, J. U. (1978). *Minority education and caste.* New York: Academic Press.

Olson, C. P. (Ed.). (1920). *Oregon laws: Showing all the laws of a general nature in force in the state of Oregon (Vol. 2).* San Francisco: The Bancroft-Whitney Company.

Onstad, S., Skre, I., Torgersen, S., & Kringlen, E. (1991). Twin concordance for DSM-III-R schizophrenia. *Acta Psychiatrica Scandinavica, 83*, 395-401.

Osborn, F. (1951). *Preface to eugenics* (rev. ed.). New York: Harper & Brothers.

Owen, M. J. (1992). Will schizophrenia become a graveyard for molecular geneticists? *Psychological Medicine, 22*, 289-293.

Paikin, H., Jacobsen, B., Schulsinger, F., Godtfredsen, K., Rosenthal, D., Wender, P. H., & Kety, S. S. (1974). Characteristics of people who refused to participate in a social and psychopathological study. In S. Mednick, F. Schulsinger, J. Higgins, & B. Bell (Eds.), *Genetics, environment and psychopathology* (pp. 293-322). New York: American Elsevier Publishing Company.

Pam, A. (1995). Biological psychiatry: Science or pseudoscience? In C. Ross & A. Pam (Eds.), *Pseudoscience in biological psychiatry: Blaming the body* (pp. 7-84). New York: John Wiley and Sons.

Pam, A., Kemker, S. S., Ross, C. A., & Golden, R. (1996). The "equal environment assumption" in MZ-DZ comparisons: An untenable premise of psychiatric genetics? *Acta Geneticae Medicae et Gemellologiae, 45*, 349-360.

Paul, D. B. (1985). Textbook treatments of the genetics of intelligence. *Quarterly Review of Biology, 60*, 317-326.

Paul, D. B. (1998). The politics of heredity: Essays on eugenics, biomedicine, and the nature-nurture debate. Albany, NY: State University of New York Press.

Pedersen, N. L., McClearn, G. E., Plomin, R., & Nesselroade, J. R. (1992). Effects of early rearing environment on twin similarity in the last half of the life span. *British Journal of Developmental Psychology, 10,* 255-267.

Pedersen, N. L., Plomin, R., McClearn, G. E., & Friberg, L. (1988). Neuroticism, extroversion, and related traits in adult twins reared apart and reared together. *Journal of Personality and Social Psychology, 55,* 950-957.

Pedersen, N. L., Plomin, R., Nesselroade, J. R., & McClearn, G. E. (1992). A quantitative genetic analysis of cognitive abilities during the second half of the life span. *Psychological Science, 3,* 346-353.

Penrose, L. S. (1973). *Outline of human genetics* (3rd ed.). New York: Crane, Russak & Company.

Peters, U. H. (2001). On Nazi psychiatry. *Psychoanalytic Review, 88,* 295-309.

Pittelli, S. J. (2002a). Genetic research on cognitive ability [Letter to the editor]. *British Journal of Psychiatry, 180,* 186.

Pittelli, S. J. (2002b). Meta-analysis and psychiatric genetics [Letter to the editor]. *American Journal of Psychiatry, 159,* 496.

Plomin, R. (1990). Trying to shoot the messenger for his message. *Behavioral and Brain Sciences, 13,* 144.

Plomin, R. (1994). The role of inheritance in behavior. *Science, 248,* 183-188.

Plomin, R. (1996). Beyond nature versus nurture. In L. Hall (Ed.), *Genetics and mental illness* (pp. 29-49). New York: Plenum Press.

Plomin, R. (1999). Genetics and cognitive ability. *Nature, 402,* (Suppl.), C25-C29.

Plomin, R. (2001). Genetics and behavior. *The Psychologist, 14,* 134-139.

Plomin, R., Corley, R., Caspi, A., Fulker, D. W., & DeFries, J. C. (1998). Adoption results for self-reported personality: Evidence for nonadditive genetic effects? *Journal of Personality and Social Psychology, 75,* 211-218.

Plomin, R., Corely, R., DeFries, J. C., & Fulker, D. W. (1990). Individual differences in television viewing in early childhood: Nature as well as nurture. *Psychological Science, 1,* 371-377.

Plomin, R., & Craig, I. (2001). Genetics, environment and cognitive abilities: Review and work in progress towards a genome scan for qualitative trait locus associations using DNA pooling. *British Journal of Psychiatry, 178,* (Suppl. 40), s41-s48.

Plomin, R., & Craig, I. (2002). Authors' reply [Letter to the editor]. *British Journal of Psychiatry, 180,* 186.

Plomin, R., & Daniels, D. (1987). Why are children in the same family so different from one another? *Behavioral and Brain Science, 10,* 1-16.

Plomin, R., & DeFries, J. C. (1985). Origins of individual differences in infancy: The Colorado Adoption Project. Orlando, FL: Academic Press.

Plomin, R., & DeFries, J. C. (1998, May). The genetics of cognitive abilities and disabilities. *Scientific American, 62*-69.

Plomin, R., DeFries, J. C., & Loehlin, J. C. (1977). Genotype-environment interaction and correlation in the analysis of human behavior. *Psychological Bulletin, 84,* 309-322.

Plomin, R., DeFries, J. C., & McClearn, G. E. (1990). *Behavioral genetics: A primer* (2nd ed.). New York: W. H. Freeman and Company.

Plomin, R., DeFries, J. C., McClearn, G. E., & Rutter, M. (1997). *Behavioral genetics* (3rd ed.). New York: W. H. Freeman and Company.

Plomin, R., Fulker, D. W., Corley, R., & DeFries, J. C. (1997). Nature, Nurture, and cognitive development from 1 to 16 years: A parent-offspring adoption study? *Psychological Science, 8,* 442-447.

Plomin, R., McClearn, G. E., Smith, D. L., Vignetti, S., Chorney, M. J., Chorney, K., Venditti, C. P., Kasarda, S., Thompson, L. A., Detterman, D. K., Daniels, J., Owen, M., & McGuffin, P. (1995). DNA markers associated with high versus low IQ: The IQ quantitative trait loci (QTL) project. *Behavior Genetics, 24,* 107-118.

Plomin, R., & Rutter, M. (1998). Child development, molecular genetics, and what to do with genes once they are found. *Child Development, 69,* 1223-1242.

Plomin, R., & Spinath, F. M. (2004). Intelligence: Genetics, genes, and genomics. *Journal of Personality and Social Psychology, 86,* 112-129.

Pope, H. G., Jonas, J. M., Cohen, B. M., & Lipinski, J. F. (1982). Failure to find evidence of schizophrenia in first-degree relatives of schizophrenic probands. *American Journal of Psychiatry, 139,* 826-828.

Popenoe, P. (1929, September). The Foster Child. *Scientific Monthly, 29,* 243-248.

Popenoe, P. (1936). Twins and criminals. *Journal of Heredity, 27,* 388-390.

Popenoe, P., & Johnson, R. (1933). *Applied eugenics* (rev. ed.). New York: Macmillan.

Portin, P., & Alanen, Y. O. (1997). A critical review of genetic studies of schizophrenia. II. Molecular genetic studies. *Acta Psychiatrica Scandinavica, 95,* 73-80.

Price, B. (1950). Primary biases in twin studies. *American Journal of Human Genetics, 2,* 293-352.

Proctor, R. N. (1988). *Racial hygiene: Medicine under the Nazis.* Cambridge, MA: Harvard University Press.

Proctor, R. N. (1992). Genomics and eugenics: How fair is the comparison? In G. Annas & S. Elias (Eds.), *Gene mapping* (pp. 57-93). New York: Oxford University Press.

Prolo, P., & Licinio, J. (2002). DRD4 and novelty seeking. In J. Benjamin, R. Ebstein, & R. Belmaker, (Eds.), *Molecular genetics and the human personality* (pp. 91-107). Washington, DC: American Psychiatric Press.

Propping, P. (1992). Abuse of genetics in Nazi Germany [Letter to the editor]. *American Journal of Human Genetics, 51,* 909-910.

Rafter, N. H. (Ed.). (1988). *White trash: The eugenic family studies 1877-1919.* Boston: Northeastern University Press.

Raine, A. (1993). *The psychopathology of crime.* San Diego: Academic Press.

Raine, A., Brennan, P., & Mednick, S. A. (1997). Interaction between birth complications and early maternal rejection in predisposing individuals to adult violence: Specificity to serious, early-onset violence. *American Journal of Psychiatry, 154,* 1265-1271.

Rainer, J. (1966). The contributions of Franz Josef Kallmann to the genetics of schizophrenia. *Behavioral Science, 11,* 413-437.

Ratner, C. (1982). Do the studies on identical twins prove that schizophrenia is genetically inherited? *International Journal of Social Psychiatry, 28,* 175-178.

Read, J., Goodman, L., Morrison, A. P., Ross, C. A., & Aderhold, V. (2004). Childhood trauma, loss and stress. In J. Read, L. Mosher, & R. Bentall (Eds.), *Models of madness: Psychological, social and biological approaches to schizophrenia* (pp. 223-252). Andover, UK: Taylor & Francis.

Reilly, P. R. (1991). *The surgical solution: A history of involuntary sterilization in the United States.* Baltimore: The Johns Hopkins University Press.

Rende, R. D., Plomin, R., & Vandenberg, S. G. (1990). Who discovered the twin method? *Behavior Genetics, 20,* 277-285.

Reuband, K. (1994). Reconstructing social change through retrospective questions: Methodological problems and prospects. In N. Schwarz & S. Sudman (Eds.), *Autobiographical memory and the validity of retrospective reports* (pp. 305-311). New York: Springer Verlag.

Rhee, S. H., & Waldman, I. D. (2002). Genetic and environmental influences on antisocial behavior: A meta-analysis of twin and adoption studies. *Psychological Bulletin, 128,* 490-529.

Richardson, K. (1998). *The origins of human potential.* London: Routledge.

Richardson, K. (2000). *The making of intelligence.* New York: Columbia University Press.

Richardson, L. F. (1913). The measurement of mental "nature" and the study of adopted children. *Eugenics Review, 4,* 391-394.

Robbins, L. C. (1963). The accuracy of parental recall of aspects of child development and of child rearing practices. *Journal of Abnormal and Social Psychology, 66,* 261-270.

Rosanoff, A. J. (1938). *Manual of psychiatry and mental hygiene* (7th ed., rewritten and enlarged). New York: John Wiley & Sons.

Rosanoff, A. J., Handy, L. M., Plesset, I. R., & Brush, S. (1934). The etiology of so-called schizophrenic psychoses. *American Journal of Psychiatry, 91*, 247-286.

Rosanoff, A. J., Handy, L. M., & Rosanoff, I. A. (1934). Criminality and delinquency in twins. *Journal of the American Institute of Criminal Law and Criminology, 24*, 923-934.

Rosanoff, A. J., & Orr, F. I. (1911). A study of heredity in insanity in the light of the mendelian theory. *Eugenics Record Office, Bulletin No. 5*. Cold Spring Harbor, New York.

Rose, R. J. (1982). Separated twins: Data and their limits [Book review]. *Science, 215*, 959-960.

Rose, R. J. (1991).Twin studies and psychosocial epidemiology. In M. Tsuang, K. Kendler, & M. Lyons (Eds.), *Genetic issues in psychosocial epidemiology* (pp. 12-32). New Brunswick, NJ: Rutgers University Press.

Rose, S. (1997). *Lifelines: Life beyond the genes*. New York: Oxford University Press.

Rosenthal, D. (1960). Confusion of identity and the frequency of schizophrenia in twins. *Archives of General Psychiatry, 3*, 297-304.

Rosenthal, D. (1961). Sex distribution and the severity of illness among samples of schizophrenic twins. *Journal of Psychiatric Research, 1*, 26-36.

Rosenthal, D. (1962a). Familial concordance by sex with respect to schizophrenia. *Psychological Bulletin, 59*, 401-421.

Rosenthal, D. (1962b). Problems of sampling and diagnosis in the major twin studies of schizophrenia. *Journal of Psychiatric Research, 1*, 116-134.

Rosenthal, D. (1963). *The Genain quadruplets*. New York: Basic Books.

Rosenthal, D. (1967). An historical and methodological review of genetic studies of schizophrenia. In J. Romano (Ed.), *The origins of schizophrenia: Proceedings of the first Rochester international conference on schizophrenia, March 29-31, 1967* (pp. 15-26). New York: Excerpta Medica Foundation.

Rosenthal, D. (1968). The heredity-environment issue in schizophrenia: Summary of the conference and present status of our knowledge. In D. Rosenthal & S. Kety (Eds.), *The transmission of schizophrenia* (pp. 413-427). New York: Pergamon Press.

Rosenthal, D. (1970a). Genetic research in the schizophrenia syndrome. In R. Cancro (Ed.), *The schizophrenic reactions* (pp. 245-258). New York: Brunner/Mazel.

Rosenthal, D. (1970b). *Genetic theory and abnormal behavior*. New York: McGraw-Hill.

Rosenthal, D. (1971a). *Genetics of psychopathology*. New York: McGraw-Hill.

Rosenthal, D. (1971b). A program of research on heredity in schizophrenia. *Behavioral Science, 16*, 191-201.

Rosenthal, D. (1972). Three adoption studies of heredity in the schizophrenic disorders. *International Journal of Mental Health, 1*, 63-75.

Rosenthal, D. (1974a). The concept of subschizophrenic disorders. In S. Mednick, et al. (Eds.), *Genetics, environment, & psychopathology* (pp. 167-176). New York: American Elsevier.

Rosenthal, D. (1974b). The genetics of schizophrenia. In S. Arieti & E. Brody (Eds.), *American handbook of psychiatry* (2nd ed., pp. 588-600). New York: Basic Books.

Rosenthal, D. (1975a). Discussion: The concept of schizophrenic disorders. In R. Fieve, D. Rosenthal, & H. Brill (Eds.), *Genetic research in psychiatry* (pp. 199-208). Baltimore: The Johns Hopkins Press.

Rosenthal, D. (1975b). The spectrum concept in schizophrenic and manic-depressive disorders. In D. Freedman (Ed.), *Biology of the major psychoses* (pp. 19-25). New York: Raven Press.

Rosenthal, D. (1978). Eugen Bleuler's thoughts and views about heredity in schizophrenia. *Schizophrenia Bulletin, 4,* 476-477.

Rosenthal, D. (1979). Genetic factors in behavioural disorders. In M. Roth & V. Cowie (Eds.), *Psychiatry, genetics and pathography: A tribute to Eliot Slater* (pp. 22-33). London: Oxford University Press.

Rosenthal, D., & Kety, S. S. (Eds.). (1968). *The transmission of schizophrenia: Proceedings of the second research conference of the foundations' fund for research in psychiatry,* Dorado, Puerto Rico, 26 June to 1 July 1967. New York: Pergamon Press.

Rosenthal, D., Wender, P. H., Kety, S. S., Schulsinger, F., Welner, J., & Østergaard, L. (1968). Schizophrenics' offspring reared in adoptive homes. In D. Rosenthal & S. Kety (Eds.), *The transmission of schizophrenia* (pp. 377-391). New York: Pergamon Press.

Rosenthal, D., Wender, P. H., Kety, S. S., Welner, J., & Schulsinger, F. (1971). The adopted-away offspring of schizophrenics. *American Journal of Psychiatry, 128,* 307-311.

Rosenthal, R. (1979). The "file drawer problem" and tolerance for null results. *Psychological Bulletin, 86,* 638-641.

Rosenthal, R., & Rosnow, R. L. (1975). *The volunteer subject.* New York: John Wiley & Sons.

Rowe, D. C. (1983). Biometrical genetic models of self-reported delinquent behavior: A twin study. *Behavior Genetics, 13,* 473-489.

Rowe, D. C. (1986). Genetic and environmental components of antisocial behavior: A study of 265 twin pairs. *Criminology, 24,* 513-532.

Rowe, D. C. (1994). The limits of family influence: Genes, experience, and behavior. New York: The Guilford Press.

Rowe, D. C. (2002). *Biology and crime.* Los Angeles: Roxbury.

Rowe, D. C., & Jacobson, K. C. (1999). In the mainstream. In R. Carson & M. Rothstein (Eds.), *Behavioral genetics: The clash of culture and biology* (pp. 12-34). Baltimore: The Johns Hopkins University Press.

Rüdin, E. (1916). *Zur Vererbung und Neuentstehung der Dementia praecox* [On the heredity and new development of Dementia praecox].Berlin: Springer Verlag OHG.

Rüdin, E. (1932). The significance of eugenics and genetics for mental hygiene. In F. Williams (Ed.), *Proceedings of the First International Congress on Mental Hygiene* (pp. 471-495). New York: International Committee for Mental Hygiene.

Rüdin, E. (1939). Bedeutung der Forschung und Mitarbeit von Neurologen und Psychiatrie im nationalsozialistischen Staat [The meaning of research and cooperation of neurologists and psychiatry in the National Socialist state]. *Zeitschrift für die Gesamte Neurologie und Psychiatrie, 165, 7-17.*

Rüdin, E. (1942). Zehn Jahre nationalsozialistischer Staat [Ten years of the National Socialist state]. *Archiv für Rassen- und Gesellschaftsbiologie, 36, 321-322.*

Rutter, M. (1996). Introduction: Concepts of antisocial behaviour, of cause, and of genetic influences. In G. Bock & J. Goode (Eds.), *Genetics of criminal and antisocial behavior* (pp. 1-20). New York: John Wiley and Sons.

Rutter, M., Silberg, J., & Simonoff, E. (1993). Wither behavioral genetics? — A developmental psychopathological perspective. In R. Plomin & G. McClearn (Eds.), *Nature, nurture, and psychology* (pp. 433-456). Washington, DC: American Psychological Association.

Samelson, F. (1979). Putting psychology on the map. In A. Buss (Ed.), *Psychology in social context* (pp. 103-167). New York: Irvington Publishers.

Sarbin, T. R. (1990). Toward the obsolescence of the schizophrenia hypothesis. *Journal of Mind and Behavior, 11, 259-283.*

Sarbin, T. R., & Mancuso, J. C. (1980). *Schizophrenia: Medical diagnosis or moral verdict?* New York: Pergamon Press.

Savan, B. (1988). Science under siege: The myth of objectivity in scientific research. Montreal: CBC Enterprises.

Scarr, S. (1968). Environmental bias in twin studies. *Eugenics Quarterly, 15, 34-40.*

Scarr, S., & Carter-Saltzman, L. (1979). Twin method: Defense of a critical assumption. *Behavior Genetics, 9, 527-542.*

Schiff, M., Duyme, M., Dumaret, A., & Tomkiewicz, S. (1982). How much *could* we boost scholastic achievement and IQ scores? A direct answer from a French adoption study. *Cognition, 12, 165-196.*

Schiff, M., & Lewontin, R. C. (1986). *Education and class: The irrelevance of IQ genetic studies.* Oxford: Clarendon Press.

Schulsinger, F. (1977). Psychopathy: Heredity and environment. In S. Mednick & K. Christiansen (Eds.), *Biosocial bases of criminal behavior* (pp. 109-125). New York: Gardner Press. (Originally published in 1972)

Schulz, B. (1934). Rassenhygienische Eheberatung [Racial hygienic marriage counseling]. *Volk und Rasse, 9*, 138-143.

Schulz, B. (1939). Über die Beteutung der empirischen Erbvorhersageforschung [On the meaning of genetic empirical prognostic research]. *Der Erbarzt, 6, (4)*, 43-44.

Schwesinger, G. C. (1952). The effect of differential parent-child relation on identical twin resemblance in personality. *Acta Geneticae Medicae et Gemellologiae, 1*, 40-47.

Segal, N. L. (1999). Entwined lives: Twins and what they tell us about human behavior. New York: Dutton.

Sharp, H. C. (1909). Vasectomy as a means of preventing procreation in defectives. *Journal of the American Medical Association, 53*, 1897-1902.

Sherrington, R., Brynjolfsson, J., Petursson, H., Potter, M., Duddleston, K., Barraclough, B., Wasmuth, J., Dobbs, M., & Gurling, H. (1988). Localization of a susceptibility locus for schizophrenia on chromosome 5. *Nature, 336*, 164-167.

Shields, J. (1954). Personality differences and neurotic traits in normal twin schoolchildren. *Eugenics Review, 45*, 213-246.

Shields, J. (1962). *Monozygotic twins brought up apart and brought up together*. London: Oxford University Press.

Shields, J. (1968). Summary of the genetic evidence. In D. Rosenthal & S. Kety (Eds.), *The transmission of schizophrenia* (pp. 95-126). New York: Pergamon Press.

Shields, J. (1978). MZA Twins: Their use and abuse. In *Twin research: Psychology and methodology* (pp. 79-93). New York: Alan R. Liss.

Shorter, E. (1997). *A history of psychiatry*. New York: John Wiley and Sons.

Siebert, A. (1999). Brain disease hypothesis for schizophrenia disconfirmed by all evidence. *Ethical Human Sciences and Services, 1*, 179-189.

Siemens, H. W. (1924a). *Race hygiene and heredity*. New York: D. Appleton.

Siemens, H. W. (1924b). *Die zwillingspathologie* [Twin pathology]. Berlin: Springer Verlag.

Siemens, H. W. (1927). The diagnosis of identity in twins. *Journal of Heredity, 18*, 201-209.

Siemens, H. W. (1937). *Grundzüge der vererbungslehre, rassenhygiene und bevölkerungspolitik* [Foundations of genetics, racial hygiene, and population policy] (8th ed.). Munich & Berlin: J. F. Lehmanns Verlag.

Simes, R. J. (1987). Confronting publication bias: A cohort design for meta-analysis. *Statistics in Medicine, 6*, 11-29.

Simon, J. L. (1997). Four comments on The Bell Curve. *Genetica, 99*, 199-205.

Sims, V. M. (1931). The influence of blood relationship and common environment on measured intelligence. *Journal of Educational Psychology, 22*, 56-65.

Slater, E. (1953). Psychotic and neurotic illnesses in twins. *Medical Research Council Special Report Series No. 278*. London: Her Majesty's Stationary Office.

Slater, E. (1968). A review of earlier evidence on genetic factors in schizophrenia. In D. Rosenthal & S. Kety (Eds.), *The transmission of schizophrenia* (pp. 15-26). New York: Pergamon Press.

Slater, E. (1971). Autobiographical sketch. In J. Shields & I. Gottesman (Eds.), *Man, mind, and heredity: Selected papers of Eliot Slater on psychiatry and genetics* (pp. 1-23). Baltimore: Johns Hopkins Press.

Slater, E., & Cowie, V. (1971). *The genetics of mental disorders.* London: Oxford University Press.

Slater, E., Hare, E., & Price, J. (1971). Marriage and fertility of psychiatric patients compared with national data. *Social Biology, 18* (Suppl.), S60-S73.

Smith, C. (1974). Concordance in twins: Methods and interpretation. *American Journal of Human Genetics, 26,* 454-466.

Smith, R. T. (1965). A comparison of socioenvironmental factors in monozygotic and dizygotic twins, testing an assumption. In S. Vandenberg (Ed.), *Methods and goals in human behavior genetics* (pp. 45-61). New York: Academic Press.

Sofair, A. N., & Kaldjian, L. C. (2000). Eugenic sterilization and a qualified Nazi analogy: The United States and Germany, 1930-1945. *Annals of Internal Medicine, 132,* 312-319.

Spitzer, R. L., & Endicott, J. (1979). Justification for separating schizotypal and borderline personality disorders. *Schizophrenia Bulletin, 6,* 95-104.

Stefansson, H., Sigurdsson, E., et al. (2002). Neuregulin 1 and susceptibility to schizophrenia. *American Journal of Human Genetics, 71,* 877-892.

Stern, C. (1962). Wilhelm Weinberg. *Genetica, 47,* 1-5.

Stocks, P. (1930). A biometric investigation of twins and their brothers and sisters. *Annals of Eugenics, 4,* 49-108.

Stoltenberg, S. F. (1997). Coming to terms with heritability. *Genetica, 99,* 89-96.

Stoolmiller, M. (1998). Correcting estimates of shared environmental variance for range restriction in adoption studies using a truncated multivariate normal model. *Behavior Genetics, 28,* 429-441.

Stoolmiller, M. (1999). Implications of the restricted range of family environments for estimates of heritability and nonshared environment in behavior-genetic adoption studies. *Psychological Bulletin, 125,* 392-409.

Straub, R. E., Jiang, Y., MacLean, C. J., Ma, Y., Webb, B. T., Myakishev, M. V., Harris-Kerr, C., Wormley, B., Sadek, H., Kadambi, B., Cesare, A. J., Gibberman, A., Wang, X., O'Neill, F. A., Walsh, D., & Kendler, K. S. (2002). Genetic variation in the 6p22.3 gene DTNBP1, the human ortholog of the mouse dysbindin gene, is associated with schizophrenia. *American Journal of Human Genetics, 71,* 337-348.

Strömgren, E. (1993). Fini Schulsinger's contribution to psychiatric research in genetic epidemiology. *Acta Psychiatrica Scandinavica* (Suppl. 370), 11-13.

Strömgren, E. (1994). Recent history of European psychiatry — Ideas, development, and personalities. *American Journal of Medical Genetics (Neuropsychiatric Genetics), 54,* 405-410.

Stumpfl, F. (1936). Die Ursprunge des Verbrechen. Dargestellt am Lebenslauf von Zwillingen [The origins of crime as represented in the resume of twins]. Leipzig: Georg Thieme.

Sullivan, P. F., & Kendler, K. S. (2003). Schizophrenia as a complex trait: Evidence from a meta-analysis of twin studies. *Archives of General Psychiatry, 60,* 1187-1192.

Szasz, T. S. (1961). The myth of mental illness: Foundations of a theory of personal conduct. New York: Hoeber-Harper.

Szasz, T. S. (1964). The moral dilemma of psychiatry: Autonomy or heteronomy? *American Journal of Psychiatry, 121,* 521-528.

Szasz, T. S. (1968). Hysteria. In D. Sills (Ed.), *International encyclopedia of the social sciences* (Vol. 7, pp. 47-52). New York: Macmillan.

Szasz, T. S. (1970). *The manufacture of madness.* New York: Delta.

Szasz, T. S. (1976). Schizophrenia: The sacred symbol of psychiatry. New York: Basic Books.

Szasz, T. S. (1987). *Insanity: The idea and its consequences.* Syracuse, NY: Syracuse University Press.

Taylor, H. F. (1980). *The IQ game: A methodological inquiry into the heredity-environment controversy.* New Brunswick, NJ: Rutgers University Press.

Taylor, P. (1996). General discussion III. In G. Bock & J. Goode (Eds.), *Genetics of criminal and antisocial behavior* (pp. 129-137). New York: John Wiley and Sons.

Teasdale, T. W. (1979). Social class correlations among adoptees and their biological adoptive parents. *Behavior Genetics, 9,* 103-114.

Tellegen, A., Lykken, D. T., Bouchard, T. J., Jr., Wilcox, K. J., Segal, N. L., & Rich, S. (1988). Personality similarity in twins reared apart and together. *Journal of Personality and Social Psychology, 54,* 1031-1039.

Terman, L. M. (1916). *The measurement of intelligence.* Boston: Houghton Mifflin.

Terman, L. M., & Merrill, M. A. (1937). *Measuring intelligence.* Boston: Houghton Mifflin.

Thompson, P. M., Cannon, T. D., Narr, K., van Erp, T., Poutanen, V. P., Huttunen, M., Lönnqvist, J., Standertskjöld-Nordenstam, C. G., Kaprio, J., Khaledy, M., Dail, R., Zoumalan, C., I., & Toga, A. W. (2001). Genetic influences on brain structure. *Nature Neuroscience, 12,* 1253-1258.

Thorndike, E. L. (1905). Measurements of twins. Archives of Philosophy, Psychology, and Scientific Methods, 1, 1-64.

Tienari, P. (1963). *Psychiatric illnesses in identical twins.* Copenhagen: Munksgaard.

Tienari, P. (1968). Schizophrenia in monozygotic male twins. In D. Rosenthal & S. Kety (Eds.), *The transmission of schizophrenia* (pp. 27-36). New York: Pergamon Press.

Tienari, P. (1971). Schizophrenia and monozygotic twins. *Psychiatria Fennica, 1971*, 97-104.

Tienari, P. (1975). Schizophrenia in Finnish male twins. *British Journal of Psychiatry Special Publication, No. 10.* M. Lader (Ed.), pp. 29-35.

Tienari, P. (1991). Interaction between genetic vulnerability and family environment: The Finnish adoptive family study of schizophrenia. *Acta Psychiatrica Scandinavica, 84*, 460-465.

Tienari, P. (1992). Implications of adoption studies on schizophrenia. *British Journal of Psychiatry, 161* (Suppl. 18), 52-58.

Tienari, P., Kaleva, M., Lahti, I., Läksy, K., Moring, J., Naarala, M., Sorri, A., Wahlberg, K. E., & Wynne, L. C. (1991). Adoption studies on schizophrenia. In C. Eggers (Ed.), *Schizophrenia and youth* (pp. 42-51). New York: Springer Verlag.

Tienari, P., Lahti, I., Sorri, A., Naarala, M., Wahlberg, K., Rönkkö, T., Moring, J., & Wynne, L. C. (1987). The Finnish adoptive family study of schizophrenia: Possible joint effects of genetic vulnerability and family interaction. In K. Halweg & M. Goldstein (Eds.), *Understanding major mental disorder: The contribution of family interaction research* (pp. 33-54). New York: Family Process Press.

Tienari, P., Sorri, A., Lahti, I., Naarala, M., Wahlberg, K., Moring, J., & Pohjola, J. (1985). Interaction of genetic and psychosocial factors in schizophrenia. *Acta Psychiatrica Scandinavica* (Suppl. No. 319, 71), 19-30.

Tienari, P., Sorri, A., Lahti, I., Naarala, M., Wahlberg, K., Moring, J., Pohjola, J., & Wynne, L. C. (1987). Genetic and psychosocial factors in schizophrenia: The Finnish adoptive family study. *Schizophrenia Bulletin, 13*, 477-484.

Tienari, P., Wynne, L. C., Läksy, K., Moring, J., Nieminen, P., Sorri, A., Lahti, I., & Wahlberg, K. E. (2003). Genetic boundaries of the schizophrenia spectrum: Evidence from the Finnish Adoptive Family Study. *American Journal of Psychiatry, 160*, 1587-1594.

Tienari, P., Wynne, L. C., Moring, J., Lahti, I., Naarala, M., Sorri, A., Wahlberg, K., Saarento, O., Seitamaa, M., Kaleva, M., & Läksy, K. (1994). The Finnish adoptive family study of schizophrenia. *British Journal of Psychiatry, 164* (Suppl. 23), 20-26.

Tienari, P., Wynne, L. C., Moring, J., Läksy, K., Nieminen, P., Sorri, A., Lahti, I., Wahlberg, K. E., Naarala, M., Kurki-Suonio, K., Saarento, O., Koistinen, P., Tarvainen, T., Hakko, H., & Miettunen, J. (2000). Finish Adoptive Family Study: Sample selection and adoptee DSM-III-R diagnoses. *Acta Psychiatrica Scandinavica, 101*, 433-443.

Tienari, P., Wynne, L. C., Sorri, A., Lahti, I., Läksy, K., Moring, J., Naarala, M., Nieminen, P., & Wahlberg, K. E. (2004). Genotype-environment interaction in schizophrenia-spectrum disorders. *British Journal of Psychiatry, 184*, 216-222.

Torrey, E. F. (1980). *Schizophrenia and civilization.* New York: Jason Aronson.

Torrey, E. F. (1990). Offspring of twins with schizophrenia [Letter to the editor]. *Archives of General Psychiatry, 47,* 976-977.

Torrey, E. F. (1995). *Surviving schizophrenia* (3rd ed.). New York: Harper Perennial.

Torrey, E. F., Bowler, A. E., Taylor, E. H., & Gottesman, I. I. (1994). *Schizophrenia and manic-depressive disorder: The biological roots of mental illness as revealed by the landmark study of identical twins.* New York: Basic Books.

Torrey, E. F., & Yolken, R. H. (2000). Familial and genetic mechanisms in schizophrenia. *Brain Research Reviews, 31,* 113-117.

Treasure, J., & Holland, A. (1995). Genetic factors in eating disorders. In G. Szmukler, C. Dare, & J. Treasure (Eds.), *Handbook of eating disorders* (pp. 65-81). New York: John Wiley & Sons.

Tsuang, M. T., & Faraone, S. V. (1990). *The genetics of mood disorders.* Baltimore: Johns Hopkins University Press.

Tsuang, M. T., & Faraone, S. V. (2000). The frustrating search for schizophrenia genes. *American Journal of Medical Genetics (Semin. Med Genet.), 97,* 1-3.

Tsuang, M. T., Stone, W. S., & Faraone, S. V. (2001). Genes, environment, and schizophrenia. *British Journal of Psychiatry, 178* (Suppl. 40), S18-S24.

Tucker, W. H. (1994). *The science and politics of racial research.* Urbana, IL: University of Illinois Press.

Turkheimer, E. (2000). Three laws of behavior genetics and what they mean. *Psychological Science, 9,* 160-164.

Uhl, G. R., & Grow, R. W. (2004).The burden of complex genetics in brain disorders. *American Journal of Psychiatry, 61,* 223-229.

Van Dyke, J. L., Rosenthal, D., & Rasmussen, P. V. (1975). Schizophrenia: Effects of inheritance and rearing on reaction time. *Canadian Journal of Behavioral Science, 7,* 223-236.

von Bracken, H. (1934). Mutual intimacy in twins. *Character and Personality, 2,* 293-309.

von Verschuer, O. (1939). Twin research from the time of Francis Galton to the present-day. *Proceedings of the Royal Society, 128 (Series B),* 62-81.

Wahlberg, K. E., Wynne, L. C., Keskitalo, P., Nieminen, P., Moring, J., Läksy, K., Sorri, A., Koistinen, P., Tarvainen, T., Miettunen, J., & Tienari, P. (2001). Long-term stability of communication deviance. *Journal of Abnormal Psychology, 110,* 443-448.

Wahlberg, K. E., Wynne, L. C., Oja, H., Keskitalo, P., Anais-Tanner, H., Koistinen, P., Tarvainen, T., Hakko, H., Lahti, I., Moring, J., Naarala, M., Sorri, A., & Tienari, P. (2000). Thought Disorder Index of Finnish adoptees and communication deviance of their adoptive parents. *Psychological Medicine, 30,* 127-136.

Wahlberg, K. E., Wynne, L. C., Oja, H., Keskitalo, P., Pykäläinen, M., Lahti, I., Moring, J., Naarala, M., Sorri, A., Seitamaa, M., Läksy, K, Kolassa, J., & Tienari, P. (1997). Gene-environment interaction in vulnerability to schizophrenia: Findings from the Finnish Adoptive Family Study of Schizophrenia. *American Journal of Psychiatry, 154,* 355-362.

Wahlsten, D. (1990). Insensitivity of the analysis of variance to heredity-environment interaction. *Behavioral and Brain Sciences, 13,* 109-120.

Wahlsten, D. (1994). The intelligence of heritability. *Canadian Psychology, 35,* 244-259.

Wahlsten, D. (1999). Single-gene influences on brain and behavior. *American Review of Psychology, 50,* 599-624.

Walker, E., Kestler, L., Bollini, A., & Hochman, K. M. (2004). Schizophrenia: Etiology and course. *Annual Review of Psychology, 55,* 401-430.

Walters, G. D., & White, T. W. (1989). Heredity and crime: Bad genes or bad research? *Criminology, 27,* 455-485.

Wasserman, D., & Wachbroit, R. (2001). Introduction: Methods, meanings, and morals. In D. Wasserman & R. Wachbroit (Eds.), *Genetics and criminal behavior* (pp. 1-21). Cambridge, UK: Cambridge University Press.

Watson, P. (1981). *Twins: An uncanny relationship?* New York: The Viking Press.

Weber, M. M. (1996). Ernst Rüdin, 1874-1952. *American Journal of Medical Genetics* (Neuropsychiatric Genetics), 67, 323-331.

Wechsler, D. (1944). *The measurement of adult intelligence* (3rd ed.). Baltimore: Williams and Wilkins.

Wechsler, D. (1958). *The measurement and appraisal of adult intelligence* (4th ed.). Baltimore: Williams and Wilkins.

Weindling, P. (1989). *Health, race, and German politics between national unification and Nazism, 1870-1945.* Cambridge, UK: Cambridge University Press.

Weindling, P. (1993). The survival of eugenics in 20th-Century Germany. *American Journal of Human Genetics, 52,* 643-649.

Weindling, P. (1999). International eugenics: Swedish sterilization in context. *Scandinavian Journal of History, 24,* 179-197.

Weingart, P. (1989). German eugenics between science and politics. *Osiris, 5 (second series),* 260-282.

Weingart, P. (1999). Science and political culture: Eugenics in comparative perspective. *Scandinavian Journal of History, 24,* 163-177.

Weinreich, M. (1946). *Hitler's professors.* New York: Yiddish Scientific Institute — Yivo.

Wender, P. H., & Klein, D. (1981). *Mind, mood, and medicine.* New York: Farrar, Straus, & Giroux.

Wender, P. H., Rosenthal, D., Kety, S. S., Schulsinger, F., & Welner, J. (1974). Crossfostering: A research strategy for clarifying the role of genetic and experiential factors in the etiology of schizophrenia. *Archives of General Psychiatry, 30,* 121-128.

Wetzell, R. F. (2000). *Inventing the criminal: A history of German criminology, 1880-1945.* Chapel Hill, NC: University of North Carolina Press.

Wheelan, L. (1951). Aggressive psychopathy in one of a pair of uniovular twins: A clinical and experimental study. *British Journal of Delinquency, 2,* 130-143.

Whitaker, R. (2002). *Mad in America: Bad science, bad medicine, and the enduring mistreatment of the mentally ill.* Cambridge, MA: Perseus.

Wiener, K. (1958). Preface. In H. W. Siemens, *General diagnosis and therapy of skin diseases.* Chicago: University of Chicago Press.

Wilson, P. T. (1934). A study of twins with special reference to heredity as a factor determining differences in environment. *Human Biology, 6,* 324-354.

Wilson, J. Q., & Herrnstein, R. J. (1985). *Crime and human nature.* New York: Simon and Schuster.

Wingfield, A. H. (1928). *Twins and orphans: The inheritance of intelligence.* London: J. M. Dent and Sons.

Wingfield, A. H. (1930). The intelligence of twins. *Eugenics Review, 22,* 183-186.

Woodworth, R. S. (1941). *Heredity and environment: A critical survey of recently published material on twins and foster children.* New York: Social Science Research Center.

World Health Organization. (1966). The use of twins in epidemiological studies. *Acta Geneticae Medicae et Gemellologiae, 15,* 109-128.

Wright, L. (1997). *Twins: And what they tell us about who we are.* New York: John Wiley & Sons.

Wright, W. (1998). *Born that way.* New York: Alfred A. Knopf.

Wyatt, W. J. (1993). Identical twins, emergenesis and environments [Letter to the editor]. *American Psychologist, 48,* 1294-1295.

Wyatt, W. J., Posey, A., Welker, W., & Seamonds, C. (1984). Natural levels of similarities between identical twins and between unrelated people. *The Skeptical Inquirer, 9,* 62-66.

Wynne, L., Singer, M., & Toohey, M. (1976). Communication of the adoptive parents of schizophrenics. In J. Jorstad & E. Ugelstad (Eds.), *Schizophrenia 75 — psychotherapy family studies, research* (pp. 413-451). Oslo: University of Oslo Press.

Yamaguchi, M. (1997, December 27). Japan's victims of sterilization stop hiding. *San Francisco Chronicle,* p. A12.

Yarrow, M. R., Campbell, J. D., & Burton, R. V. (1970). Recollections of childhood: A study of the retrospective method. *Monographs of the Society for Research in Child Development, 35* (5, Serial No. 138).

Yeh, M., Morley, K. I., & Hall, W. D. (2004). The policy and ethical implications of genetic research on attention deficit hyperactivity disorder. *Australian and New Zealand Journal of Psychiatry, 38,* 10-19.

Zerbin-Rüdin, E. (1972). Genetic research and the theory of schizophrenia. *International Journal of Mental Health, 1,* 42-62.

Zerbin-Rüdin, E., & Kendler, K. S. (1996). Ernst Rüdin (1874-1952) and his Genealolgic-Demographic Department in Munich (1917-1986): An introduction to their family studies of schizophrenia. *American Journal of Medical Genetics (Neuropsychiatric Genetics), 67,* 332-337.

Index of Names

Ainslie, R. C., 15, 349
Albee, G. W., 340, 349
Allen G. E., 275
Allen, M. G., 170
Allport, G. W., 105–106, 128–129, 349
Alper, J. S., 97, 331, 350
Andreasen, N. C., 158, 195, 211, 324–325, 328, 335, 349, 367

Badner, J. A., 359
Banu, G., 39, 371
Barkley, R. A., 6, 350
Beckmann, H., 167, 358
Beckwith, J., 97, 120, 331, 333, 350
Benjamin, L. S., 225, 227–228, 249, 350, 378
Berry, C., 82
Bertelsen, A., 34–35, 44, 48, 187–193, 360
Biederman, J., 31, 44, 327, 357, 364
Billings, P. R., 97, 350
Binding, R., 37
Binet, A., 108, 305, 312–313, 316–317
Blatt, S., 220–221, 225, 228, 236, 238–239, 245, 248, 256, 370
Bleuler, E., 160, 195, 219, 247
Bleuler, M., 71, 231
Bohman, M., 289–295, 301, 351
Bowler, A. E., 49, 386

Boyle, M., 156–157, 160, 166, 168–170, 200–201, 225, 236, 258–259, 291, 352
Brennan, P., 281, 302, 353, 373, 378
Brigham, C. C., 313–315, 317, 353
Brunner, H. G., 275, 353, 367, 379
Burt, C., 59, 108–109, 118, 122, 308, 358, 361, 363, 365

Cadoret, R. J., 289, 353
Cannon, T. D., 167, 353, 384
Cardon, L. R., 149, 354
Carey, G., 68, 286, 354
Carter, N. D., 24–25, 69, 88–89, 92–94, 354, 381
Carter-Saltzman, L., 69, 88–89, 92–94, 381
Cassill, K., 106, 354
Cassou, B., 201, 205, 207, 354
Ceci, S. J., 134, 311, 354, 375
Cheverud, J. M., 147–148, 354
Christiansen, K. O., 281–283, 285–287, 296, 298, 354, 362, 373, 381
Chumakov, I., 324, 354
Churchill, W., 82
Cloninger, C. R., 283, 285, 287, 324, 351, 354
Colbert, T. C., 323, 355
Conklin, E. G., 25

391

SUBJECT INDEX

A

aborigines, 318
Adoptees method, 199, 251, 254–255, 267
adoption studies, 1–6, 9, 60–63, 111, 150, 158,
  161, 193, 195–197, 202, 206–207, 210–212,
  224, 228, 251, 262, 265, 270, 272, 275–276,
  281, 287–289, 294, 303–304, 308–311, 319,
  321–322, 325–327, 337–338, 344, 363–
  364, 367, 370, 373, 378–379, 383, 385
  of criminal and antisocial behavior, 4,
    280–281, 303, 372–373, 381, 384

  of IQ, 61, 118, 147, 305–306, 308, 310, 312,
    317, 319, 345, 381

  of schizophrenia, 3, 6, 9, 27, 45–48, 51–52,
    54–56, 58, 61–63, 151, 155, 157–161, 163–
    165, 167–168, 172, 175–176, 184–187,
    190–191, 193, 195–196, 202, 204, 206–
    207, 211, 214–215, 217–219, 222, 224,
    226–227, 230, 235–238, 242, 244, 249,
    255–256, 261–262, 264–273, 294, 297,
    322, 325, 330, 349–351, 353, 355, 358–
    360, 362–370, 372, 374, 377–380, 382–
    383, 385, 388–389

American, 7, 15–16, 18, 23–25, 37, 39, 42–45,
  48–49, 64–65, 92, 103, 109, 118, 152, 156,
  158, 162, 164, 195–199, 204, 206, 209, 212–
  213, 215, 219–221, 224–225, 233–234, 238,
  240, 242, 251, 257, 262–263, 271–272,
  275–279, 284, 289, 302, 306, 311–312, 316–
  317, 324, 345, 349–350, 353–354, 356–367,
  370–373, 375–389
  Eugenics Society, 48

American Adoptees' Family studies Daphne
  (and Barbara), 100
anorexia, 84
Antisocial
  behavior (see also 'criminality'), 4, 275,
    279–281, 283–284, 286, 288, 292, 295,

302–303, 353–354, 356, 360, 364–365,
372–373, 378, 380–381, 384

  personality, 289, 355

Arabs, 318
Association studies, 323–324
Attention-Deficit   Hyperactivity   Disorder
  (ADHD)
  genes, 6, 142, 326–327, 329, 334, 350, 364,
    369

Auschwitz, 42–43, 49, 60, 64

B

Barbara, 100, 341
behavior genetics, 1, 5, 8, 63, 81, 87, 137, 143–
  144, 196, 295, 331, 334, 336, 350, 355, 358,
  383, 386
Bias
  in IQ tests, 93

Black, 346, 350
  and white populations, 139

Board of Eugenics, 208
British (see also English and United Kingdom)
  schizophrenia twin study, 12, 23, 39, 45,
    48, 59–60, 108, 156, 185, 286, 353, 355–
    356, 358, 360–362, 369, 376, 385–386, 388

C

Canadian, 24, 147, 372, 386–387
castration, 266
Cohort effects, 106, 134
Colorado Adoption Project (CAP), 130, 211,
  310, 377
contact (between 'separated' twins), 112
co-twin control method, 27, 43, 60
criminal and antisocial behavior, 4, 280–281,
  283, 286, 302–303, 372–373, 381, 384
culture/al, 36, 105–106, 115, 127, 135, 152, 173,
  226, 314, 318, 346